Logical Empiricism

Logical Empiricism

HISTORICAL

&

CONTEMPORARY

PERSPECTIVES

Edited by
Paolo Parrini, *University of Florence*
Wesley C. Salmon, *University of Pittsburgh*
Merrilee H. Salmon, *University of Pittsburgh*

UNIVERSITY OF PITTSBURGH PRESS

Published by the University of Pittsburgh Press, Pittsburgh, Pa., 15260

Copyright © 2003, University of Pittsburgh Press

10 9 8 7 6 5 4 3 2 1

ISBN 0-8229-4194-5

This book is dedicated to the memory of

Wesley C. Salmon

1925–2001

Contents

Preface   ix

Introduction   1
Paolo Parrini and Wesley Salmon

I. Turning Points and Fundamental Controversies *11*

A Turning Point in Philosophy: Carnap-Cassirer-Heidegger   13
Michael Friedman

Carnap's "Elimination of Metaphysics through Logical Analysis of
Language": A Retrospective Consideration of the Relationship
between Continental and Analytic Philosophy   30
Gottfried Gabriel

Schlick and Husserl on the Essence of Knowledge   43
Roberta Lanfredini

Carnap versus Gödel on Syntax and Tolerance   57
S. Awodey and A. W. Carus

II. On the Origins and Development of the Vienna Circle *65*

On the Austrian Roots of Logical Empiricism: The Case of the First
Vienna Circle   67
Thomas Uebel

On the *International Encyclopedia,* the Neurath-Carnap Disputes, and
the Second World War   94
George Reisch

Carl Gustav Hempel: Pragmatic Empiricist   109
Gereon Wolters

III. The Riddle of Wittgenstein 123

The Methods of the *Tractatus:* Beyond Positivism and Metaphysics?   125
  David G. Stern

IV. Philosophy of Physics 157

Two Roads from Kant: Cassirer, Reichenbach, and General Relativity   159
  T. A. Ryckman

Vienna Indeterminism II: From Exner to Frank and von Mises   194
  Michael Stöltzner

V. The Mind-Body Problem 231

The Mind-Body Problem in the Origin of Logical Empiricism:
Herbert Feigl and Psychophysical Parallelism   233
  Michael Heidelberger

Logical Positivism and the Mind-Body Problem   263
  Jaegwon Kim

VI. Scientific Rationality 279

Kinds of Probabilism   281
  Maria Carla Galavotti

Smooth Lines in Confirmation Theory: Carnap, Hempel, and
the Moderns   304
  Martin Carrier

Changing Conceptions of Rationality: From Logical Empiricism
to Postpositivism   325
  Gürol Irzik

VII. Nonlinguistic Empiricism 347

Reason and Perception In Defense of a Non-Linguistic Version of
Empiricism   349
  Paolo Parrini

Commit It Then to the Flames ...   375
  Wesley C. Salmon

Preface

Under the joint sponsorship of the Florence Center for the History and Philosophy of Science and the Pittsburgh Center for Philosophy of Science, a group of scholars met in Florence in November of 1999 to discuss historical and contemporary aspects of the philosophical movement known as logical empiricism. Organizers of the meeting were Professors Maria Carla Galavotti, Alessandro Pagnini, Paolo Parrini, Wesley Salmon, Raffaella Simili, and Gereon Wolters.

The highly original papers presented at the meeting greatly illuminated some previously unexplored features of logical empiricism. Moreover, the way in which the works of the various authors in this fast-growing area of research complemented and reinforced one another gave the conference an unusual degree of cohesion. Paolo Parrini and Wesley Salmon undertook the task of preparing a volume of collected papers based on the conference. The volume was nearly complete before Wes died. All of the contributors had submitted drafts of their papers, the organization of the volume had been decided and the introduction written. The University of Pittsburgh Press agreed to publish the work, and in the ensuing months, after I joined with Paolo Parrini to complete the volume, authors were able to revise their works and bring references up to date. We believe that the resulting work is an exciting contribution to the understanding of twentieth-century scientific philosophy.

Generous and indispensable financial support for the meeting in Florence was provided by the National Science Foundation, the Harvey and Leslie Wagner Endowment, the Consiglio Nazionale della Ricerca (C.N.R), the Ministero dell'Università e della Ricerca Scientifica e Tecnologica, and the Università di Firenze. The editors are grateful for that support. They also thank the Center for Philosophy of Science, and its Director J. G. Lennox, the Department of History and Philosophy of Science, Cynthia Miller, Director of the University of Pittsburgh Press, and an anonymous reader for the Press for their advice and support.

Merrilee H. Salmon
University of Pittsburgh

Logical Empiricism

Introduction

PAOLO PARRINI

University of Florence

WESLEY C. SALMON

University of Pittsburgh

We live in an age in which the word "positivism" is used mainly as a term of abuse, both by philosophers and by workers in many other disciplines in the arts and sciences. It is gleefully declared that we live in a "postpositivist age." In a certain sense, this claim is obviously true; the logical positivism that flourished in the Vienna Circle of the 1920s and 1930s has clearly been superseded. Even Hans Reichenbach, who led the group of scientific philosophers in Berlin before Hitler took power in 1933, published his opposition to this brand of logical positivism in *Experience and Prediction* (1938). Nevertheless, two considerations underlie the recent usage of "logical empiricism" to include both the Vienna and the Berlin groups. First, the widespread misunderstanding of "positivism" strongly suggests the wisdom of abandoning this term. Second, the underlying unity of method and spirit that has developed from these early roots firmly supports a comprehensive label. Thus, in this introduction, we follow the precedent of contemporary Viennese scientific philosophers in using the term "logical empiricism" to denote a philosophical movement that grew from nineteenth-century roots early in the twentieth century and that progressed to the scientific philosophy we pursue at the dawn of the twenty-first century. The present book is devoted to a deep investigation of the historical development and present status of this tradition. The reference to contemporary perspectives in the title emphasizes the fact that the majority of the essays involve theoretical considerations as well as historical scholarship.

For decades, logical empiricism has been characterized by the opposition between two distinct philosophical outlooks, which have frequently been called "Continental philosophy" and "analytic philosophy." This terminology is unfortunate for various reasons. First, the English Channel is not a philosophical distinction; it is a geographical fact of nature. To talk of the opposition between "Continental" and "analytic" philosophy is a category mistake.[1] Second, many of the most important roots of "analytic" philosophy grew in Continental soil, especially Vienna and Berlin, where logical empiricism was born.[2] Another major root is, however, in British soil; the empiricist tradition of John Locke, David Hume, John Stuart Mill, and Bertrand Russell also contributes importantly to "analytic" philosophy. Thus, the distinction is geographically unsound. Third, what is generally regarded as "analytic philosophy" split into two distinct branches. In the Anglo-American world, much of "analytic philosophy" took the form of "ordinary language analysis," an approach often at odds with the scientific orientation of logical empiricism. Very roughly, it might be said, logical empiricism found inspiration in the early work of Ludwig Wittgenstein (the *Tractatus Logico-Philosophicus*), while ordinary language analysis related quite directly to his later works (especially *Philosophical Investigations*).[3]

Although we cannot avoid the "Continental/analytic" terminology, a major purpose of this book is to override this inaccurate and inappropriate distinction by examining in detail the development of important twentieth-century movements in philosophy. There are, of course, certain philosophers who are almost universally taken as outstanding representatives of so-called Continental philosophy—for example, Edmund Husserl and Martin Heidegger. Certain others—for example, Moritz Schlick and Rudolf Carnap—obviously represent so-called analytic philosophy (of the scientifically oriented variety).

The essays in part 1, "Turning Points and Fundamental Controversies," lay out some of the main historical features of the philosophical developments to which this book is devoted. They furnish the big picture. Michael Friedman takes us back to Immanuel Kant in order to exhibit the emergence of two distinct versions of *neo*-Kantian philosophy. One of these developed into "Continental philosophy," the other into "analytic philosophy." Gottfried Gabriel pursues a similar theme. Both of these essays deal extensively with the conflict between Carnap and Heidegger. Roberta Lanfredini analyzes the opposition between Schlick and Husserl with regard to the form versus content issue in scientific knowledge. Paolo Parrini's essay in part 7 deals with the closely associated issue of the relationship between reason and perception, a vital problem for post-Kantian philosophy.

Not all roots of "analytic" philosophy grew on Continental soil. The work of Bertrand Russell was greatly admired by logical empiricists on the Continent.[4]

Paolo Parrini and Wesley Salmon

Recall that the opening epigram of Carnap's *Logische Aufbau* is Russell's famous "supreme maxim of scientific philosophizing"—namely, "Wherever possible, logical constructions are to be substituted for inferred entities." A key element of the logical empiricist tradition is the criterion of cognitive significance; its intent is the elimination of traditional metaphysics.[5] This aim has been expressed in a great variety of ways ever since the days of the medieval philosopher William of Ockham. Both Hume and Russell share this goal.[6]

In general, logical empiricists took the term "logical" very seriously. They were strongly influenced by Gottlob Frege and by Whitehead and Russell. Alfred Tarski and many other Polish logicians were influential. Kurt Gödel firmly established results of great significance. S. Awodey and A. W. Carus deal with a dispute between Carnap and Gödel. Like the Carnap-Heidegger controversy, the Carnap-Gödel debate centers on deep metaphysical issues, in particular, the reality of abstract entities. These authors maintain that Gödel's objections do not constitute a decisive refutation of Carnap's views.

Part 1 deals essentially with external relations between logical empiricism and other philosophical movements; part 2, "On the Origin and Development of Logical Empiricism," is internally oriented. Thomas Uebel urges us to take account of the *anti*-Kantian—as opposed to *neo*-Kantian—orientation of "the first Vienna Circle" and its contribution to the later Vienna Circle. The philosopher-scientists of the first Vienna Circle (especially Hans Hahn, Otto Neurath, and Philipp Frank) were positively influenced by such other philosopher-scientists as Ernst Mach and Henri Poincaré. The influence of Kant was overwhelmingly negative. Uebel takes the admission of the German scientific philosopher Carnap to the group as the dividing line between the first Vienna Circle and the later one. George Reisch investigates a deep division within logical positivism—namely, the conflict between Carnap and Neurath. Gereon Wolters, citing an important unpublished statement by Carl. G. Hempel, provides further elaboration of this internal division.

Logical empiricists were influenced by many philosophers who cannot be classified as members of their movement. Wittgenstein is an outstanding example. The logical atomism of Russell and Wittgenstein was taken very seriously by the "second Vienna Circle." Stern deals in depth with Wittgenstein's *Tractatus* and with the exceedingly wide variety of interpretations it has enjoyed over many decades since its publication. The unique influence of the *Tractatus* deserves to be singled out in a part of its own—part 3, "The Riddle of Wittgenstein."

The essays in part 4 take up some of the main themes of part 1, but they do so in the context of technical physics. In particular, the essay of T. A. Ryckman is closely associated with Friedman's; Michael Stöltzner's essay bears a close relation to Uebel's. According to Friedman and Ryckman, Ernst Cassirer is a focal point.

Friedman concludes that Cassirer is an appropriate starting point for understanding the relation between "Continental philosophy" and "analytic philosophy"; Ryckman takes up the issue at that point and studies in detail the conflicting paths taken by Cassirer and Reichenbach (whose relationship to Einstein was closer than that of any other logical empiricist).[7] Our attention is drawn to their conflicting interpretations of the general theory of relativity. Individuals with an interest in history and philosophy of physics, and with some sophistication in physics, would find it profitable to read the contributions of Friedman and Ryckman in direct succession.

Uebel and Stöltzner discuss developments in Vienna, where the influence of Ernst Mach was strong. Unlike Ryckman, Stöltzner focuses on microphysics—that is, the kinetic theory of gases and quantum mechanics. He shows that, before the advent of quantum mechanics as developed by Schrödinger and Heisenberg, an indeterministic world picture was emerging. The work of Ludwig Boltzmann on kinetic theory showed that the second law of thermodynamics is a statistical law that appears deterministic only because of the large numbers of molecules in any sample of gas we might study. In cases where much smaller numbers were involved, the second law might actually be violated. Brownian motion, which had been known for most of the nineteenth century, seemed to be a case in point. While some physicists, for example, Max Planck, resisted an indeterministic interpretation, others, like Philipp Frank, came to accept indeterminism. On this issue, Stöltzner argues, there was no conflict between Mach and Boltzmann. Reichenbach and Richard von Mises strongly supported the indeterministic approach even though they had serious disagreements regarding the concept of probability itself.[8] The arrival of quantum mechanics greatly strengthened the indeterministic position. Stöltzner introduces Franz Serafin Exner as a pivotal participant in the transition from deterministic to indeterministic physics in Vienna. "Exner's synthesis between Mach and Boltzman paved the way to accept genuine indeterminism in physics without any reference to quantum mechanics" (this volume, 199). Individuals with strong interests in history and philosophy of physics, and a modicum of sophistication in physics, would do well to read Uebel's and Stöltzner's essays in direct succession.

Physics was not the only science to which logical empiricists directed their attention; psychology was also a subject of major interest. Indeed, it may be said that Cartesian mind-body dualism posed serious challenges. The essays in part 5, "The Mind-Body Problem," deal with problems in this area. Michael Heidelberger argues against the view that the contemporary debate of the mind-body problem emerged independently in the United States and Australia in the late 1950s. Going back to Gustav Theodor Fechner in the middle of the nineteenth century, he shows how the articulation of the doctrine of psychophysical parallelism "marks a turning point in the history of experimental and quantitative psychology, but also . . . in the history of the mind-body debate" (this volume, 236). The ensuing discussions of psychophysical parallelism led eventually to the neutral monism of Mach and Russell. Moreover, in the early decades of the twentieth century, Schlick and Carnap struggled with exactly the same problems. Just as contemporary philosophy of physics grew from nineteenth-century roots, so also did contemporary philosophy of psychology.

Jaegwon Kim takes up the mind-body problem in the work of logical positivists in the first three decades of the twentieth century—in particular, Schlick, Carnap, Hempel.[9] After pointing to the early phenomenalistic theories of the world, especially in Carnap's *Logische Aufbau*, he discusses the logical behaviorism and physicalism that followed. While logical behaviorism construes the realm of the mental in terms of overt behavior, physicalism admits neurological facts as well. This latter move opens the door to mind-brain identity theories. Kim then points out that in Carnap's work in the 1930s we find a clear anticipation of contemporary functionalist theories. Carnap construes mental events in terms of *dispositions* to exhibit certain types of behavior in certain circumstances, but he also voices the theory that the dispositions themselves may, on the basis of advances in neuroscience, be identified with neurological states. Kim concludes, "Reading the literature again, I was impressed by the metaphysical depth and sophistication in the positivist philosophers, especially Carnap" (this volume, 277).

Scientific rationality, the subject of part 6, was obviously a major focal point of interest for the entire movement of logical empiricism. Carnap and Reichenbach, in particular, saw rationality as intimately connected to probability. Reichenbach's doctoral dissertation (1915) was devoted to this concept,[10] as were several of his later works—most important, *Wahrscheinlichkeitslehre* (1935), *Experience and Prediction* (1938), and *The Theory of Probability* (1949).[11] Carnap turned his attention to probability in the 1940s, and his work of this period culminated in *Logical Foundations of Probability* (1950) and *The Continuum of Inductive Methods* (1952). As Maria Carla Galavotti explains, Reichenbach steadfastly maintained until his death in 1953 that the frequency interpretation is the only legitimate interpretation of the probability calculus. He distanced his conception from that of another logical empiricist, Richard von Mises, who also admitted only a frequency interpretation. Carnap, too, supported the frequency interpretation, but, unlike the others, he firmly denied that it is the only satisfactory concept. Instead, he developed in great technical detail a logical interpretation that he identified with degree of confirmation.

During the 1920s, as Galavotti points out, two important figures who were not identified with logical empiricism began developing a subjective interpretation

—Frank Ramsey in England and Bruno de Finetti in Italy. Both Ramsey and de Finetti made important direct connections between probability and rational decision theory. Reichenbach never took the subjective approach seriously. Carnap, in his later works on probability, was strongly influenced by subjectivists, though he never actually adopted any thoroughgoing form of subjectivism.

Martin Carrier discusses nonprobabilistic theories of confirmation—in particular, the qualitative approaches of Hempel and Clark Glymour. Although Hempel's 1945 theory came to grief on a number of issues—including some substantial difficulties exposed by Carnap (1950)—Carrier shows how Hempel's fundamental notion of confirmation by instantiation was explicitly adopted in a much improved form in Glymour's bootstrapping theory. Carrier explains how the Hempel-Glymour approach contradicts the hypothetico-deductive method as well as the Bayesian theory. Carrier emphasizes both the continuity and the progressive nature of the logical empiricist movement.

Gürol Irzik brings Thomas Kuhn's famous challenge to the logical empiricist concept of scientific rationality into the limelight, and he considers with great care the real versus the merely perceived conflict between Kuhn and Carnap. He exposes common misconceptions of conflict between Kuhn and Carnap showing —roughly speaking—that Kuhn was less of a relativist and Carnap more of a relativist than they have traditionally been pictured. He exposes the inaccuracy of charges of irrationalism leveled against Kuhn, especially those appearing soon after *The Structure of Scientific Revolutions*. He shows how both Carnap and Kuhn differ from Karl Popper regarding the degree of normativity of principles of scientific methodology.

The final section, "Nonlinguistic Empiricism," contains the essays by two of the editors. Paolo Parrini takes denial of the existence of synthetic a priori knowledge as the basic tenet of empiricism. (His coeditor Wesley Salmon agrees.) Looking at Kantian and neo-Kantian defenses of synthetic a priori knowledge, as well as arguments of empiricists against such knowledge, he concludes that the key epistemological issue is the relation of dependence or independence between reason and perception. According to Kant, reason and perception are not independent, because reason imposes general conditions on perceptual knowledge. The role of the synthetic a priori is precisely to characterize these conditions. The problem for the empiricist is to explain how perceptual knowledge can exist without prior constraints of reason.

The early logical empiricists failed to deal adequately with this problem, fearing, perhaps, that careful examination of perception would lead to psychologism —a position they staunchly rejected. Failure to address this issue opened the door to critiques of empiricism by such philosophers as Kuhn, Feyerabend, and Hanson.

Paolo Parrini and Wesley Salmon

Answering such challenges has not been easy. Yet, if one desires to defend empiricism, one must come to terms with this basic issue. Because of the standard account of scientific theories that emerged among logical empiricists, the issue turned into a linguistic one. According to this account, we have two separate languages, the theoretical language and the language of perception. Scientific theories are conceived as axiomatic systems, which, as uninterpreted systems, say nothing about the world. A link to the world is achieved via correspondence rules that connect the perceptual terms to the terms of the theory. The strongest attack against this conception of scientific theories was made by those who claim that all scientific terms, including the so-called observational terms, are theory-laden.[12]

Parrini argues that the fundamental epistemological problem of perception cannot be resolved strictly within a linguistic context. A relationship between language and reality requires something akin to ostensive definition. Citing Schlick's *Allgemeine Erkenntnislehre*, he points out that the kind of knowledge involved is not acquaintance *(kennen)*—that, for Schlick, is not knowledge at all—but rather recognition *(wiedererkennen)*. This requires that humans recognize different objects as being of the same kind—it amounts to the discovery of identity within diversity. How this is achieved is a matter for psychology. The crucial epistemological point is that we are able to do it. This is a matter of fact. The answer avoids psychologism because it leaves the how question to psychology. It turns out, then, that the question of the relation between reason and perception cannot be settled a priori. The answer to this question takes us out of the realm of language and provides the basis for a nonlinguistic form of empiricism. It answers a question that logical empiricists largely left unanswered—namely, How can we establish the independence of perception from reason, thereby sustaining the conclusion that there is no synthetic a priori knowledge?

Wesley Salmon's essay discusses the grounds on which the notorious criterion of cognitive significance was quite generally rejected by logical empiricists and suggests that it occurred mainly for the wrong reasons. He points out that Alonzo Church's famous refutation of A. J. Ayer is actually *totally irrelevant* to the significance criterion; instead, it refutes Ayer's characterization of verification. Salmon's chief complaint—directed at several criticisms—is that confirmation (in the context of verifiability to some degree) was construed exclusively in terms of deductive relations. This is the underlying problem in Ayer's case. Deductive chauvinism strikes again!

One frequent criticism directed against the criterion is that, if applied to itself, it classifies itself as meaningless. This criticism is entirely misguided. The criterion does not pretend to be a cognitive statement; it is intended as a regulation or a rule of procedure. Accordingly, the nonapplicability of empirical evidence is irrelevant;

fundamental rules of this sort require vindication in terms of pragmatic considerations, not validation on the basis of empirical evidence. The justifiability of the criterion hinges on issues of intellectual responsibility. The thesis that the verifiability principle must be considered, not as a cognitive statement, but as rule of method and that, as such, had to be defended on the basis of considerations regarding the type of intellectual culture they were trying to promote—responsible, free, intersubjective, etc.—was defended in Italy by the philosopher Giulio Preti (1911–72) in his numerous essays on logical empiricism. Salmon maintains that knowledge of unobservable objects (frequently designated by the unfortunate term "theoretical entities"—another category mistake) can be achieved by analogical and causal reasoning that depends on the nature of the physical world. Both Salmon and Parrini attempt to show how a viable empiricism is dependent on how the world actually works.

᠄

Logical empiricism dominated epistemology—in philosophy of science specifically and in philosophy more generally—throughout the twentieth century. It has been embraced by some philosophers and rejected by others, but it has served as the main point of reference throughout this time. Its most severe crisis arose around midcentury from the work of such postpositivist philosophers as Kuhn, Hanson, Feyerabend, and Lakatos. Even if we consider their attacks decisive—a point of view we do not consider conclusively established—we believe that empiricism survives at the beginning of the twenty-first century in its nonlinguistic form. In the past thirty years, we have become progressively more aware that the crisis of logical empiricism cannot be considered a crisis for empiricism as such. We hold, moreover, that this nonlinguistic form of empiricism need not be as restrictive as van Fraassen's nonlinguistic empiricism; our view sanctions empirical knowledge of unobservable objects. For logical empiricists and those who hold opposing views, the papers collected in this volume should be proof of the influence exercised by the philosophical spirit that animated the protagonists of logical empiricism.

Notes

1. This point is articulated clearly in Gottfried Gabriel's contribution to this volume.
2. This point is elaborated in Gereon Wolters's contribution to this volume.
3. David Stern's essay on the *Tractatus* searchingly investigates Wittgenstein's influence on "analytic philosophy."
4. We should not forget the role played by the English philosopher Alfred North Whitehead in the authorship of *Principia Mathematica*. Whitehead, not Russell, is the first author of this work.

Paolo Parrini and Wesley Salmon

5. Maria Carla Galavotti's essay exhibits a deep relationship between the criterion of cognitive significance and the work of Carnap and Reichenbach on the concept of probability.

6. David Hume provides the point of departure for Wesley Salmon's essay on the verifiability principle in this volume.

7. Schlick and Reichenbach held closely related interpretations of general relativity, and they interacted constructively on this issue.

8. See Maria Carla Galavotti's essay in part 6 for further information regarding this conflict.

9. The above-mentioned distinction between "scientific analytic philosophy" and "ordinary language analysis" became especially significant in philosophy of psychology. The former approach takes account of empirical work in psychology; the latter attempts to eliminate confusions in our discourse on mind and matter. Kim's essay brings out the distinction between these approaches in his analysis of the distinction between two senses of "definition" found in these two schools of thought.

10. In his "Autobiographical Sketches for Academic Purposes," in M. Reichenbach and R. S. Cohen, eds., 1978, vol. 1 (Dordrecht: Reidel), Reichenbach says, "I came to explore the validity of the laws of probability for reality. . . . Although I had used Kantian philosophy to dress up the solution I had offered for the problems, I am still convinced that the basic idea of this work is very essential; it is one of the foundations of my present conception of the problems of probability" (1). At the time he wrote his dissertation, he was familiar with the ideas of Johannes von Kries.

11. This is an English translation of the 1935 book with much added material.

12. In the mid-1940s, I (W.C.S.) was employed as a laboratory technician at the University of Chicago Metallurgical Laboratories (later Argonne National Laboratory). One of my tasks was to build a galvanometer pier; the instrument we were using was extremely sensitive to external vibrations, such as those caused by trucks passing on the street. My goal was to build a structure that would dampen such vibrations. I began by getting a large wooden box, which I filled almost completely with sand. On the sand, I piled a layer of concrete blocks, and on top of that, a thick layer of newspapers. I continued to make several alternate layers of concrete blocks and newspapers. On the top layer of concrete blocks I put some sort of tabletop (perhaps a piece of plywood; I can't recall). The galvanometer stood on top of the whole thing. While the term "galvanometer" is obviously theoretical, I cannot find any interesting sense in which "wooden box," "sand," "concrete block," "newspaper," and "plywood" are theory laden.

Introduction

I

Turning Points and
Fundamental Controversies

A Turning Point in Philosophy

Carnap-Cassirer-Heidegger

MICHAEL FRIEDMAN
Stanford University

In the early 1990s I began a project focused on two episodes in the development of early-twentieth-century philosophy: the famous Davos disputation between Martin Heidegger and Ernst Cassirer in 1929, and Rudolf Carnap's notorious criticism of Heidegger (and the statement "nothingness itself nothings *[das Nichts selbst nichtet]"*) in his paper "Überwindung der Metaphysik durch logische Analyse der Sprache," published in 1932.[1] I was surprised and fascinated to learn (from Thomas Uebel) that Carnap had attended the Davos disputation; and, when I looked into the relevant parts of Carnap's diary where he reports on this period, I learned that he had met Heidegger at Davos, taken a very serious interest in *Being and Time* when he returned to Vienna, and had in fact written (and delivered) drafts of "Überwindung der Metaphysik" directly in the wake of this experience. I also found, not surprisingly, that the issues at stake in Carnap's criticism (for both him and Heidegger) were charged with social and political significance reflecting the deep and pervasive cultural struggles of the late Weimar period. Indeed, shortly after the Nazi seizure of power in 1933 (during which, as is well known, Heidegger assumed the rectorship at Freiburg while publicly embracing the new regime) both Cassirer and Carnap emigrated to the New World, and Heidegger was the only active philosopher of the first rank to remain on the Continent.

It occurred to me, then, that the Davos encounter between Carnap, Cassirer, and Heidegger had particular importance for our understanding of the ensuing split between what we now call the analytic and Continental philosophical traditions. Before this encounter there was no such split—at least within the German-speaking world. Logical empiricism, Husserlian phenomenology, neo-Kantianism, and Heidegger's new existential-hermeneutical variant of phenomenology were rather engaged in a fascinating series of philosophical exchanges and struggles—all addressed to the revolutionary changes that were then sweeping both the *Naturwissenschaften* and the *Geisteswissenschaften*. The differing philosophical movements of course disagreed with and opposed one another about the interpretation and significance of these revolutionary changes, but they still spoke the same language and actively engaged one another on a common set of philosophical problems. Moreover, since the Davos disputation itself concerned the fate of neo-Kantianism and the proper interpretation of Kant himself (with Heidegger taking as his main target the Marburg School of neo-Kantianism with which Cassirer was closely associated), it further occurred to me that examining the differing ways in which all three philosophers (Carnap, Cassirer, and Heidegger) evolved in sharply diverging directions from a common neo-Kantian heritage might greatly illuminate the nature and sources of the analytic/Continental divide. This is the "turning point in philosophy" I refer to in the title.

In the early 1990s I completed approximately the first half of this project, covering the encounter at Davos and the relationship between Carnap and Heidegger in the wake of this encounter. I published an abbreviated version of this work as "Overcoming Metaphysics: Carnap and Heidegger" in 1996.[2] I have now completed a monograph further developing this phase of the project and also adding an extended treatment of Cassirer, including his relationship to both Heidegger and Carnap.[3] What I want to do here is present some of the newer ideas from the second part of the monograph, especially concerning the relationship between Carnap and Cassirer. In order properly to frame this material I will first briefly summarize some of the main themes from the first part.

At the time of the Davos disputation Cassirer and Heidegger were arguably the two leading philosophers in Germany. Cassirer was the most eminent active Kant scholar and the editor of the then standard edition of Kant's works; he had just completed his own magnum opus, *The Philosophy of Symbolic Forms*. Heidegger had recently published *Being and Time* and was in the process of taking Edmund Husserl's place as the leader of the phenomenological movement. The interpretation of Kant Heidegger presented there (in explicit opposition to Marburg neo-Kantianism) aimed to show that the *Critique of Pure Reason* does not present a theory of mathematical natural scientific knowledge. The real contribution of the *Critique* is rather to work out, for the first time, the problem of the laying of the ground for metaphysics. On this reading, Kant argues (in remarkable agreement with the main argument of *Being and Time*) that metaphysics can only be grounded in a prior analysis of the nature of *finite* human reason. Moreover, Kant's introduction of the so-called transcendental schematism of the understanding has the effect of dissolving both sensibility and the understanding in a "common root"— namely, the transcendental imagination, whose ultimate basis (again in remarkable agreement with the argument of *Being and Time*) is temporality. And this implies, finally, that the traditional basis of Western metaphysics in logos, *Geist*, or reason is definitively destroyed. Cassirer, for his part, strongly opposes such a renunciation of reason. Although man as the "symbolic animal" must begin with the transcendental imagination and thus with finitude, it is also clear, as Kant himself has shown, that the finite human creature can nevertheless break free from finitude into the realm of objectively valid, necessary and eternal truths both in moral experience and in mathematical natural science.[4]

Carnap's criticism of Heidegger, although not expressed in the same language, brings up closely related issues. The main problem with "nothingness itself nothings," for Carnap, is a violation of the logical form of the concept of nothing. Heidegger uses the concept both as a substantive and as a verb, whereas modern logic has shown that it is neither—its logical form is constituted solely by existential quantification and negation. Yet Carnap clearly recognizes, at the same time, that this kind of criticism would not affect Heidegger himself in the slightest; for the real issue between the two lies in the circumstance that Heidegger denies while Carnap affirms the philosophical centrality of logic and the exact sciences. Carnap accordingly refers explicitly to such Heideggerian passages as the following:

> Nothingness is the source of negation, not vice versa. If the power of the understanding in the field of questions concerning nothingness and being is thus broken, then the fate of the dominion of "logic" within philosophy is also decided therewith. The idea of "logic" itself dissolves in a vortex of more original questioning. . . .
>
> The supposed soberness and superiority of science becomes ridiculous if it does not take nothingness seriously. Only because nothingness is manifest can science make what exists itself into an object of investigation. Only if science takes its existence from metaphysics can it always reclaim anew its essential task, which does not consist in the accumulation and ordering of objects of acquaintance but in the ever to be newly accomplished disclosure of the entire expanse of truth of nature and history. . . .
>
> Therefore no rigor of a science can attain the seriousness of metaphysics. Philosophy can never be measured by the standard of the idea of science.[5]

Moreover, Heidegger, in what looks like a response to Carnap in 1943, reiterates the same point in even stronger terms:

The suspicion directed against "logic," whose conclusive degeneration may be seen in logistic [modern mathematical logic], arises from the knowledge of that thinking that finds its source in the truth of being, but not in the consideration of the objectivity of what exists. Exact thinking is never the most rigorous thinking, if rigor *[Strenge]* receives its essence otherwise from the mode of strenuousness *[Anstrengung]* with which knowledge always maintains the relation to what is essential in what exists. Exact thinking ties itself down solely in calculation with what exists and serves this [end] exclusively.[6]

It is clear, then, that Heidegger and Carnap are actually in remarkable agreement. "Metaphysical" thought of the type Heidegger is trying to awaken is possible only the basis of a prior overthrow of the authority and primacy of logic and the exact sciences. The difference is that Heidegger eagerly embraces such an overthrow, whereas Carnap is determined to resist it at all costs.

The philosophical issues between Carnap and Heidegger, and also between Heidegger and Cassirer, are thus based, in the end, on a stark and profound disagreement over the nature and centrality of logic, mathematics, and the mathematical exact sciences. For both Carnap and Cassirer these are of central philosophical significance indeed, whereas Heidegger is self-consciously looking for a source of philosophical significance lying at a much "deeper" and less "rational" level. In tracing out the roots of this fundamental disagreement, it turns out, we need to return to the issues about neo-Kantianism and the "transcendental schematism of the understanding" raised in the Cassirer-Heidegger disputation at Davos.

The first point to notice is that all three men—Carnap, Cassirer, and Heidegger —were philosophically trained within the neo-Kantian tradition that dominated the German-speaking world at the end of the nineteenth and beginning of the twentieth century. At the time, there were two main distinguishable versions: the Marburg School, founded by Hermann Cohen and continued by Paul Natorp and (at least until about 1920) Cassirer himself, and the Southwest School, founded by Wilhelm Windelband and systematically developed by Heinrich Rickert. Heidegger completed his habilitation under Rickert at Freiburg (before the latter succeeded Windelband at Heidelberg). Carnap, for his part, studied Kant at Jena with Bruno Bauch—another student of Rickert's from Freiburg—and, in fact, wrote his doctoral dissertation under Bauch. It is clear, moreover, that Carnap carefully studied both versions of neo-Kantianism, including the writings of Natorp, Cassirer, and Rickert.

Common to both versions is a certain conception of epistemology and the object of knowledge inherited from Kant. Our knowledge or true judgments should not be construed as representing objects or entities that exist independently of our judgments, whether these independent entities are the "transcendent" objects of the metaphysical realist existing somehow "behind" our sense experience or the naked, unconceptualized sense experience itself beloved of the empiricist. Rather, following Kant's "Copernican Revolution," the object of knowledge does not exist independently of our judgments at all. On the contrary, this object is first created or "constituted" when the unconceptualized data of sense are organized or framed within the a priori logical structures of judgment itself. In this way, the initially unconceptualized data of sense are brought under a priori "categories" and thus first become capable of empirical objectivity.

Yet there is a crucially important difference between this neo-Kantian account of the object of knowledge and judgment and Kant's original account. For Kant, we cannot explain how the object of knowledge becomes possible on the basis of the a priori logical structures of judgment alone. We need additional a priori structures that mediate between the pure forms of judgment comprising what Kant calls general logic and the unconceptualized manifold of impressions supplied by the senses. These mediating structures are the pure forms of sensible intuition, space and time. Thus the pure logical forms of judgment only become categories when they are "schematized"—when they are given a determinate spatiotemporal content in relation to the pure forms of sensible intuition. The pure logical form of a categorical judgment, for example, becomes the category of *substance* when it is schematized in terms of the temporal representation of permanence; the pure logical form of a hypothetical judgment becomes the category of *causality* when it is schematized in terms of the temporal representation of succession; and so on. For Kant, then, pure formal logic (general logic) must, if it is to play an epistemological role, be supplemented by what he calls transcendental logic—with the theory of how logical forms become schematized in terms of pure spatiotemporal representations belonging to the independent faculty of pure intuition. And it is precisely this account of schematization, in fact, that forms the heart of the transcendental analytic of the *Critique of Pure Reason*—the so-called metaphysical and transcendental deductions of the categories.

But both versions of neo-Kantianism, following the tradition of post-Kantian idealism more generally, entirely reject the idea of an independent faculty of pure intuition. The a priori formal structures in virtue of which the object of knowledge becomes possible must therefore derive from the logical faculty of the understanding and from this faculty alone. Since space and time no longer function as independent forms of pure sensibility, the constitution of experience described by "transcendental logic" must now proceed on the basis of purely conceptual—*and thus essentially non-spatiotemporal*—a priori structures. And it is this last feature of their epistemological conception, moreover, that associates the neo-Kantians with the nineteenth-century tradition of "pure logic *[reine Logik]*" represented by Herbart, Bolzano, and Lotze, a tradition culminating with the polemic against psychologism of Husserl's *Logical Investigations* (1900).

How, then, do we explain the constitution of the object of empirical knowledge within this revised form of Kantianism? Here the two different traditions —Marburg and Southwest—fundamentally diverge. On the so-called genetic conception of knowledge favored by the Marburg School, the object of empirical knowledge is explained as the never completed "X" toward which the methodological progress of mathematical natural science is converging. In Cassirer's sophisticated presentation of this conception in *Substance and Function* (1910), for example, pure mathematics is represented by the totality of what we now call relational structures (the number series, the structure of Euclidean space, and so on) described by the modern logical theory of relations developed in Bertrand Russell's *Principles of Mathematics* (1903). In mathematical natural science, however, we develop a particular ordered sequence of such structures—representing the historical-methodological evolution of this science—which is never complete but yet converging. The empirical object itself is then simply defined as the ideal limit structure (or limit theory) toward which this historical-methodological sequence is converging. The Marburg School thus advocates what we might call a "logicization" of the object of empirical knowledge, and it is no wonder, then, that this view becomes known as "logical idealism."

In the tradition of the Southwest School, by contrast, logic is identified with traditional Aristotelian syllogistic and is sharply and explicitly distinguished from mathematics. Moreover, we also follow Kant, in this tradition, in separating the logical forms of judgment, on the one side, from the unconceptualized manifold of sense-impressions, on the other. (In the Marburg tradition there is no such unconceptualized manifold.) Yet, since we have deliberately rejected the mathematical intermediary between these two sides developed by Kant himself—the pure forms of sensible intuition—overwhelming problems arise within the Southwest School in explaining how the pure forms of judgment can possibly serve to constitute the object of empirical knowledge. These problems become especially sharp and explicit in the work of Emil Lask, a brilliant student of Rickert's who perished in the Great War in 1915. What Lask does, in essence, is entirely reject the project of Kant's metaphysical deduction aiming to derive the categories constitutive of experience from the logical forms of judgment. For Lask, what is fundamental is the concrete, already categorized real object of experience itself, and the subject matter of formal logic only arises subsequently in an artificial process of abstraction, by which the originally unitary categorized object is broken down into form and matter, subject and predicate, and so on. Moreover, since this comes about due to a fundamental weakness or peculiarity of our human understanding—our inability to grasp the unitary categorized object as a unity—the entire realm of "pure logic" is nothing but an artifact of our subjectivity possessing no explanatory power whatsoever.

Turning Points and Fundamental Controversies

Now Carnap, in the *Aufbau,* can be seen as continuing, and radicalizing, the Marburg epistemological tradition. Carnap makes this kinship particularly explicit, in fact, when he first introduces the basic relation of what he calls his "constitutional system":

> Cassirer *([Substanzbegr]* 292ff.) has shown that a science having the goal of determining the individual through lawful interconnections *[Gesetzseszusammenhänge]* without its individuality being lost must apply, not class ("species") concepts, but rather *relational concepts;* for the latter can lead to the formation of series and thereby to the establishing of order-systems. It hereby also results that relations are necessary as first posits, since one can in fact easily make the transition from relations to classes, whereas the contrary procedure is only possible in a very limited measure.
>
> The merit of having discovered the necessary basis of the constitutional system thereby belongs to two entirely different, and often mutually hostile, philosophical tendencies. *Positivism* has stressed that the sole *material* for cognition lies in the undigested *[unverarbeitet]* experiential *given;* here is to be sought the *basic elements* of the constitutional system. *Transcendental idealism,* however, especially the neo-Kantian tendency (Rickert, Cassirer, Bauch), has rightly emphasized that these elements do not suffice; *order-posits [Ordnungssetzungen]* must be added, our "basic relations."[7]

Carnap then develops, by an elaborate series of logical constructions, a sequence of definitions ordered in the type-hierarchy of Whitehead's and Russell's *Principia Mathematica* (1910–13), proceeding from the private or *autopsychological* realm of a single cognitive subject, to a world of public external objects constituting the *physical* realm, and finally to the intersubjective and thus cultural realities belonging to the *heteropsychological* realm. In this way, all objects of empirical science (starting with empirical psychology and ending with sociology and history) are step-wise constructed or "constituted" by logical means, and this conception, as Carnap himself notes, is indeed closely analogous to that of the Marburg School:

> Constitutional theory and *transcendental idealism* agree in representing the following position: all objects of cognition are constituted (in idealistic language, are "generated in thought"); and, moreover, the constituted objects are only objects of cognition *qua* logical forms constructed in a determinate way. This holds ultimately also for the basic elements of the constitutional system. They are, to be sure, taken as basis as unanalyzed unities, but they are then furnished with various properties and analyzed into (quasi-) constituents (§116); first hereby, and thus also first as constituted objects, do they become objects of cognition properly speaking—and, indeed, objects of psychology.[8]

Indeed, there is one important respect in which Carnap's conception is even more radical than that of the Marburg School. Cassirer's *Substance and Function,* for example, retains a significant element of dualism between pure thought and empirical reality—the contrast between the pure relational structures of pure logic and mathematics, on the one side, and the historical sequence of successor theories representing the methodological progress of empirical natural science, on the other. For Carnap, by contrast, empirical reality simply *is* a particular logical structure, a type-theoretic structure (representing the epistemological progress of an initial cognitive subject) erected on the basis of a single primitive nonlogical relation—a structure, moreover, in which each empirical object is *completely* defined or constructed at a definite finite level within the hierarchy of logical types.

Just as Carnap, in the *Aufbau*, can be seen as attempting to realize the philosophical ambitions of the Marburg School using the new mathematical logic of *Principia Mathematica*, Heidegger can be seen as attempting to resolve the outstanding problems of the Southwest School using the new phenomenological method due to Husserl. Thus Heidegger, in *Being in Time*, explicitly rejects the "Copernican Revolution" favored by both neo-Kantian schools in favor a "direct realist" conception of truth and relation to an object derived from the Husserlian notion of "identification" and the work of Emil Lask. Objects, in their "disclosedness," are prior to all valid judgments and thus prior, in particular, to the notion of logical form. Indeed, Dasein's most fundamental relation to the world is not cognitive at all, but rather one of either "authentic" or "inauthentic" existence, in which Dasein's own peculiar mode of being (that is, "being-in-the-world") is itself either disclosed or covered over. In this way, Heidegger further radicalizes Husserlian phenomenology by incorporating the historically oriented *Lebensphilosophie* of Wilhelm Dilthey, according to which the true "living subject" (in contrast to Husserlian "pure consciousness") is both concrete and fundamentally historical. Heidegger's version of "direct realism" is thus only possible on the basis of what he calls the "historicity" of Dasein, and so all truth, in the end, must be seen as historically relative. Just as pure formal logic, for Lask, has lost all explanatory power even to begin the constitution of any actual empirical object, pure formal logic, from Heidegger's point of view, is similarly irrelevant to the "existential analytic" of Dasein. This analytic must rather rest on a deeper, existential-hermeneutic analysis wherein both logic in particular and the notion of valid judgment in general emerge as decidedly secondary and "derivative" phenomena. Logical objectivity itself must be subordinated in principle to the analytic of Dasein and its historicity.

I hope to have now given you a sense of how the fundamental disagreement between Carnap and Heidegger traces back to the deep fissures that had meanwhile appeared (and were philosophically expressed in diverging directions in the two leading schools of neo-Kantianism) in Kant's original conception of how the pure logical activities of the understanding meet the given empirical data supplied by the senses. I would now like to discuss, against this background, how Cassirer's further development, throughout the 1920s, of his philosophy of symbolic forms is similarly framed within, and ultimately severely challenged by, this same post-Kantian predicament.

As we have seen, Cassirer's *Substance and Function* (1910) is perhaps the most sophisticated expression of the classical Marburg tradition. Kant's "transcendental method" is seen as beginning with the "fact of science"—the fact, that is, of the existence of mathematical exact science—and then seeking, by a regressive argument, for the conditions of the possibility of this fact. No longer exclusively tied to the Newtonian mathematical physics that provided Kant himself with his only model, however, science is now seen as an essentially dynamical process in which the fundamental mathematical structures employed in physics are continually revised without end. There is, nonetheless, a necessary element of convergence in this process of continual revision; and it is precisely this convergence that now underwrites the objectivity of the entire developmental sequence and also serves to "constitute" the empirical correlate to which it "corresponds." This is now simply defined, according to the genetic conception of knowledge, as the ideal limit structure toward which the sequence in question is converging.

In the late teens and early twenties, however, Cassirer became convinced that this classical Marburg conception of knowledge is much too narrow. And he attempted to extend it, accordingly, to include the place of mathematical natural science in particular within a much wider conception of the development of human culture as a whole. Science, along with language, art, myth, religion, and so on is now seen as just one "symbolic form" among others, with no exclusive claim to objective validity. On the contrary, the notion of objective validity itself can now be properly understood only when we appreciate the differing claims to such validity expressed in the different symbolic forms, and when we then further grasp how the *totality* of symbolic forms are interrelated. The task of transcendental philosophy is now no longer simply oriented to the fact of science, but rather to the much more general "fact of culture" as a whole.

It is important to be clear about the precise nature of Cassirer's break with neo-Kantianism here. For the Marburg neo-Kantians, like Kant himself, in no way limited philosophy to the study of scientific knowledge. Just as Kant himself had done, for example, Cohen wrote works on ethics, aesthetics, and religion as well. Indeed, because of their overarching focus on the genetic conception of knowledge, it was especially natural for the Marburg School to incorporate ethics, aesthetics, and religion within a broadly Kantian framework. For Kant, scientific knowledge is framed by what he calls the regulative use of reason, through which reason is guided but not constrained by the idea of the unconditioned in pursuing the never to be attained goal of an ideally complete scientific description of the world. In the *Critique of Judgment* (1790) Kant then incorporates aesthetics within this framework as a distinctive exercise of "reflective judgement" (which is closely connected with the regulative use of reason) and secures the unity of practical and theoretical reason by arguing that the idea of the unconditioned, which is wholly indeterminate as far as theoretical reason is unconcerned, first receives determinate meaning and content through the moral law as a product of pure practical reason. The full realization of the moral law in the "idea of the highest good" thereby secures the unity of theoretical, practical, and aesthetic judgment (and religion as well) as a "focus imaginarius" or "infinitely distant point." Ethics, aesthetic, and religion thus emerge via a teleological extension, as it were, of the never ending genetic procedure characteristic of natural scientific knowledge.

The philosophy of symbolic forms is distinguished from traditional Kantian and neo-Kantian attempts to incorporate both scientific and nonscientific modes of thought within a single philosophical framework by its emphasis on more *primitive* forms of world-presentation—on the ordinary perceptual awareness of the world expressed, for Cassirer, primarily in natural language, and, above all, on the mythical view of the world lying at the most primitive level of all. These more primitive forms do not arise through any kind of extension or completion of the scientific form. On the contrary, they lie at a deeper level of spiritual life, which then gives rise to more sophisticated forms by a developmental process taking its starting point from this given basis. From mythical thought develops religion and art, from natural language develops theoretical science. In place of a teleological structure, therefore, we have what has been aptly called a "centrifugal" structure, as the more primitive forms give birth to the more sophisticated forms arranged around a common origin and center.[9] And it is precisely here, in particular, that Cassirer appeals to "romantic" philosophical tendencies lying outside the Kantian and neo-Kantian traditions—to speculations about the origins of language and human culture in Vico and Herder, to the pioneering work in the comparative study of languages and cultures by Wilhelm von Humboldt, to the *naturphilosophische* and aesthetic ideals of Goethe. It is here that he supplements Kant with an historical dialectic self-consciously derived from Hegel (the title of the third volume is *The Phenomenology of Knowledge*, which, as Cassirer carefully explains, is taken from Hegel rather than modern Husserlian phenomenology), and it is here that he comes to terms with the contemporary *Lebensphilosophie* of Dilthey, Bergson, Scheler, and Simmel.

Cassirer's dialectical development of more sophisticated symbolic forms from more primitive forms runs through three fundamental types or "functions" of symbolic meaning. The expressive function *(Ausdrucksfunktion)* of symbolic meaning characterizes the most primitive level of mythical thought. Here we experience events in the world around us as charged with affective and emotional significance, as desirable or hateful, comforting or threatening. We thereby experience reality as a fleeting complex of events bound together by their affective and emotional "physiognomic" characters, rather than as a world of stable and enduring substances manifesting themselves from various points of view and on various occasions. In mythical thought, therefore, there is no essential distinction between appearance and reality. This distinction only emerges at the next stage, that of the representative function *(Darstellungsfunktion)* of symbolic meaning, which then has the task of precipitating out of the original mythical flux of "physiognomic" characters a world of stable and enduring substances, distinguishable and reidentifiable as such. And it is primarily through the medium of natural language, according to Cassirer, that we construct what he calls the "intuitive world" of ordinary sense perception, on the basis of what he calls intuitive space and intuitive time. We distinguish the enduring thing-substance from its variable manifestations from different points of view (in intuitive space) and on different occasions (in intuitive time), and we thereby arrive at a fundamental distinction between truth and falsehood, appearance and reality, which is expressed in its most developed form, for Cassirer, in the propositional copula.

This distinction between appearance and reality, as expressed in the propositional copula, then leads naturally to a new task of thought, the task of theoretical science, of systematic inquiry into the realm of truths. Here we encounter the third and final function of symbolic meaning, the *significative* function *(Bedeutungsfunktion)*, which is exhibited most clearly, according to Cassirer, in the "pure category of relation." For it is precisely here, in the theoretical view of the world, that the pure relational concepts characteristic of modern mathematics, logic, and mathematical physics are finally freed from the bounds of sensible intuition. Mathematical space and time arise from intuitive space and time, for example, when we abstract from all demonstrative relation to a "here and now" and consider instead the single system of relations in which all possible "here-and-now" points are embedded; the mathematical system of the natural numbers arises when we abstract from all concrete applications of counting and consider instead the single potentially infinite progression wherein all possible applications of counting are comprehended, and so on. The result, in the end, is the world of modern mathematical physics described in Cassirer's earlier scientific works—a pure system of mathematical relations where, in particular, the intuitive concept of substantial thing has finally been replaced by the relational-functional concept of universal law.

In constructing this mathematical-physical world of pure significance, moreover, we need a fundamentally new type of symbolic instrument—namely, that which Leibniz had much earlier envisioned in the form of a "universal characteristic":

This task would not be achievable if thought, in posing it, did not simultaneously create a new *organ* for it. It can no longer remain with those formations that the intuitive world brings to it ready made, as it were, but it must rather make the transition to *constructing [aufzubauen]* a realm of symbols in complete freedom, in pure self-activity. . . . A complex of *signs* now underlies the system of relations and of conceptual significance—a complex of signs that is built up in such a way that the interconnections holding between the individual elements of the latter system can be surveyed and read off from it. . . . Alongside of the *"Scientia generalis"* a *"Characteristica generalis"* is required. The work of *language* continues in this characteristic, but it simultaneously enters into a new logical dimension. For the signs of the characteristic have divested themselves of everything merely expressive, and, indeed, of everything intuitively-representative: they have become pure "signs of significance *[Bedeutungszeichen]*." We are thereby presented with a new type of "objective" meaning relation that is specifically different from every kind of "relation to an object" subsisting in perception or in empirical intuition.[10]

The language of mathematical-physical theory thereby transcends the stages of expressive and representative meaning exhibited in the mythical and intuitive worlds, and we thereby finally attain the stage of "purely symbolic" significative meaning. All "picturing," that is, must be replaced by the purely logical *coordination (Zuordnung)* described in Helmholtz's theory of signs, and this project, as first clearly envisioned by Leibniz, finds its most precise and exact fulfillment in modern mathematical logic.

By here endorsing a purely formal characterization of mathematical-physical representation in the tradition of Leibniz's "universal characteristic" and Helmholtz's theory of signs, Cassirer has moved quite far indeed from the original Kantian point of view and has come extremely close, in fact, to the position of Carnap and the logical positivists. The difference, however, is that this kind of mathematical-physical representation (what Cassirer calls "significative" or "purely symbolic" meaning) is, for Cassirer, only one type of objective symbolic meaning among others. And it can only be properly understood, according to Cassirer, when it is placed in the developmental or dialectical process we have just described—as something that necessarily evolves from the more primitive stages of expressive and representative meaning. Indeed, it is only in this way, according to Cassirer, that we can properly understand the respective claims to objective validity and truth within both the *Naturwissenschaften* and the *Geisteswissenschaften*—which Cassirer himself prefers to refer to as the *Kulturwissenschaften*.

It is precisely here that we encounter the sole critical remarks on Carnap's philosophy that Cassirer ever published, which occur in *Zur Logik der Kulturwissenschaften* (1942). Cassirer argues that expressive perception (*Ausdruckswahrnehmen*), based on the expressive function of symbolic meaning, is just as fundamental as thing perception *(Dingwahrnehmung)*, based on the representative function of symbolic meaning. Neither can be reduced to the other, both are "primary phenomena *[Urphänomene]*." Moreover, whereas the physical sciences take their evidential base from the sphere of thing perception, the cultural sciences take theirs from the sphere of expressive perception and, more specifically, from the fundamental experience of other human beings as fellow selves sharing a common intersubjective world of "cultural meanings." And it is at this point that Cassirer explicitly challenges the "physicalism" of the Vienna Circle, as articulated, in particular, in Carnap's *Pseudoproblems in Philosophy* (1928) and *The Unity of Science* (1932). What Cassirer want to oppose, of course, is the idea (originally developed in the *Aufbau*) that the domain of the "heteropsychological" must be constructed or constituted from the domain of the physical, from the purely physical phenomena of bodily motion, sign production, and the like.

After citing a claim from Carnap to the effect that the language of physics is the only intersubjective language, Cassirer suggests that this conclusion, if strictly carried through, would entail the elimination of the genuinely cultural sciences. The only escape from this dilemma is to acknowledge, on the contrary, the independence and autonomy of *all* symbolic functions: "We must strive, without reservation or epistemological dogma, to understand each type of language in its own particular character—the language of science, the language of art, of religion, and so on; we must determine how much each contributes to the construction of a 'common world.'"[11] Acknowledging the autonomy of the expressive function, in particular, then allows us to grant the cultural sciences their own authority and autonomy. When we see that all "cultural meaning" is based, in the end, on the equally fundamental "primary phenomenon" of expression, there is no longer any reason to question the legitimacy of the cultural sciences from the point of view of the physical sciences, which are themselves based on the "primary phenomenon" of representation.

Now Carnap, in the *Aufbau*, has no intention of questioning the legitimacy of the cultural sciences. These sciences acquire their own, *relatively* autonomous domain immediately after the construction of the heteropsychological realm. At the same time, however, Carnap does maintain the privileged position of the physical sciences, and he does say, in particular, that only the world of physics provides for the possibility of a "univocal, consistent intersubjectivization." But Carnap does not base this claim on any "naive empiricist" prejudice privileging "thing perception" over "expressive perception." Carnap has no such philosophical views on the ultimate nature of perception at all. Rather, the language of physics is privileged because its exactness and precision of mathematical representation make it an exemplary vehicle for the kind of purely formal (purely structural) meaning Carnap hopes to capture in his fundamental method of logical construction: *"Every scientific statement can in principle be so transformed that it is only a structural statement. But this transformation is not only possible, but required. For science wants to speak about the objective; however, everything that does not belong to structure but to the material, everything that is ostended concretely, is in the end subjective. In physics we easily recognize this desubjectivization, which has transformed almost all physical concepts into pure structure concepts."*[12] Mathematical physics is privileged, for Carnap, because it provides the most highly developed example of what Cassirer calls the *significative* function of symbolic meaning.

Cassirer himself of course agrees completely with this idea. He, too, holds that modern mathematical physics, as expressed especially within the language of modern mathematics and mathematical logic, presents us with the most highly developed form of the significative function of symbolic meaning. But the whole point of the philosophy of symbolic forms is that objectivity as such, intersubjective validity and communicability, is by no means confined to the significative function. Physical science has its own characteristic type of objectivity, expressed in universally valid mathematical laws holding for all times and all places. In the cultural sciences, however, we have access to a different but analogous type of objectivity, expressed in our ability continually to interpret and reinterpret human products or "works" from our own particular point of view within the historical development of culture:

> The constancy required for this purpose is not that of properties or laws, but rather that of meanings *[Bedeutungen]*. The more culture develops, the more particular domains [there are] in which it disperses itself, the more richly and variously this world of meanings forms itself. We live in the words of language, in the forms of poetry and plastic art, in the forms of music, in the structures of religious representation and religious belief. And only within these forms do we "know" one another.... The goal [of cultural science] is not the universality of laws, but neither is it the [mere] individuality of facts and phenomenon. In contrast to both it establishes its own ideal of knowledge. What it wishes to know is the *totality of forms* in which human life is realized. These forms are infinitely differentiated, but they do not lack unified structure. For it is ultimately the "same" human being that we always continually encounter in the development of culture, in thousands of manifestations and in thousands of masks.[13]

The constancy and objectivity in question, however, are not given, as it were, in the facts and phenomena of human cultural history itself. They are rather *made* by the philosophical history of culture, as it *interprets* those phenomena from precisely this point of view.

The problem for Cassirer, however, is to explain more fully how the totality of symbolic forms in fact has a "unified structure" making possible a *common* claim to universal validity. In what precise sense is it the "same" human being encountered in *all* the symbolic functions? This problem is especially challenging and urgent for Cassirer, because he continues to maintain, in good Marburg style, that the clearest and, as it were, "highest" form of universal validity is given by the language of mathematical exact science:

> And with this transition [to pure significative meaning] the realm of proper or rigorous "science" first opens up. In *its* symbolic signs and concepts everything that possesses mere expressive value is extinguished. Here there is to be no longer any individual subject, but only the thing itself is to "speak." . . . However, what the formula [of this language] lacks in closeness to life and in individual fullness —this is now made up, on the other side, by its universality, by its scope and its universal validity. In this universality not only individual but also national differences are overcome. The plural concept of "languages" no longer holds sway: it is pushed aside and replaced by the thought of the *characteristica universalis,* which now enters the scene as *"lingua universalis."*
>
> And now we hereby first stand at the birthplace of mathematical knowledge and mathematical natural scientific knowledge. From the standpoint of our general problem we can say that this knowledge begins at the point where thought breaks through the veil of language—not, however, in order to appear entirely unveiled, to appear devoid of all symbolic clothing, but rather in order to enter into an in principle different symbolic form.[14]

This commitment to the originally Leibnizean ideal of a truly universal, transnational, and transhistorical system of communication, exemplified by the logical-mathematical language of exact scientific thought, is enthusiastically embraced by Carnap as well. It constitutes, in fact, the very core and basis of Carnap's whole philosophical orientation. The crucial difference between the two is that, whereas Carnap limits such truly universal intersubjective communicability to precisely that which is expressible in rigorous logical notation, Cassirer wants to extend it to all the other symbolic forms as well.

But there are deep systematic difficulties standing in the way of this ambition. In the original Kantian architectonic, what Kant himself calls pure general logic (traditional Aristotelian formal logic), frames the entire system at the highest level, in that the traditional logical theory of concepts, judgments, and inferences supplies the formal systematic scaffolding on which Kant's comprehensive synthesis is constructed. The logical forms of concepts and judgments, when schematized by the faculty of sensibility, generate both the table of categories and the system of principles, which in turn underlie Kant's "constitutive" theory of human sensible experience of the phenomenal world as made possible in pure mathematics and pure natural science. These same logical forms, considered independently of the faculty of sensibility, then generate the concept of the noumenon, which remains merely "problematic," however, from a theoretical point of view. Moreover, the basic logical forms of (syllogistic) inference, again considered independently of sensibility, generate the idea of the unconditioned, which also remains "indeterminate" from a theoretical point of view, although, nonetheless, it possesses positive guiding force in the "regulative use of reason." This same faculty of reason, finally, when applied to the determination of the will, also generates the moral law as a product of pure *practical* reason. Here the idea of the unconditioned receives positive "constitutive" content, which can then set a definite teleological goal (the ideal of the highest good) as the highest guiding principle of *all* regulative activity, both practical and theoretical.

The most fundamental problem created by the post-Kantian rejection of the central Kantian distinction between sensible and intellectual faculties, then, lies in the destruction of this intricate architectonic. It is no longer possible, in particular, to view pure formal logic as the most clearly and uncontroversially universal form of human thinking *and* as the framework for a comprehensive philosophy of the whole of our intellectual and cultural life. Cassirer's herculean efforts to construct a similarly comprehensive philosophy of symbolic forms make this problem especially clear. For, whereas Cassirer perseveres in the idea that pure formal logic (in its modern, post-Kantian guise as a sufficient and adequate language for all of mathematics and exact science) provides us with the paradigm of truly universal intersubjective communicability as well, he also wants to maintain a complementary but still universal intersubjective validity in the essentially nonmathematical cultural sciences. He never satisfactorily explains, however, how these two characteristically different types of validity are related, and he never makes clear, in particular, how truly universal, transcultural validity is possible outside the mathematical exact sciences. Thus, if Cassirer cannot make good on the idea of an underlying unity for the totality of symbolic forms, it appears that we are finally left with the fundamental philosophical dilemma presented by Carnap and Heidegger after all. We can either, with Carnap, hold fast to formal logic as the ideal of universal validity and confine ourselves, accordingly, to the philosophy of the mathematical exact sciences, or we can, with Heidegger, cut ourselves off from logic and "exact thinking" generally, with the result that we ultimately renounce the ideal of truly universal validity itself.

If I am not mistaken, then, it is precisely this dilemma that lies at the heart of the twentieth-century opposition between analytic and Continental philosophical traditions, which thus rests, from a philosophical point of view, on the systematic cracks that had meanwhile appeared in the original Kantian architectonic. And, whereas the philosophies of Carnap and Heidegger represent the best efforts we have seen at rigorously working out the two diverging directions arising from these systematic cracks—the divergence, more generally, between the "scientific" and the "humanistic" strands in Kant's original synthesis—it is Cassirer, more than any other thinker of the twentieth century, who nevertheless tries to hold these two strands together. As I have just argued, Cassirer does not completely pull of this most delicate and difficult of balancing acts. Those interested in finally beginning a reconciliation of the analytic and Continental traditions, however, can find no better starting point than the rich treasure of ideas, ambitions, and analyses stored in his astonishingly comprehensive body of philosophical work.

Notes

1. R. Carnap, "Überwindung der Metaphysik durch logische Analyse der Sprache," *Erkenntnis* 2 (1932): 219–41; translated as "The Elimination of Metaphysics through Logical Analysis of Language," in *Logical Positivism*, ed. A. J. Ayer (New York: Free Press, 1959).

2. M. Friedman, "Overcoming Metaphysics: Carnap and Heidegger," in *Origins of Logical Empiricism*, ed. R. Giere and A. Richardson (Minneapolis: University of Minnesota Press, 1986). © University of Minnesota Press. I am indebted to the University of Minnesota Press for permission to reprint parts of this paper here.

3. M. Friedman, *A Parting of the Ways Carnap, Cassirer, and Heidegger* (Chicago: Open Court, 2000). © Open Court Publishing Company. I am indebted to Open Court Publishing Company for permission to reprint parts of this volume here.

4. See M. Heidegger, *Kant und das Problem der Metaphysik, Gesamtausgabe*, vol. 3 (Frankfurt: Klostermann, 1991); English version is *Kant and the Problem of Metaphysics* (Bloomington: Indiana University Press, 1990).

5. M. Heidegger, *Was ist Metaphysik?* (Bonn: Friedrich Cohen, 1929), 14, 18; translated as "What Is Metaphysics?" in *Basic Writings*, ed. D. Krell (New York: Harper and Row, 1977), 107, 111–12.

6. M. Heidegger, "Nachwort," in *Was ist Metaphysik?* 4th ed. (Frankfurt: Klostermann, 1943), 104; translated as "Postscript" to "What Is Metaphysics?" in *Existence and Being*, ed. W. Brock (Chicago: Henry Regnery, 1949), 356.

7. R. Carnap, *Der logische Aufbau der Welt* (Berlin: Weltkreis, 1928), §75; translated from the second (1961) edition as *The Logical Structure of the World* (Berkeley and Los Angeles: University of California Press, 1967). The reference is to Cassirer's *Substance and Function*.

8. Carnap, *Der logische Aufbau der Welt*, §177.

9. For this terminology, see J. M. Krois, *Cassirer: Symbolic Forms and History* (New Haven: Yale University Press, 1987).

10. E. Cassirer, *Philosophie der symbolischen Formen, dritter Teil: Phänomenologie der Erkenntnis* (Berlin: Bruno Cassirer, 1929), 331–32; translated as *The Philosophy of Symbolic Forms, Volume Three: The Phenomenology of Knowledge* (New Haven: Yale University Press, 1957), 285.

11. E. Cassirer, *Zur Logik der Kulturwissenschaften* (Göteborg: Högskolas Arsskrift 47, 1942), 48; translated as *The Logic of the Humanities* (New Haven: Yale University Press, 1961), 97.

12. Carnap, *Der logische Aufbau der Welt*, §16.

13. Cassirer, *Zur Logik der Kulturwissenschaften*, 84; *Logic of Humanities*, 143‒44.

14. Cassirer, *Philosophie der symbolischen Formen, dritter Teil*, 394; *Philosophy of Symbolic Forms, Volume Three*, 339.

Carnap's "Elimination of Metaphysics through Logical Analysis of Language"

A Retrospective Consideration of the Relationship between Continental and Analytic Philosophy

GOTTFRIED GABRIEL

Friedrich Schiller University

Rudolf Carnap is a classic proponent of the ideal language school within analytic philosophy. He has divided opinion more sharply than other representatives of this tradition and thus contributed decisively to the ongoing separation of analytic and Continental philosophy. The essay "The Elimination of Metaphysics through Logical Analysis of Language" (Carnap 1931, 1959), in particular, contributed to the polarization because it made Martin Heidegger, *the* classic author of Continental philosophy, the target of exercises in a logically inspired criticism of metaphysics.

The following essay reconsiders the relationship between analytic and Continental philosophy, using the Carnap-Heidegger controversy as an example. We should bear in mind, however, that the roles of analytic and Continental philosophy have in the meantime been strangely reversed. Whereas the Continental tradition struggles above all with the deconstruction of supposed remnants of old metaphysics, a new metaphysics is celebrating its reemergence in logically ingenious theories of analytic philosophy. The order of the day is not an elimination of metaphysics, but its new foundation through the logical analysis of language. We would suppose that this development would have the disapproval of both Carnap and Heidegger. For example, Carnap would surely have accused the metaphysics of possible worlds (and his philosophical grandchild David Lewis) of confusing internal and external existence, whereas Heidegger would have criticized such metaphysics as a kind of forgetfulness of Being of a presence-at-hand ontology *(Vorhanden-heitsontologie).*

There are more profound reasons for assuming such agreement between Carnap and Heidegger. For it is not so much the attitude toward metaphysics itself, but their views as to what remains for philosophy to do following the end of metaphysics that constitutes the opposition between Carnap and Heidegger.[1] This opposition shows in different forms of linguistic presentation, and, indeed, it is primarily through linguistic differences that analytic and Continental philosophers can be recognized today. Before pursuing this idea, we discuss the philosophical-historical setting of Carnap's and Heidegger's thinking to display their common ground.

To begin, we should remember that the distinction between analytic and Continental philosophy is problematic in two ways. Firstly, the two designations are askew, because "analytic" is a methodological determination, whereas "Continental" is geographical. Secondly, the implicit geographical division, according to which all analytic authors are assumed to be Anglo-Saxon, doesn't work. The ranks of analytic philosophy include, along with Carnap, Frege, Russell, Moore, Wittgenstein, Ryle, and Austin. Frege, Wittgenstein, and Carnap not only come from the Continent but also received their essential intellectual formation there (Frege and Carnap in Jena; Wittgenstein in Vienna).

With regard to Carnap and Heidegger, their philosophical beginnings at least are congruous, albeit with different emphasis. Both received their initial education in the context of neo-Kantianism: Carnap under Bauch in Jena; Heidegger under Rickert in Freiburg. (During his time in Freiburg, Carnap also heard Rickert.) Carnap was influenced decisively by Frege, and Heidegger in a corresponding manner by Husserl. Both concurred with the antipsychologistic logism of their respective teachers in the form of a theory of validity (which goes back to Lotze). Both, however, also acknowledged the other tradition: the early Heidegger (1912, 20) refers to Frege, and the young Carnap to Husserl.[2] Whereas for Carnap the distinction between validity and genesis of statements remained decisive throughout his life, Heidegger, under the influence of life philosophy (Dilthey, Nietzsche), distanced himself from linking the concept of truth to statements, that is from the propositional concept of truth, essential for the logical tradition since Aristotle.

This difference concerning the nature of truth drives Carnap and Heidegger in differing directions. But this difference did not arise through Carnap's having paid no attention to life philosophy. Instead, here too the initial situation for Carnap and Heidegger is identical. Both experienced the contest between neo-Kantianism and life philosophy at the beginning of the twentieth century. Central to this conflict was the question regarding the relationship between logic and life and—

proceeding from this question—the determination of the task of philosophy. The result was that the two neo-Kantians treated the challenge of life philosophy in different ways. For Heidegger the course was set by Husserl, and for Carnap by Wittgenstein. Their paths parted for good with Heidegger's inaugural lecture in Freiburg, "What Is Metaphysics?" (Heidegger 1929). Carnap had previously taken part in the famous dispute between Cassirer and Heidegger in Davos. Although the independence of Heidegger's thinking had also impressed Carnap, he remained philosophically bound to the rationalistically oriented neo-Kantianism as represented by Cassirer. By taking his examples of meaningless metaphysical statements from Heidegger's inaugural lecture in particular, he sent a signal, for, as he emphasized, he might "just as well have selected passages from any other of the numerous metaphysicians of the present or the past" (Carnap 1959, 69 n. 2).

The list of metaphysicians adduced by Carnap as examples (Fichte, Schelling, Hegel, Bergson, and Heidegger) displays a large degree of agreement with the usual enumeration of Continental authors, with one important exception. Nietzsche, who in today's controversy is viewed (by both sides) as a model author in the Continental tradition, is spared in Carnap's criticism. This fact has not been appreciated sufficiently until now, not least because with his move to the United States, Carnap had cut or at least suppressed his own Continental roots in orienting himself to the new philosophical circumstances, especially to the pragmatism found there.[3] In his intellectual biography only weak traces of the Continental tradition are recognizable, and even these have been ignored.

For several years a historical reconsideration of the origins of analytic philosophy that seeks to revise its "forgetfulness of the Continent" has been in progress. After Frege had found his fixed place in neo-Kantianism (cf. Gabriel 2002), Carnap's relationship with this tradition, which dominated German philosophy at the turn of the century, has been largely investigated (cf. in particular Richardson 1998). A part of Carnap's Continental roots have thus been revealed.

Friedman has presented the first thorough historical study of the relationship between Carnap and Heidegger, in which he attempts to show how, by dealing with the same basic ideas of neo-Kantianism, two traditions of philosophy were able to develop that have led to the current opposition of analytic and Continental philosophy (Friedman 1996). One factor, however, that Friedman does not take into account is the role of life philosophy.[4] It is only by considering this that the picture is completed and the actual differences are brought into the open. The final section of Carnap's (1931) essay, "Metaphysics as Expression of an Attitude towards Life *(Lebensgefühl),*" provides important clues here. The expression "Lebensgefühl" is a central term for W. Dilthey. Presumably Carnap adopted the term not directly from Dilthey but from his student Herman Nohl, whom Carnap had heard in

Jena.[5] It is revealing that, apart from Bauch and Frege, Nohl is the only one of his teachers in Jena that Carnap mentions by name in his *Intellectual Autobiography*. And this mention is not limited to academic reasons. Carnap writes:

> I remember with special pleasure and gratitude the seminars of Hermann [Herman—G.G.] Nohl (at that time a young instructor in Jena), in philosophy, education, and psychology, even when the topic, for example, Hegel's *Rechtsphilosophie*, was often somewhat remote from my main interests. My friends and I were particularly attracted by Nohl because he took a personal interest in the lives and thoughts of his students, in contrast to most of the professors in Germany at that time, and because in his seminars and in private talks he tried to give us a deeper understanding of philosophers on the basis of their attitude toward life (*"Lebensgefühl"*) and their cultural background. (Carnap 1963, 4)

The personal element addressed here has its place in Carnap's life itself. The "friends" (such as the later pedagogue Wilhelm Flitner) were committed members of the German youth movement with whom Carnap bounded through the Jena woods. The "deeper understanding of philosophers" mentioned here amounts to discerning their respective Lebensgefühl as the driving force of their differing metaphysics. The basis of such an assessment, which Carnap adopted from Nohl, is Dilthey's Weltanschauung doctrine. Before turning to this as the actual "point" of Carnap's "Elimination of metaphysics," I would first like to expand on the methodological framework as set out in Carnap's essay (1931).

The basic features of the criticism of metaphysics developed by Carnap, namely the linking of formal logic with the principle of verifiability are not new: they had already been worked out in Wittgenstein's *Tractatus*, to which Carnap explicitly refers. The familiar problems with the formulation of the empiricist criterion of meaning are not to be elaborated here. Difficulties already result from the fact that the meaning of a sentence is supposed to be determined by its truth conditions (cf. Carnap 1959, 62). Meaning is hence linked to the form of statements and is reduced to propositional meaning. Accordingly, normative statements are considered meaningless by Carnap: "It is altogether impossible to make a statement that expresses a value judgment" (77). Now the criterion of meaning is not to be understood as a merely descriptive criterion of distinction, but as a normative criterion of exclusion and hence falls prey to its own verdict of meaninglessness.[6]

In comparing Carnap and Wittgenstein, an important terminological difference is to be noted. For Wittgenstein the only statements (propositions) that have sense are those that describe logically possible facts (in the sense of existent or nonexistent states of affairs). Already the statements of logic lack sense. This determination is not to be understood pejoratively, but merely as characterizing their status as logically true—that is, as tautologies that say nothing about the world. Carnap, however, understands these statements as being meaningful because they are "true solely by virtue of their form" and follows Kant in defining them as analytic statements (Carnap 1959, 76). Accordingly they are valid a priori. Alongside analytic statements Carnap recognizes empirical statements corresponding to Kant's synthetic a posteriori statements. As distinct from his Jena dissertation "Der Raum" (1921), Carnap now denies the possibility of a priori synthetic statements.

Carnap speaks of meaningless statements *(sinnlose Sätze)* when words without meaning occur in them, or when they are not correctly formed according to logical syntax. The *meaningless* statements in Carnap's terminology thus correspond to the *nonsensical (unsinnigen)* propositions in Wittgenstein's terminology.[7] The thesis that words without meaning occur in metaphysical texts is old. It is stated emphatically, for example, by Hume. New, however, is the view that there exist statements in which the logical syntax is violated, though they accord with historical-grammatical syntax (cf. Carnap 1959, 69). This insight goes back to Frege and was first developed into the basis of a criticism in principle of metaphysics by Wittgenstein in the *Tractatus*. The basic features of this criticism are adopted by Carnap. The consequences he sees are admittedly quite different from those of Wittgenstein.

Whereas Wittgenstein's farewell to metaphysics was not without sadness, Carnap cheerfully issues the command for philosophy of science to "clean up" and allows philosophy to be absorbed by the logic of science. This "way out" is ruled out for Wittgenstein because for him a metalogic that attempted to "say" once again what can only "show" itself is impossible. The categorial discourse that sets out the logic of our language is compelled to overstep the limits of this logic. This discourse itself breaks the rules of syntactic well-formedness that it seeks to explicate. Hence not only are the statements of traditional metaphysics nonsensical, but so too are those statements in which this criticism is formulated, in particular, therefore, the statements of the *Tractatus* itself.

This view has consequences for the form of presentation of Wittgenstein's texts, a form that must be called literary rather than logical. Wittgenstein's *Tractatus* presents logic as literature. It is no wonder that through to the present day the logical and scientific faction within analytic philosophy has preferred to appeal to Carnap and has its difficulties with Wittgenstein. It is the logical tradition of propositionalism that binds knowledge to the (true) proposition and causes the methodical function of forms of presentation to be misunderstood.

The logic taken as a basis by both Wittgenstein and Carnap is the propositional and predicate logic developed by Frege in his *Begriffsschrift*. In this logic the traditional subject-predicate structure of propositions is replaced by an argument-function structure, through which a completely new and far-reaching analysis of language is made possible. Within the framework of such an analysis Frege had in particular proposed a logical distinction between four categorially different uses of the verb "to be": predication (subsumption), identity, subordination (of concepts), and existence.

Carnap makes use of these distinctions in his criticism of Heidegger. In doing so he presents some exemplary sentences of Heidegger's, which—so as to heighten the rhetorical "effect"—are contracted into the following passage:

> What is to be investigated is being only and—*nothing* else; being alone and further —*nothing*; solely being, and beyond being—*nothing. What about this Nothing?* . . . *Does the Nothing exist only because the Not, i.e. the Negation, exists?* Or is it the other way around? *Does Negation and the Not exist only because the Nothing exists?* . . . We assert: *the Nothing is prior to the Not and the Negation.* . . . Where do we seek the Nothing? How do we find the Nothing? . . . We know the Nothing. . . . *Anxiety reveals the Nothing.* . . . That for which and because of which we were anxious, was "really"—nothing. Indeed: the Nothing itself—as such—was present. . . . *What about this Nothing?—The Nothing itself nothings.* (Carnap 1959, 69)

In Carnap's view, Heidegger makes the logical mistake "of employing the word 'nothing' as a noun *(Gegenstandsname)*, because it is customary in ordinary language to use it in this form in order to construct a negative existential statement" (70). The negative existential statement "It is not the case that there exists something which has a certain property" is also expressed by the sentence that *nothing* has this property. It is through the objectification of this use of "nothing" that the meaningless talk of "the nothing" comes about. Part of the responsibility for this lies in the confusion of the uses of "to be" in the sense of predication and in the sense of existence.

Now Carnap in no way fails to recognize that something important can be addressed in metaphysics. He disputes, however, that it can be represented in the form of meaningful statements. Apart from this, Carnap admits that language still has functions other than making statements. Alongside a cognitive function it assumes an emotional one. This serves in particular to give expression of the attitude toward life *(Ausdruck des Lebensgefühls)*. It is in precisely this function that Carnap sees metaphysics, which, however, attempts to clothe something in the form of statements that cannot be said. A legitimate need underlies metaphysics; however, the adequate expression of the attitude toward life is not metaphysics but art: "In the case of metaphysics we find this situation: through the form of its works it pretends to be something that it is not. The form in question is that of a system of statements which are apparently related as premises and conclusions,

Gabriel / Carnap's "Logical Analysis of Language"

that is, the form of a theory. . . . The metaphysician believes that he travels in territory in which truth and falsehood are at stake. In reality, however, he has not asserted anything, but only expressed something, like an artist" (Carnap 1959, 79). As the historical source for his surrogate thesis ("Metaphysicians are musicians without musical ability") Carnap adduces Nietzsche as that metaphysician "who perhaps had artistic talent to the highest degree" (80) and was hence able to give expression to the Lebensgefühl in the form of poetry (in *Zarathustra*). At this point a surprising contiguity shows up between the positions of Carnap and Heidegger that leads us back to the theme of forms of presentation.

If we consider the historical stock of philosophical forms of presentation, we find the complete spectrum between the poles of science and poetry. The question is always to which does one orient oneself. Carnap orients himself methodically toward science—that is, toward the justification of statements. With him philosophy is absorbed by the logic of science; it no longer has contents of its own. These contents are passed onto poetry, where they find the form appropriate to them. With Carnap, so to speak, Frege's *Begriffsschrift* lies on the desk and Nietzsche's *Zarathustra* on the bedside table. For the intermediate form of a "concept-poetry" (*Begriffsdichtung* in the sense of Friedrich Adolf Lange), there is no place on either. The result is a problematic dichotomy of cognition and feeling. Apart from this dichotomous accentuation Heidegger seems to proceed from the same finding of a conflict between the form and content of metaphysics. But since it is the contents that matter to him, he departs from the scientific form and consistently approaches (as Nietzsche did) the form of poetry. Carnap and Heidegger, as well as the philosophical traditions founded by the two, have a common point of departure but proceed from there in opposite directions and thus arrive at diametrically contrary forms of philosophy.

What Carnap announces in "The Elimination of Metaphysics" (Carnap 1959, 62), namely a detailed exposition of a "metalogical" theory of syntactically meaningful languages, is presented by him in *The Logical Syntax of Language* (Carnap 1934, 1967). It should be noted that this exposition of the (later) so-called linguistic framework represents a transformation of traditional category theories along the lines of the linguistic turn. Whereas in Aristotle we are concerned with categories of being, and in Kant with categories of thinking, the analyses of analytic philosophy of language pertain to the categories of language. Considered in terms of philosophical history, the following line of development results: ontology— epistemology—philosophy of language. Carnap's analyses differ from related endeavors within analytic philosophy in that they do not lead to the establishment of a single categorial framework but conceive of different frameworks as being theoretically possible.

In the course of its development—starting with *Der Logische Aufbau der Welt* (Carnap 1928) via *Die logische Syntax der Sprache* (Carnap 1934) through to *Meaning and Necessity* (Carnap 1956)—there were indeed shifts in accentuation in Carnap's thinking, namely from a more epistemological (in the sense of neo-Kantianism), via a formal logical, through to a semantic analysis, but an attitude that Carnap (1934, 44f.; 1967, 51f.) himself formulated as his "principle of tolerance" prevailed constantly. This attitude is characterized by a *conventionalist* apprehension of languages within the framework of the theory of science.

Carnap also spoke later of the "principle of conventionality of language forms" (Carnap 1963, 55). The basic idea lies in resolving scientific-theoretical disputes over *content*, such as the foundational dispute between logicists and intuitionalists in mathematics, through the description of different language *forms*: "*In logic, there are no morals.* Everyone is at liberty to build up his own logic, i.e. his own form of language, as he wishes. All that is required of him is that, if he wishes to discuss it, he must state his methods clearly, and give syntactical rules instead of philosophical arguments" (Carnap 1967, 52). The principle of tolerance is thus part of Carnap's endeavor to eliminate so-called metaphysical pseudoproblems from the sciences. It formulates a metatheoretical standpoint that amounts to replacing ontology with logical syntax.

Carnap also applied the principle of tolerance in modified form to the dispute between nominalists and Platonists in semantics. In doing so he distinguished between "internal" questions of existence that must be answered relative to a specific language form (a "linguistic framework") and "external" questions of existence that are concerned with reality as such. External questions of existence continued for him to be pseudoproblems, whereas with admitting language forms that provide the framework for internal questions of existence he recommended the principle of tolerance. The criterion for the admission of linguistic forms was to be *scientific* utility alone (Carnap 1950, 1956).

Carnap's distinction between internal and external questions, which has Kantian roots, has for a long time not been taken seriously enough in analytic philosophy.[8] This is shown by the discussions of scientific realism that are troubled by not having distinguished (from the beginning) between internal and external realism. Yet Carnap had already made clear that science presupposes an internal realism—an empirical realism in the Kantian sense—but that every attempt to found, inductively as it were, an external metaphysical realism on a scientific basis is an impossible undertaking, because it involves a transcendent use of experience.[9] The analytic students did not understand their teacher's Continental inheritance.

If, for a moment, one ignores the fact that Carnap restricts philosophy from the outset to the explication of scientific language forms, then one could say that he is in agreement with Heidegger at least in the criticism of the presence-at-hand *ontology;* he adheres, however, to a presence-at-hand *syntax.*

The subject of language forms is the key to understanding the conflict between Carnap and Heidegger. With the rejection of certain language forms the expressive possibilities of philosophy are curtailed. Carnap acts tolerant, but his tolerance extends only so far as it is possible to translate the language form into a *logical* syntax. He accuses Heidegger of adopting "many peculiarities of the Hegelian idiom *(Sprachform)* along with their logical faults" (Carnap 1959, 75). Heidegger reacted to this objection in his "Epilogue to *What Is Metaphysics?*" His answer shows that he had understood the point precisely. "The suspicion against 'logic,' of which logistics may be considered a consistently developed degeneration,[10] emerges from the knowledge of that thinking which finds its source in the experience of the truth of Being [*Sein*], but not in the consideration of the objectivity of the being [*des Seienden*]. Exact thinking is never the strictest thinking" (Heidegger 1976a, 308).

Heidegger denies the logic oriented toward propositional thinking the right to establish logical linguistic forms as the possible forms of thought altogether. He sees metaphysics as being at work precisely in logistics, and this in the sense of a reifying presence-at-hand ontology. (With his self-critique in the *Philosophical Investigations* of the one-sided ontology of objects in the *Tractatus,* Wittgenstein later agreed with him indirectly.) Even though Heidegger has a different metaphysics from Carnap in mind, he, too, is concerned with an "*Überwindung* of metaphysics" (he uses this formulation several times). Yet he does not want to *eliminate* metaphysics; he wants to *overcome* it. Heidegger wants to direct thinking away from the question of being as the being *(Seiendes)* toward the question of Being *(Sein)* itself (Heidegger 1976b, esp. 367f.).

The comparison of Heidegger and Carnap ought to have made clear that the difference between Continental and analytic philosophy is above all a matter of the style of thought that manifests in linguistic style. The linguistic style is a matter of rhetoric, and it is not difficult to distinguish authors according to whether their rhetoric orients itself toward poetry or logic. Such an orientation is not simply a matter of personal taste; rather, the apprehension of philosophy itself comes to bear therein—namely as to whether poetic metaphors or logical analysis take on the guiding function in philosophical speech.

It remains to be asked whether analytic and Continental philosophy can be "reconciled" with one another against the background of their shared past. What is meant by this cannot be a colorful "postmodern" mixture of styles, or even the leveling of the differences existing between the forms of presentation. What matters far more is a thorough analysis of the respective peculiarities from the point of view of their functions. If the one-sided orientation toward the *logic* of linguistic forms could be overcome and extended in favor of a more comprehensive *rhetorics* of linguistic forms, then the distinction between analytic and Continental philosophy will prove to be historically explicable but systematically mistaken. For the time being, however, something would already have been gained if the "fears of contact" were broken down further. In this spirit I would like to conclude by bringing out some elements of Continental thinking with Carnap.

If one takes Kant's understanding of metaphysics as a basis, then the appropriate place for the metaphysical ideas (of Freedom, God, and Immortality) is in practical philosophy. Such a shift in location is also undertaken by Carnap in that he traces the theoretical hypostasizations of metaphysics back to practical needs. Carnap's moral point of view is comparable to that of a Kantian socialism, as had been developed in the Marburg neo-Kantianism. The essential difference lies in that, for Carnap, there is no practical reason that could do justice to this need *discursively*. To this extent he follows Nietzsche. Morality without justifiability amounts to moral decisionism.

We are, however, also involved with practical, albeit not *morally* practical, decisions in the approach to the sciences. Carnap's best-known example is the decision between idealism and realism (in the question of the existence of an external world), which is classified only as theoretically, but not as practically, irrelevant. Carnap also deals with other metaphysical questions in this way. At the official theoretical front door of his philosophy he turns them away, but at the same time he keeps open a back door to the courtyard of practical decision making. This theory-practice dualism is the result of his "crossing" of neo-Kantianism and life philosophy.[11]

The influence of Kant shows itself with Carnap not only in the adoption of several distinctions (such as those between genesis and validity, and between analytic and synthetic statements), but also in matters of theory construction. Carnap was very well aware that the development of scientific theories takes place from the points of view of unity and fruitfulness (Kant's *Zweckmäßigkeit*). Such points of view are, however, external in kind. The metaphysical suspicion must then apply to them, too. Here once more Carnap backs out of the difficulties by declaring these matters to be practical ones relating to application.

As we have already seen, Carnap uses the expression "metaphysics" in a somewhat indeterminate manner. In doing so he refers to authors and questions as examples and not generally to everything that has been traditionally counted as metaphysics. In a remark to the English text of "The Elimination of Metaphysics" he notes that the expression "metaphysics" is used "for the field of alleged knowledge of the essence of things which transcends the realm of empirically founded, inductive science. Metaphysics in this sense includes systems like those of Fichte,

Schelling, Hegel, Bergson, Heidegger. But it does not include endeavors towards a synthesis and generalization of the results of the various sciences" (Carnap 1959, 80). With the addition (in the last sentence), the *regulative* epistemological function (in the Kantian sense) of metaphysical ideas is obviously being addressed.

Naess has rightly noted that, despite the life-philosophical reinterpretation of the concern of metaphysics, Carnap and Dilthey have one thing in common with the metaphysical systems: "They are 'totalizing' views of reality, in Dilthey's sense; they 'set' certain values and represent decisions" (Naess 1968, 46). Last but not least, this is already expressed linguistically in the program of a "scientific *Weltauffassung.*" The replacement of the Diltheyan expression "Weltanschauung" (cf. Carnap 1959, 79) with "Weltauffassung" is nothing more than a word-political measure.[12] The difficulty remains of determining how the categorial discourse is to be understood that attempts to speak in a "totalizing" manner about the whole of meaningful language and is hence compelled to go beyond the bounds that it determines as being such. It does not have to become "poetic" for this reason alone, but it will not be able to abstain from metaphors and other "rhetorical" forms of presentation. As a result, philosophy has at least one foot in the poetry camp; it *is* concept-poetry. It is at this point, I think, that the discussion between Carnap and Heidegger, between analytic and Continental philosophy, should be taken up once again.

Translated by Andrew Inkpin

Notes

1. On the tacit common ground between Heidegger and Carnap cf. Kambartel 1968, 195–97; also Rentsch 1985, 140–46, 192f.

2. Carnap 1928. See the references to Husserl in the index of names and bibliography.

3. It should be borne in mind that the older pragmatism itself had links with the Continent. Lotze in particular left traces of influence. It would be worth examining whether the holism of Quine goes back *indirectly* to the Hegelianism of Lotze's metaphysics.

4. In his new book (Friedman 2000), he does, however, draw attention to this.

5. The influence of Dilthey on Carnap was first pointed out by Naess (1968, 41–48): "Metaphysics as the Expression of an Attitude to Life: Dilthey, Nohl and Carnap"; cf. also Patzig 1966, 100.

6. Cf. Carnap's self-critique in Carnap 1963, 45f.

7. At one point Carnap erroneously calls contradictory—i.e., logically false—sentences absurd *(unsinnig)* (Carnap 1959, 71, but cf. correctly 76).

8. Cf. on this distinction Krauth 1970, especially chapter 7; also Norton 1977. On the relationship to Continental philosophy cf. Parrini 1994, esp. 274 n. 4 and the literature listed there.

9. Contrary to Carnap, H. Reichenbach, Carnap's companion in promoting a "scientific *Weltauffassung,*" was convinced at times that he could give an inductive justification of realism. For a critical evaluation cf. Klein (2000, sect. 2.10–2.12).

10. "Logistics" *(Logistik)* was at the time the usual term, also used by Carnap, for mathematical logic.

11. This connection already becomes clear in Carnap 1928. Cf. in particular the conclusion §§179–83.

12. It probably goes back to Neurath. Cf. Stadler 1997, 372.

References

Carnap, R. 1928. *Der logische Aufbau der Welt.* Berlin: Weltkreis-Verlag. 2d ed. Hamburg: Meiner, 1961.

———. 1931. "Überwindung der Metaphysik durch logische Analyse der Sprache." *Erkenntnis* 2:220–41.

———. 1934. *Logische Syntax der Sprache.* Vienna: Julius Springer.

———. 1950. "Empiricism, Semantics, and Ontology." *Revue Internationale de Philosophie* 4:20–40. Reprinted in Carnap 1956, 205–21.

———. 1956. *Meaning and Necessity.* 2d ed. Chicago: University of Chicago Press.

———. 1959. "The Elimination of Metaphysics through Logical Analysis of Language." In *Logical Positivism,* ed. A. J. Ayer, 60–81. Glencoe, Ill.: Free Press. Translation of Carnap (1931).

———. 1963. "Intellectual Autobiography." In *The Philosophy of Rudolf Carnap,* ed. P. A. Schilpp, 1-84. The Library of Living Philosophers. Vol. 11. La Salle, Ill.: Open Court.

———. 1967. *The Logical Syntax of Language.* London: Routledge & Kegan Paul.

Friedman, M. 1996. "Overcoming Metaphysics: Carnap and Heidegger." In *Origins of Logical Empiricism,* ed. R. N. Giere and A. W. Richardson, 45–79. Minneapolis: University of Minnesota Press.

———. 2000. *A Parting of the Ways: Carnap, Cassirer, and Heidegger.* La Salle, Ill.: Open Court.

Gabriel, G. 2002. "Frege, Lotze, and the Continental Roots of Early Analytic Philosophy." In *From Frege to Wittgenstein: Perspectives on Early Analytic Philosophy,* ed. E. H. Reck, 39–51. Oxford: University Press.

Heidegger, M. 1912. "Neuere Forschungen über Logik." In *Frühe Schriften (Gesamtausgabe,* vol. 1), 17–43. Frankfurt am Main: Klostermann 1978.

———. 1929. *Was ist Metaphysik?* Bonn: Friedrich Cohen. In *Wegmarken (Gesamtausgabe,* vol. 9), 109–22. Frankfurt am Main: Klostermann 1976.

———. 1976a. Afterword to *Was ist Metaphysik?* In *Wegmarken (Gesamtausgabe,* vol. 9), 303–12. Frankfurt am Main: Klostermann.

———. 1976b. Introduction to *Was ist Metaphysik?* In *Wegmarken (Gesamtausgabe,* vol. 9), 365–83. Frankfurt am Main: Klostermann.

Kambartel, F. 1968. *Erfahrung und Struktur: Bausteine zu einer Kritik des Empirismus und Formalismus.* Frankfurt am Main: Suhrkamp.

Klein, C. 2000. *Konventionalismus und Realismus: Zur erkenntnistheoretischen Relevanz der empirischen Unterbestimmtheit von Theorien.* Paderborn: Mentis.

Krauth, L. 1970. *Die Philosophie Carnaps.* Vienna and New York: Springer.

Naess, A. 1968. *Four Modern Philosophers: Carnap, Wittgenstein, Heidegger, Sartre.* Chicago: University of Chicago Press.

Norton, B. G. 1977. *Linguistic Frameworks and Ontology: A Re-Examination of Carnap's Metaphilosophy.* New York: Mouton.

Parrini, P. 1994. "With Carnap, Beyond Carnap: Metaphysics, Science, and the Realism/Instrumentalism Controversy." In *Logic, Language, and the Structure of Scientific Theories: Proceedings of the Carnap-Reichenbach Centennial*, ed. W. Salmon and G. Wolters, 255–77. Pittsburgh and Konstanz: University of Pittsburgh Press and Universitätsverlag Konstanz.

Patzig, G. 1966. Afterword to Carnap, *Scheinprobleme in der Philosophie*, 85–136. Frankfurt am Main: Suhrkamp.

Rentsch, T. 1985. *Heidegger und Wittgenstein. Existential- und Sprachanalysen zu den Grundlagen philosophischer Anthropologie.* Stuttgart: Klett-Kotta.

Richardson, A. W. 1998. *Carnap's Construction of the World: The Aufbau and the Emergence of Logical Empiricism.* Cambridge: Cambridge University Press.

Stadler, F. 1997. *Studien zum Wiener Kreis: Ursprung, Entwicklung und Wirkung des Logischen Empirismus im Kontext.* Frankfurt am Main: Suhrkamp.

Schlick and Husserl on the Essence of Knowledge

ROBERTA LANFREDINI

University of Florence

Preface

The relationship between Schlick and Husserl is pervasive. The number of Schlick's references to phenomenology (explicit or understood) is equal to—if not greater than—the number of his references to Kant, Mach, and Avenarius. Furthermore, such references are invariably polemical, and sometimes scornful, as when Schlick states that he feels "almost disgust" ([1918] 1974, 136) about the *ideation* meant as a solution to the problem of the relation between the psychological configurations (or configurations of thought) and logical configurations that exist independently from the real world.[1]

In the first edition of the *Allgemeine Erkenntnislehre*, Schlick's critical comments on the phenomenological method were even more numerous than in the second edition. Moreover, Husserl, in his preface to the second edition of volume 2 of *Logische Untersuchungen*, written in 1921, attacks Schlick with caustic observations —as Schlick himself describes it. Besides, Husserl inadvertently quotes Schlick's text incorrectly, referring to the *Allgemeine Erkenntnistheorie*, which probably caused Schlick a certain amount of irritation. Apart from these details, Husserl's accusations against Schlick are serious. The principal one is that he misunderstood all the fundamental theses of phenomenology, perhaps because he did not read these theses with sufficient attention. If Schlick's tone is considered scornful,

Husserl's tone is that of one who has been offended. So there is a deep sentiment of dislike between the two. Going beyond psychological considerations, on the one hand, I believe in one point Husserl is right: Schlick, perhaps annoyed by the contorted and involved Husserlian language, was unable to understand the real sense of the majority of phenomenological theses to which he constantly refers. On the other hand (this is above all what I really want to emphasize), Schlick, despite the crystalline clarity of his argumentation, has not sufficiently recognized the problematic depth in philosophical terms of two questions, which were crucial for Husserl: (1) What is experience? (2) What role does experience play in the cognitive process?

Expressibility and Comprehensibility of Content

To show this defect in Schlick's grasp of the problem, my essay focuses on two fundamental questions: (1) How do we know and what do we know—that is, what is the object of knowledge? (2) What, in general, is knowledge for Schlick and what is knowledge for Husserl?

Let's start with the first question: the ways and the objects of knowledge. A crucial question for Schlick, as well as for logical empiricism in general, is the total insignificance of the concept of content (in particular, intuitive content) for both a theory and a practice of knowledge. Besides, it is this thesis (related to a notable distinction between *kennen* and *erkennen*) that marks—on Schlick's explicit admission—his radical detachment with respect to Husserl's phenomenology.

Why is the concept of content unimportant for a theory of adequate and, above all, scientifically informed knowledge? Schlick's answer: because the content is inexpressible and indescribable—that is, linguistically untouchable. "The difference between structure and material, between form and content is, roughly speaking, the difference between that which can be expressed and that which cannot be expressed" ([1938] 1979, 291).

This thesis, however, is not completely clear. It seems to hover between the two following interpretations: first, that we can identify the concept of expressibility with the concept of communicability; and, second, that we can identify the concept of expressibility with the concept of comprehensibility. If by the term "expression" we mean "communication," the thesis of the inexpressibility of content is accepted by both Husserl and Schlick. The content indicated by the word "fear" cannot be communicated but must be learned by the experience of being afraid. Likewise, we cannot "extract" from my consciousness the greenness of a color that I see and put it into the consciousness of another person.

According to Husserl, in communication we essentially use signs *(Anzeichen)*, which are physical objects in all their effects (sound or writing). When we communicate, what we communicate is not directly the content of our assertion, but it is something that accompanies a content in a report, which Husserl says is unnecessary (signs can differ; in fact, they vary in different languages) ([1899–1900] 1970). An individual mental experience cannot be passed on to the intuition of another. When common language attributes some kind of perception felt by others (when, for example, to "see" their indignation or to "feel" their pain), such a way of expressing oneself should not be understood in the literal sense. The listener perceives that the person speaking shows some mental experiences, but he does not experience them directly. He does not have what can be called an "internal" perception, but only an "external" perception. The difference that separates an adequate intuition, a mental experience personally experienced, from an inadequate one that is simply presumed, is for Husserl insurmountable.

So Husserl does not deny the existence of a communicative (or informative) function of language. However, on this point he resolutely disagrees with Schlick. Such a function is essentially distinguished from, and is completely inessential to the expressive function of language, a function that every individual develops in his incommunicable, isolated life. In it, and only in it, is a real sense of expression established; only in this do we experience the comprehension of meaning. So, not only do we express the content, but such a content or, more particularly, the union of mental experiences that form it, is the essence of expression. Only through an analysis of such a content is it possible to understand the semantic property (such as, to make sense; to refer to an object, a state of things, or an event) contained in an expression.

On the contrary, for Schlick, expression and communication do not constitute two distinct functions of language, but they are perfectly assimilable. In other words, for Schlick, the essence of expression is communication: "We may, therefore, regard communicability as a criterion of expressibility, i.e., of structure. . . . No fact can be an 'expression' except by agreement. Nothing expresses anything by itself" ([1938] 1979, 292, 304–5). In short, for Husserl, signs (writing and sounds) are able to cause in the interlocutor significant mental experiences that constitute the content of expression. Without the presence of these contents, there is no sense in talking about comprehension of a proposition. That does not mean, however, that in communication there is a kind of "transportation" from one consciousness to another. On the contrary, in communication the speaker makes known to the listener or reader the existence, in the speaker, of certain mental experiences, certain significant intentions. But making it known does not mean having directly experienced those mental contents. According to Schlick, as we have seen, the mental experience, the content, has no role either in expression or in communication.

Looking back at what we have already said, let us consider the famous example of the person who is born blind. Let us consider the perception of green. Let us look at the ineffable quality of greenness, which makes the essence of the content. This quality is accessible only to beings endowed with eyesight and power of color perception. It could not possibly be conveyed to a person born blind. Shall we conclude, Schlick asks, that such a person could not understand any of our statements about the color, that they must be quite meaningless to him because he can never possess the green content? Schlick's answer to this question is no. We can communicate to the person born blind, as we can to a person who can see, the meaning of "green." Nevertheless, what we communicate is not the content—the greenness. "Since content is essentially incommunicable by language, it cannot be conveyed to a seeing man any more or any better than to a blind one" ([1938] 1979, 295).[2] What can be communicated (or expressed, which is for Schlick the same) is the fact that something exists that we call green, and it is something possessing a certain structure or belonging to a certain system of internal relations.

I can give a particular description of this green leaf lying on my desk by placing the color in a certain order. I assert, for instance, that it is a bright green, or a rich green, or a bluish green, trying to describe the green by comparing it to other colors. Evidently it belongs to the intrinsic nature of our green that it occupies a definite position in a range of colors and in a scale of brightness, and this position is determined by relations of similarity and dissimilarity to the other elements—in this case shades—of the whole system. In this sense every quality has a certain definite logical structure. "In this way every quality (for instance, the qualities of sensation: sound, smell, heat, etc., as well as colour) is interconnected with all others by internal relations which determine its place in the system of qualities" ([1938] 1979, 294). In such a perspective the difference between a colorblind person and a normal one is a purely formal property. There is a greater variety in the perception of a normal individual, or, what is the same, the system of color is more complicated in a normal person than in a blind one. The internal relations are not as simple, and this is a difference of structure.

In *The Idea of Phenomenology* Husserl's example is perfectly analogous to Schlick's, with respect to a person born deaf ([1950] 1966). A person born deaf knows that sounds exist and sounds make harmonies. But he cannot understand how sounds do this, how musical operas are possible. He cannot represent a thing of that kind, he cannot look at it, and looking at it he cannot understand *the how*. No kind of physical or psychological theory about color can add anything to this "pure vision," which, according to Husserl, establishes the sense of color.

At this point we seem to have arrived at a dead end. With Schlick we have found the identification of expression with communication on one side and the identification of expression (or meaning) with structure and form on the other.

With Husserl, in contrast, we have the distinction between communication and expression on one side, and the identification of the expression with content—that is, with certain mental experiences—on the other. For the former, the content coincides with the inexpressible; consequently, it is necessarily outside an authentic cognitive ambit. The other writes that the experiences that confer meaning coincide with what is expressible, and so they are an integrating part of knowledge.

The Structure of Content

In truth the rigorous separation between Schlick and Husserl can be analyzed further. The problem, I think, is to establish what we mean by content, a concept that has a more complex structure than Schlick has supposed. For Schlick, meaning and content are two rigorously different notions.

> The essential feature of expression is order. . . . It is the kind of thing with which logic is concerned, and we may, therefore, call it logical order, or simply structure.
> If it is true that verbal sentences, the propositions of our spoken language, can communicate nothing but the logical structure of the green colour, then they seem to be unable to express the most important thing about it, namely that ineffable quality of greenness which appears to constitute its very nature, its true essence, in short, its content. (Schlick [1938] 1979, 290, 295)

This thesis could be accepted without reservation by Husserl, if we only identify the concept of content with that of intuitive filling. The metaphor used by Schlick to give an example of the concept of content is that of filling an empty frame with the material furnished by an individual's experience, which has a clear phenomenological matrix.

But what Schlick doesn't seem to remember or understand is that the intuitive filling is for Husserl only an aspect of the content. One of the most important distinctions in phenomenology is that between significant intention (that is an empty intention) and filling intention (that is an intuitive intention, which is perceptive or imaginative). Both types of intentions meet in the general concept of content. In a certain sense, for Husserl as for Schlick, content (in the sense of intuitive content) and meaning are rigorously distinct. But it is important to underline that for Husserl the formal content, the meaning, the comprehension, the structure, are in any case part of a more general notion of content—that is, intention of a subject. We shall later analyze the importance of this consideration.

Not only the concept of content seems to be more complex than maintained by Schlick. The concept of intuition also deserves to be examined. Schlick often speaks, with polemical reference about phenomenology, of perception of content. "I can perceive a green leaf; I say that I perceive it if (among other things) the content 'green' is there, but it would be nonsense to say that I perceive this content" ([1918] 1974, 319). The content (in the sense of intuitive content) is simply there. If we insist on using a verb, the word "enjoying" would present itself: it is the nearest equivalent to the German *erleben*. "Here we uncover the great error committed by the philosophy of intuition: the confusing of acquaintance (*Kennen*) with knowledge (*Erkennen*)" ([1918] 1974, 83). Nevertheless, once again, the concept of living or experiencing does not exhaust the concept of intuition. In fact, in the *Logical Investigations* Husserl continuously repeats the distinction between experiencing (*erleben*) the content and apprehending (*auffassen*) or perceiving a property or an object ([1899–1900] 1970, 2:1, §14).

In the first case we refer to experience of immanent contents. In the second case we refer to perception (or imagination) of a transcendent object. We experience (or enjoy, as Schlick says) acoustic sensations, but we hear (we perceive) the singer's song. One of Husserl's fundamental criticisms, addressed to his teacher Brentano, is that he did not distinguish adequately between simply having contents and the apprehension of those contents. It is ambiguous to talk about sense data. Sensation and sensory property (or quality) are two entirely distinct concepts; sensation is immanent content, while property and, in general, object are transcendent. The same form can present itself from different perspectives, the same color with different illuminations, the same sound as near or far away. So, contrary to what Schlick says—"when I gaze at a red surface, the red is part of the content of my consciousness" ([1918] 1974, 102)—red is not a part of the content of my consciousness. It is this thesis that allows Husserl to go beyond the phenomenalistic notion of consciousness (for example, that of classic empiricism and, in particular, that of David Hume). For phenomenalists, consciousness is a place, and the objects that we would call intentional are not the things meant through immanent data, but the immanent data in themselves—that is, sensations, ideas, perceptions, images, and so on. On the contrary, for Husserl, consciousness is not a closed place and the object is nothing that is inside a consciousness, as in a sack, almost as if consciousness were an empty form that is the same everywhere, an empty sack that is always identical, in which we put at one time this and at another time that. The same metaphor is used by Schlick in *Form and Content* and in *General Theory of Knowledge*.

> "It is really too primitive a picture to compare consciousness to a box in which objects could be put and taken out." ([1938] 1979, 326)
> We are accustomed to say—and this is only metaphorical . . . that whatever I

imagine or feel or sense is "in" my consciousness. The word "in" has only a figu-
rative meaning; for it is certain that consciousness is not a receptacle—nor is it
indeed comparable to a receptacle, which in itself always remains the same and
which can be filled by ever changing "contents." ([1918] 1974, 122–23)

Nevertheless, we should not be deluded: the use of the same metaphor does not
correspond to a convergence of intents, as we shall soon see.

For now it is enough to emphasize how affirming that content is simply there
and simply present means supporting a simplistic and undifferentiated vision of
the concept of intuition. We experience contents, but we mean objects that go beyond these contents. Another clarification: for Schlick, content (in the sense of intuitive content) has nothing to do with knowledge. There is still more: "the most
fundamental mistake of philosophy of all times" is to identify knowledge with "immediate awareness" or with "intuition" ([1938] 1979, 318). Once again, the polemic
reference is to phenomenology. But Husserl has never identified knowledge and
intuition. When I hear a sound or see a color, it is not with these acts of hearing
and seeing that I come to understand what a sound is or a color is. Knowing and
having acquaintance with are not at all synonymous.

According to Schlick, "knowledge is expression: there is, consequently, no inexpressible knowledge" ([1938] 1979, 315–16). For Schlick, knowledge is a result of
an act of comparing, recognizing, naming: "I have to recognise the colour as that
particular one I was taught to call 'blue'" ([1938] 1979, 322).

The terminology that Husserl uses is more or less identical: knowledge means
to recognize, identify the intended object in an empty act of meaning with the object that is given to us in an act of perception ([1899–1900] 1970, 2:2). This does
not mean, as Schlick hints, to distinguish two types of knowledge: one conceptual
and one intuitive. On the contrary, knowledge is the result of an integration between two components. This is exactly the reason why it is possible to distinguish
in phenomenological terms the determination or discrimination of a perceptive
object (the result of a simple act of perception) from the identification of an object (the result of the conjunction of a meaning act and a perceptive act). This important distinction does not seem present in Schlick's philosophical perspective.

In a passage from General Theory of Knowledge, Schlick distinguishes unelaborated perception or sensation from sensation associated with an apperception
process, identified at that time with perception or with having experience of something. "Pure unelaborated perception or sensation is mere acquaintance (Kennen).
If this is what one has in mind, it is entirely wrong to speak of a 'perceptual knowledge.' Sensations give us no knowledge whatsoever of things, but only an acquaintance with them" ([1918] 1974, 87). Perception in a proper sense is, for Schlick,

knowledge, even if "of the most primitive kind." Why knowledge? "For I am not left with a mere sense impression; on the contrary, the latter is at once incorporated into the range of my previous experience, recognised as being of such and such a sort. Consequently if we restrict the expression 'perception' to apperceived sense impression, then indeed, but only then, may we speak of perceptual knowledge" (87).

"But this concept of intuitive knowledge—Schlick immediately tries to make clear—has not the least connection with the one found in Bergson and Husserl" (88). In reality Husserl's analysis is finer, for he distinguishes sensation from perception, on one side, and perception from conceptual elaboration, on the other. Perception, for Husserl, is not a judgment. It is not, in itself, knowledge of the object. This distinction marks the important difference between an epistemic conception and a nonepistemic conception of perception. Husserl accepts the second type of conception. What exactly makes this agreement important is that it can be used to correct Kant's famous principle according to which "intuition without conception is blind, and conception without intuition is empty." While the second part of the affirmation is without doubt true, the first needs a correction: it is not the concept that allows intuition to see. Perception has already, in itself, an organization, a structure which does not have any reference to conceptualization.

Intuition, Experience, and Knowledge

In sum, the famous distinction between *kennen* and *erkennen* is not at all denied by Husserl. Certainly for Husserl, as for Schlick, *kennen* is not knowledge. Husserl agrees with what Schlick says, that "so long as an object is not compared with anything, is not incorporated in some way into a conceptual system, until then it is not known. In intuition the objects are only given not included," and that "without apperceptive or conceptual elaboration there is no knowledge" (83, 88).

Through the mere experiencing, the mere intuition, it is not possible to have knowledge of anything. Yet there is a crucial difference between the perspectives of Schlick and Husserl. Schlick declares explicitly that "intuition and conceptual knowledge do not both strive for the same goal, they move in opposite directions" (82). For Husserl, in contrast, intuition (perception or imagination) is an integrating part in the process of knowledge. It is the fusion of two elements: one is purely structural or formal and the other intuitive. In general terms: authentic knowledge cannot do without a qualitative factor (which, in modern philosophy of the mind, falls under the term *qualia*). Even if the identification with a sensation of red is not a sufficient condition for the determination of an intuitive act, it is in any case a necessary condition. Because we have an identification of a red object we have to experience "what it is like" (using Thomas Nagel's expression [Nagel 1974; see also Jackson 1982, 1986]) to see something red. We have knowledge when we operate or work on recognition. But recognition implies necessarily an act of intuition—that is, an act of perception or imagination.

On the contrary for Schlick (it is not just by chance that this thesis has been silently taken up by many contemporary materialists, such as Churchland [1985, 1989]) the qualitative element is irreducibly outside of an adequate theory of knowledge. When I look at the blue sky and lose myself entirely in the contemplation of it without thinking, then I am enjoying the blue, I am in a state of pure intuition. The blue fills my mind completely. But that does not mean knowing what blue really is. The meaning of the word "blue" is entirely included in the structure of the intuitive content. So the inexpressibility of the blue does not operate effectively in the comprehension, or in the knowledge of the blue.

Considering all this we have arrived at the center of our topic. What is experience for Schlick? Which role does such an experience play in our knowledge of the world that surrounds us? I believe that the real element of the deep disagreement between Schlick and Husserl resides both in a different philosophical conception of the notion of experience and in the different philosophical (and not only programmatic) value that the two authors attribute to the concept of experience with reference to a general theory of knowledge.

With reference to the first question we have now seen that Schlick has the tendency to furnish an undifferentiated and extreme report on the concept of intuition. He speaks almost always about intuition as "an exceptionally close relation between subject and object" ([1918] 1974, 81). The borderline case of this is the mystic relation between conscience and God. In reality this is a caricature of the concept of intuition. To start we have seen that intuition has an internal structure. Schlick says:

> In knowing there are always two terms: something that is known and that as which it is known. In the case of intuition, on the other hand, we do not put two objects into relation with one another, we confront just one object, the one intuited. Thus an essentially different process is involved; intuition has no similarity whatever to cognition. When I give myself fully to an intuitive content of my consciousness, say a red patch I see before me, or when in behaving I submerge myself fully in the feeling of activity, I experience through intuition the red or the activity. But have I really come to know the essence of the red or of the activity? Not at all. (82)

This is an oversimplified conception of intuition. In intuition there is also a clear distinction between *what* we perceive and *how* we perceive it. In the case of perception *the how* corresponds to a determined perspective. The object of perception is, for its essence, something that never reveals itself entirely to us, but only reveals itself in partial appearances.

So intuition doesn't mean to give myself fully to an intuitive content of my consciousness, say a red patch I see before me, but to perceive (or to imagine) an object in certain completely objective ways.

Now we go on to the second question: what role does experience have in the process of knowledge. Without doubt at the programmatic level, Schlick, as is well known, attributes an enormous value to experience and, indirectly, to intuition. Intuition for Schlick is "the indescribable that precedes everything else"; "it is the ineffable and ever present foundation of all else"(321, 322). So it is also the foundation of knowledge, even if it is not, in itself, knowledge. His formulation of the principle of verification, the idea that science is founded on concrete definitions, his passionate participation in a polemic about protocol sentences, the distinction between analytic judgments (a priori) and synthetic judgments (a posteriori), together with the future distinction between observational and theoretical language, are all the elements that point to the recognition of the essential role that experience plays in the cognitive process. Not only this: for Schlick, it is also the experience—or the content—that supplies the "empty frame" with a *univocal* determination, which is the indication of a particular object. Moreover, this is exactly what happens in the interpretation of a hypothetical deductive system.

> How is the empty structure of a hypothetical deductive system actually filled with meaning? What is the stuff which must be added to the empty frame in order to make a science of it? There seems to be but one possible answer to this question, namely: the purely formal structure must be filled with content—it could not be anything else, because there is nothing else.
>
> All the different individuals communicate to each other the structural forms, the patterns . . . but each one has to find out for himself their applicability to the world, each one has to consult his own experience, thereby giving the symbols a unique meaning and filling the structure with content, as a child may colour drawings of which only the outlines are given. ([1938] 1979, 331, 334)

Above all that is what happens in measurement, as well as in the particular type of knowledge that is essential to connect our theories to the world, which is knowledge of facts in contrast to explanatory knowledge.

Nevertheless, according to Schlick, every observer fills the empty frame of every hypothetical deductive system with its own content. Such filling up, which Schlick called *interpretation,* is an essentially and radically private and subjective act. In this sense, experience (or, better, the intrinsically qualitative element that distinguishes it), as crucial as it is, is outside the circle of our assertions. For Husserl, in contrast, experience is an integral part of knowledge. Without it no determinate knowledge is possible. The first thing that must be clear for Husserl when we approach phenomenology is that this is a priori logic of phenomena, well founded on evidence and that every kind of knowledge must necessarily start with the data. So Husserl's perspective is rigorously empiricistic and positivistic. Also for Husserl, as we have seen, the comprehension of a proposition does not need the presence of intuitive content, but only a presence of an "empty" intentional act (the frame to which Schlick also makes reference). But knowledge for Husserl is not simple comprehension.

Understanding effectively which role experience plays in Schlick's theory of knowledge means answering the second question that we placed at the beginning: what is knowledge for Schlick and what is knowledge for Husserl? In fact, on a general level, the perspectives of the two authors converge on many fundamental points. To start with, both adopted a rigorously antipsychologistic point of view. The theory of knowledge has nothing to do with the individuation and analysis of the nature of mental processes. The task of theory of knowledge, for Schlick as for Husserl, is to ask oneself on which universal basis knowledge becomes necessary. "He inquires into the universal grounds on which valid knowledge in general is possible—an inquiry that clearly differs basically from one that addresses itself to the mental processes by which knowledge develops over time in one or another individual" ([1918] 1974, 3).

Moreover, both authors seem interested, in Kant's approach on reflection, more in the problem of *how* than of *what* we know, more in the individuation of the possibilities of knowing than in the objects of knowledge (what Husserl calls the "enigma of transcendence"). For Schlick and Husserl, as we have seen, "to know is to recognise *(wiedererkennen)* or rediscover *(wiederfinden)*" (15). Recognize and rediscover what? Which are the things that are confronted in a cognitive process? In everyday life we refer to representations. But representations for Schlick are "quite vague and blurred" (17). Schlick rejects in successive stages in *General Theory of Knowledge* the fluctuating, indeterminate, transient character of the intuitive configurations to the determination of concepts, leading from the qualitative analysis to the quantitative analysis. "The kind of knowledge that meets the needs of pre-scientific thought and practical life cannot find legitimate employment in a science that demands at all times the greatest possible rigor and the highest degree of certitude" (19).

The first step that knowledge, in its real sense, must take is to substitute the notion of concept for that of representation. "What is a concept? A concept is to be distinguished from an intuitive image above all by the fact that it is completely determined and has nothing uncertain about it. . . . Thus a concept is not an image. It is not a real mental structure of any sort. Indeed it is not real at all, but imaginary—something that we assume in place of images with strictly determined content" (20). The gnoseological meaning of the conceptual function is in *designation,* which for Schlick means *coordination.* (The essence of the concept consists simply in its being a sign that we coordinate to an object; the essence of judgment consists in its being a sign for relations among objects). The essence of thought for Schlick is the univocal association of objects and symbols. The original link between concept and object, judgment and fact, is for Schlick a more conventional bond. "The task of a sign is to be a representative of that which is designated, to act in its place in some respect or other. Wherever it is impossible or inconvenient to operate with the objects themselves, we replace them with signs which can be manipulated more easily and as desired" (59).

This link has the same function as a catalog in a library. "The catalogue is simply an ordered collection of signs each of which corresponds to a volume in the library" (59). With one clarification: truth and knowledge are for Schlick distinct concepts. Since the truth consists simply in a univocal designation, "it would be in principle child's play to arrive at perfect truth . . . to designate all things in the world simply by inventing individual signs for each of them, and then committing to memory the meaning of each sign" (66). But "knowledge is more, much more, than mere truth. . . . Knowledge means unique coordination with the help of certain definite symbols, namely those that have already found application elsewhere" (59). Clarification is important, but it does not change the substance of our speech: an element of stipulation and convention is absolutely crucial for understanding Schlick's theory of knowledge at all its levels.

If the word for order for Schlick is *designation,* or *coordination,* for Husserl it is *constitution.* Constitution presupposes for its essence the presence of subjectivity in the determination of the object of knowledge. This is the ultimate sense of the theory of Husserlian intentionality, and it is a gnoseological sense, not a psychological one.

Conclusion

The principle according to which the essence of knowledge is coordination presumes on the one side the existence and independence of the two poles of coordination. On the other side, it is exactly this independence that is brought into the discussion by Husserl. To understand knowledge as the logic of intentional acts means that the intended object has an exclusive correlation with its meaning. So the intentional analysis is only an analysis of the capacity and limits of what we can mean (that is, think or experience). For Husserl, a world outside our world is an "effective absurdity."

In this sense, the general characteristic of intentionality coincides with its transcendental capacity. One of the principal foundations of such a perspective is the existential independence of the intentional relation: when we present an object in an intentional act, it does not necessarily mean that it really exists. This is evident in the case of an object that we think about or imagine. It is also true, according to Husserl, for a perceived object: perceptive hallucination, until it exists, is effectively fully illusory, exactly because in it the object is as if it existed "in person" *(leibhaft)*.

So the ontological plan understood by the two philosophical perspectives are radically different. Nevertheless, there is a still more fundamental difference between Schlick and Husserl, which refers to the value of the concept of life. For Schlick, life, like intuition, is radically outside every cognitive capacity: "intuition is enjoyment, enjoyment is life, not knowledge" ([1938] 1979, 323).

On the contrary, Husserl writes, it is the concept of life that characterizes the sphere of subjectivity that establishes the object to which it is intentionally addressed. The intentional act is an intentional experience, the subject experiences immanent circumstances, and through them, approaches those that are transcendent. So the life that Husserl speaks about is the life of sense; it is that dynamical structure (extremely complex) in which sense organizes itself to constitute the object in its different ways, cognitive and noncognitive. It is exactly the attempt to isolate the way in which the life of subjectivity expresses itself, and making these ways a fundamental part of a philosophical project, that marks the deepest and most incurable disagreement between Husserl's phenomenology and Schlick's proposal of a theory of knowledge.

Notes

1. In *General Theory of Knowledge,* Schlick merely criticized Husserl's concepts of ideation and evidence as unclear and psychologistic. In such a perspective, Schlick's concept of analyticity, for example, is not different from that of Kant. In fact, the definition is based on the subject-predicate distinction. Later on, Schlick seems to have been particularly impressed by Wittgenstein's concept of tautology and thus he was able to more adequately distinguish formal from material truth (see [1932] 1979). The dispute now is much more about the concept of analyticity. The problem is: what kind of sentence is "every color has some extension" or "every tone has some strength" or "every spatial perception has some perspective"? For Schlick, as for Wittgenstein, propositions like these do not express our experience but are in some sense tautologies, and their denials are contradictions. So the simultaneous presence of two colors at the same place in the visual field is logically impossible, since it is ruled out by the *logical structure* of color. But there is an important dis-

tinction in Husserl's perspective: the distinction between the *non-sense (unsinning)* and the *without sense (sinnlos)*. The difference is very important and it is a difference between a formal contradiction and a material contradiction, a distinction that is absent in Schlick's philosophy. See Simons (1992).

2. "What we shall have to claim is that there exists a necessary deficiency of expressive power which (being necessary) afflicts every possible natural language, and which makes it impossible for language (language *per se* that is) to penetrate, as it were, below the level of recording discriminations and failures to discriminate, to the level of phenomenal description. . . . It might be argued at this point that the reason why language is necessarily incapable of being used to describe the *qualia* of experience is that there are simply no such things: our experience possesses structure but no content" (Harrison 1973, 5–6).

References

Churchland, P. M. 1985. "Reduction, Qualia and the Direct Introspection of Brain States." *Journal of Philosophy* 82:8–28.

———. 1989. *A Neurocomputational Perspective: The Nature of Mind in the Structure of Science.* Cambridge: MIT Press.

Harrison, B. 1973. *Form and Content.* Oxford: Basil Blackwell.

Husserl, E. 1899–1900. *Logische Untersuchungen.* Vol. 1, *Prolegomena zur reinen Logik;* vol. 2/1, *Untersuchungen zur Phänomenologie und Theorie der Erkenntnis;* vol. 2/2, *Elemente einer Phänomenologisch Aufklaurg der Erkenntnis.* 1913. 2nd ed. Halle: Niemeyer.

———. 1970. *Logical Investigations.* Trans. J. N. Findley. London: Routledge & Kegan Paul. Translation of Husserl (1913).

Husserl, E. 1950. *Die Ideen der Phänomenologie.* Den Haag: Martinus Nijhoff.

———. 1966. *The Idea of Phenomenology.* Trans. W. T. Halston and E. G. Nakhnikian. Den Haag: Martinus Nijhoff.

Jackson, J. 1982. "Epiphenomenal Qualia." *Philosophical Quarterly* 32:127–36.

———. 1986. "What Mary Didn't Know." *Journal of Philosophy* 83:291–95.

Nagel, T. 1974. "What Is It Like to Be a Bat?" *Philosophical Review* 83:435–50.

Schlick, M. 1918. *Allgemeine Erkenntnislehre.* Berlin: Springer-Verlag.

———. 1974. *General Theory of Knowledge.* Trans. E. Blumberg. New York: Springer-Verlag. Translation of Schlick (1925).

———. 1932. "Gibt es ein materiales Apriori?" In *Wissenschaftlicher Iahresbericht der Philosophischen Gesellschaft an der Universität zu Wien. Ortsgruppe Wien der Kant-Gesellschaft für das Vereinsjahr 1931–32,* 55–65. Wien: Gerold.

———. 1979. "Is There a Factual a Priori?" In *Philosophical Papers II, 1925–1936.* Trans. W. Sellars, 161–70. Dordrecht: Reidel.

———. 1938. *Form and Content: An Introduction to Philosophical Thinking.* In *Gesammelte Aufsatze, 1926–1936,* 151–249. Wien: Gerold. (Written in English.)

———. 1979. "Form and Content: An Introduction to Philosophical Thinking." In *Philosophical Papers II, 1925–1936,* 285–369. Dordrecht: Reidel.

Simons, P. 1992. "Wittgenstein, Schlick and the a Priori." In *Philosophy and Logic in Central Europe, from Bolzano to Tarski,* 361–76. Dordrecht: Kluwer.

Carnap vs. Gödel on
Syntax and Tolerance

S. AWODEY

Carnegie Mellon University

A. W. CARUS

University of Chicago

In the 1950s Gödel wrote several drafts of a paper to refute what he called Carnap's "syntactic interpretation" of mathematics (Gödel *1953/9). Some of these drafts were published in 1995 and have already stimulated extensive philosophical commentary.[1] Much of this attention has focused on Gödel's view that Carnap's overall framework, based on the "principle of tolerance," is self-undermining.[2] Gödel's argument is accepted to varying degrees; most commentators, like Goldfarb and Ricketts (1992), have held that Carnap's view can only be upheld in a weakened or diluted (and rather empty) form. Carnap's own later views differ (as Gödel himself acknowledges) significantly from the "syntactic interpretation" Gödel attributes to Carnap; it would be worth confronting Gödel's arguments with the *actual* views of the later Carnap.[3] And although it has been acknowledged that the *Syntax* view *also* differs in critical respects from Gödel's portrayal of it (Ricketts 1994; Goldfarb and Ricketts 1992), it seems not to have been noticed that Gödel's argument contains a mistake. Carnap's *Syntax* view need make no concessions whatever to this argument, therefore, whatever other objections may be raised against it and whatever the later Carnap might have said in response to Gödel's paper had it been published at the time.

⌐

Gödel summarizes the syntactic program he attributes to Carnap in two assertions, that *mathematics can be interpreted to be syntax of language* and that *mathematical sentences have no content*. The investigations in Carnap's *Logical Syntax of Language*, he claimed, as well as those of the Hilbert school, had shown the following about these two points:

> (1) *Mathematics can be interpreted to be syntax of language* only if the terms "language" or "syntax" or "interpreting" are taken in a very generalized or attenuated sense, or if only a small part of what is commonly regarded as "mathematics" is acknowledged as such. . . . (2) *Mathematical sentences have no content* only if the term "content" is taken from the beginning in a sense acceptable only to empiricists and not well founded even from the empirical standpoint. Thereby these results become unfit to . . . support . . . the philosophical views in question (such as nominalism or conventionalism). (Gödel *1953/9-III, 337)

Regarding the second of these assertions, Gödel maintains that the examination of the syntactical viewpoint "leads to the conclusion that there *do* exist mathematical objects and facts which are exactly as objective (i.e., independent of our conventions or constructions) as physical or psychological objects and facts" (337). Carnap, of course, would have regarded such questions (asked outside the context of a particular linguistic framework) about the existence or nonexistence of any objects, whether physical or mathematical, as empty of cognitive significance. To that extent, he would have been happy to agree with Gödel that there is no difference between the two kinds of objects.

But it is the *first* of the two assertions Gödel makes in the above quotation that has excited all the recent commentary: his assertion that the "syntactic interpretation of mathematics," as maintained by Carnap and other members of the Vienna Circle, can be *proved false*.[4] Carnap would have found Gödel's argument very interesting but would not have accepted it.[5] To understand why, it will be best to treat this rather subtle point with Carnap-like, perhaps almost pedantic, thoroughness. Gödel's argument can be paraphrased in the following four steps:

(i) For mathematics to be interpreted as syntax of language—and thus empty of empirical content—it must be proved that no syntactic (that is, purely linguistic) stipulation can possibly have empirical consequences; otherwise, mathematics is in danger of making claims about the empirical world on purely arbitrary, definitional (however convenient or practical) grounds.

(ii) But even the choice of a very weak language framework (as restricted as primitive recursive arithmetic) has the consequence, by Gödel's own second incompleteness theorem, that the consistency of our chosen language cannot be proven with its own resources.

(iii) Any proof that our chosen language is consistent, then, presupposes the consistency of the stronger metalanguage required for the proof, so the attempt to prove consistency—at any level—incurs an infinite regress, and we cannot completely exclude the possibility that the chosen language is inconsistent, and thus has empirical consequences (as it would imply not only every mathematical sentence, but every empirical sentence).

(iv) Conclusion: The requirement of step (i) cannot be met, so mathematics cannot be syntax of language.

Carnap would have pointed out, though, that Gödel's assertion in step (i), that the consistency of any stipulated language must be *provable,* is not implied by what Gödel calls the "syntactic interpretation" of mathematics (SIM). He is right to point out that the SIM entails the consistency of any language stipulated, but this is not the same as *provable* consistency. Carnap could have agreed with Gödel on "if P then Q," in other words, but Gödel uses "if P then *provably* Q" as the basis for his argument. This stronger assertion rests on an apparent non sequitur in step (i); let us examine the argument more carefully. Gödel begins with the correct statement that

(A)   SIM implies that mathematics is empirically vacuous.

He also reminds us, correctly, that

(B)   If a stipulated language for mathematics is inconsistent, then it may have empirical consequences.

From this, by contraposition, it follows that

(B')   If a stipulated language for mathematics is to be empirically vacuous, it must be consistent.

But from A and B' it follows only that

(C)   SIM requires the consistency of any stipulated language for mathematics.

It does not follow, as Gödel suggests in (i), that

(D)   SIM requires the *provable* consistency of any stipulated language for mathematics.

In short, where Gödel says "the rules of syntax must be demonstrably consistent, since from an inconsistency *every* proposition follows" (337), he should correctly say "the rules of syntax must be consistent, since from an inconsistency *every* proposition follows."

This small difference is an important one, as can perhaps be seen more clearly by considering Gödel's argument (shown on the left-hand side below) in conjunction with an analogous one—which might easily have occurred to Carnap if he had written his own reply to Gödel—shown on the right-hand side:

(i)   For mathematics to be syntax of language, it must be proved that no stipulation can have empirical consequences.

(i')   For space-time to be flat, it must be shown that such and such conditions (indicating curvature) do not obtain in any region of the universe.

(ii)   But even a very weak language cannot be proved consistent without further assumptions. Any proof that our chosen language is consistent, then, presupposes the consistency of the stronger metalanguage required for the proof.

(ii')   But the required observations may be affected by the presence of curvature; for example, measurements may be distorted or instruments become unreliable. Moreover, the universe may be infinite, or there may be regions that are in principle inaccessible to us.

(iii)   We can never be certain that the chosen language is consistent and has no empirical consequences.

(iii')   We can therefore never be certain that our observations are conclusive and that the required conditions obtain.

(iv)   Conclusion: The requirement of step (i) cannot be met, so mathematics cannot be syntax of language.

(iv')   Conclusion: The requirement of step (i') cannot be met, so space-time cannot be flat.

The argument on the right differs from that on the left in that it concerns an empirical question. But it has the same logical form as Gödel's argument (on the left), and it is no less sound. In both cases, the conclusion is unwarranted.

The erroneous claim in the right-hand argument is "it must be *shown* that . . ." in (i'). It would be correct to say "it must be *the case* that . . . ," for space-time to be flat, such and such conditions do not obtain. It would also be correct to say "for space-time to be *shown* to be flat, it must be *shown* that such and such conditions do not obtain." In Gödel's argument, likewise, though, "if P then Q" is correct, and "if provably P then provably Q" is correct, neither of these is equivalent to Gödel's "if P then provably Q."

Gödel's argument, therefore, does not refute the possibility that mathematics can be interpreted as syntax of language, as he claims. However, a slight modification of this argument does show something else of equal interest. Though we saw that (D) above is unwarranted, it would be correct to say

(E)   A *proof* of SIM requires the *provable* consistency of any stipulated language for mathematics.

As Gödel correctly argues, the consistency of any stipulated language cannot be proved. If Q is false, then "if P then Q" is true only if P is false. So by (E) there can be no proof of SIM. *We cannot prove that mathematics is syntax of language by mathematical reasoning.* But this result, far from undermining SIM, is in complete harmony with it. For the syntactic view implies the vacuity of mathematics, which would surely be violated if that viewpoint could itself be *proved* mathematically, as such a result would itself be a nontrivial mathematical proposition.

Gödel appears to have regarded Carnap's "syntactic interpretation" as a *claim*, like the differing claims made about mathematics by logicism, intuitionism, or formalism. But just as the syntactic view reconceived of these positions as *proposals* for the language of science (Carnap 1934, 42–45, 253–56), it is itself to be understood as a proposal. It is precisely the proposal, in fact, that considerations regarding the form of language for scientific knowledge not be regarded as claims. There are, of course, facts about languages. But the syntax view proposes that these need not be taken as having a bearing on facts about the world (assertions made in a language). Such linguistic facts, it is proposed, are to be taken as practical advantages or disadvantages of the language in question for various possible uses to which it might be put.

As a proposal for a way of understanding questions regarding the form to be given the overall language of science, the syntactic view (and the principle of tolerance) falls within the scope of the questions (and answers) it is proposed to govern —it is self-applicable. This is not an argument for its *truth* (this would be to confuse what the later Carnap called internal and external questions), but it does address the question of its *status*. (In the way that the status of the "verification principle"—as an "elucidation" in Wittgenstein's sense—had worried the Vienna Circle.) The syntactic view and its principle of tolerance are a proposed set of constitutive or defining principles of a way of understanding proposals about the form of the scientific language.

The self-applicability of the syntactic view amounts to this: it belongs to the class of statements (those informally called "proposals for language frameworks") to which it applies. This circularity is not vicious; the proposal conforms to the rule it proposes, but not by stipulative fiat of that rule itself. This is perhaps easier to see if we contrast it with other explications. The concept of differentiable function, for instance, left intuitive by Leibniz, Newton, and Euler, was explicated in the well-known way by Cauchy and Weierstrass. In this case, what required clarification was not a property of statements but a property of functions on R^n, which some functions possess and others do not. It does not make sense to ask whether the Weierstrass definition of differentiable function is differentiable. There is no problem of self-applicability.

In contrast to this case, consider Tarski's explication of the concept of truth. Here we can, without contradiction, ask "Is Tarski's truth definition true?" This question is answerable, however, only if it is asked with specific reference to a particular semantic system containing a truth definition. Without reference to a particular language system, it no more makes sense to ask this question than it does to ask whether numbers (or uncountable sets, or electromagnetic fields) exist. If truth is defined for a language in Tarski's way, then the truth of this definition, expressed in an appropriate meta-metalanguage, follows trivially from the fact that it is the definition (just as the statement "a real-valued function $y = f(x)$ for which the limit of $\{f(x) - f(x + d)\}/d$ as d approaches o exists is differentiable at x" is true in the framework of classical analysis). The truth definition applies to itself, but trivially. (And it is important to note that its truth says nothing whatever about its quality or usefulness as an explication of truth compared to other possible explications.)

What Gödel calls "interpretations of mathematics" are generally not self-applicable even in this trivial sense. They stand outside the discourse they purport to be about; they are not themselves intended as mathematical statements. But what sort of statements are they? Not, in any case, ones that have a clear significance within the discourse they themselves constitute. The syntactic view has the advantage that it is consistent with itself in this respect; it has a clear status (it is analytic) within the discourse it proposes. The stipulation that analyticity is a matter of constitutive stipulation makes itself analytic within the framework it constitutes.[6] And although this self-applicability does nothing to *justify* the syntactic interpretation, the fact that it does apply consistently to itself would have seemed to Carnap a clear advantage over other "interpretations" of mathematics—which, lacking significance within the framework they constitute, are forced back on a claim to some other, transcendental or absolute, status.

In the above pseudo-argument about space-time, let us assume the truth of (ii') —that is, suppose we cannot determine whether curvature exists. Then we have two choices. Either "space-time is flat" becomes cognitively meaningless (as no facts can be brought to bear on it) or it becomes analytic. In the latter case, we *define* the curvature of space-time as flat, just as one might define the notion of being "at rest" relative to the position of the earth for the purpose of describing the motions of the heavenly bodies.

A realist about the physical world rejects the idea that the flatness of space-time could be a matter of definition. For him, there is a true curvature of space-time, however unknowable it may be now or in principle. In mathematics, too, the Gödelian realist would hold the notion of truth to be absolute and not dependent on a particular choice of linguistic framework. The two forms of realism are analogous; they do not recognize the language relativity of claims or assertions. They refuse to distinguish between (language-constituting) *proposals* and (language-relative) *claims.*

So Gödel did not *refute* the syntactic program, as he claimed. The syntactic view was never *asserted* as a claim. It was a proposal for a way of understanding questions about the form of the scientific language. The merits of such a proposal can, of course, be discussed. But the question whether we should adopt this proposal is a practical question; the proposal is not the kind of sentence to which the concepts "proof" or "disproof" apply. This is not to say that no questions relevant to the merits of such a proposal can be proved; Gödel's incompleteness results are an obvious example. Our historical experience with mathematics is also relevant. But such empirical and rational considerations—even Gödel's theorems, for all their importance—do not by themselves uniquely determine a way of understanding the role and significance of mathematics in our knowledge.

Notes

1. By, among others, Goldfarb (1995), Goldfarb and Ricketts (1992), Richardson (1994), Ricketts (1994, 1996), Parsons (1995), Friedman (1999), and Potter (2000).

2. Especially Friedman (1999), Richardson (1994), Parsons (1997), and Potter (2000).

3. This we undertake in a separate paper (Awodey and Carus 2003). Carnap's later view is still surprisingly unknown, despite the informative discussions of Stein (1992), Bird (1993), and Jeffrey (1994).

4. Following the quotation at the beginning of section 1 above, Gödel says, "if the terms occurring are taken in their ordinary sense, then assertion 1 ("*mathematics can be interpreted to be syntax of language*") is disprovable" (Gödel *1953/9, 337).

5. In draft III, published as the first part of Gödel (*1953/9), he formulates the argument as follows: "a rule about the truth of sentences can be called *syntactical* only if it is clear from its formulation, or if it somehow can be known beforehand, that it does not imply the truth or falsehood of any 'factual' sentences (i.e. one whose truth, owing to the semantical rules of the language, depends on extra-linguistic facts). This requirement not only follows from the concept of a convention about the use of symbols, but also from the fact that it is the lack of content of mathematics upon which its a priori admissibility despite strict empiricism is to be based. The requirement under discussion implies that the rules of syntax must be demonstrably consistent, since from an inconsistency *every* proposition follows, all factual propositions included" (339). Similarly, in the same version: "*To eliminate intuition or empirical induction by positing the mathematical axioms to be true by convention is not possible.* For, before any convention can be made, mathematical axioms of the same power or empirical findings with a similar content are necessary already in order to prove the consistency of the envisaged convention. A consistency proof, however, is indispensable because it belongs to the concept of a convention that one knows it does not imply any propositions which can be falsified by observation (which, in the case of mathematical "conventions," is equivalent with consistency)" (347).

6. There is a kind of analogy in this respect between the *Syntax* approach to proving consistency and the self-applicability of the syntactic interpretation. As Goldfarb and Rick-

etts put it, "Is Carnap's position infected with a vicious circularity here? We think not. To be sure, there is a regress, but it is not obviously circular or vicious unless one thinks that some foundational work must be done by the syntactical description of a language. If no such task is at issue, then the upshot is simply that we can never make the conventional nature of mathematics fully explicit in any framework. The structure of Carnap's view is then coherent. Given the distinction between issues within a linguistic framework and issues between linguistic frameworks—a distinction that is always central to Carnap's thought—then the position is not circular so much as self-supporting at each level. If the mathematical part of a framework is analytic, then it's analytic; and so invoking mathematical truths at the level of the metalanguage is perfectly acceptable, since they flow from the adoption of the metalanguage" (Goldfarb and Ricketts 1992, 71).

References

Awodey, S., and A. W. Carus. 2003. "How Carnap Could Have Replied to Gödel." In *Carnap Brought Home: The View from Jena,* ed. S. Awodey and C. Klein (LaSalle, Ill.: Open Court).

Bird, G. H. 1993. "Carnap and Quine: Internal and External Questions." *Erkenntnis* 42:41–64.

Carnap, R. 1934. *Logische Syntax der Sprache.* Vienna: Springer.

Friedman, M. 1999. "Tolerance and Analyticity in Carnap's Philosophy of Mathematics." In *Reconsidering Logical Positivism,* 198–234. Cambridge: Cambridge University Press.

Gödel, K. *1953/9, version III. "Is Mathematics Syntax of Language," in his *Collected Works,* 3:334–55. Oxford: Oxford University Press.

Goldfarb, W. 1995 "Introductory Note to *1953/9," in K. Gödel *Collected Works,* 3:324–33. Oxford: Oxford University Press.

Goldfarb, W., and T. Ricketts. 1992. "Carnap and the Philosophy of Mathematics." In *Science and Subjectivity: The Vienna Circle and Twentieth Century Philosophy,* ed. D. Bell and W. Vossenkuhl, 61–78. Berlin: Akademie-Verlag.

Jeffrey, R. 1994. "Carnap's Voluntarism." In *Logic, Methodology, and Philosophy of Science IX,* ed. D. Prawitz, B. Skyrms, and D. Westerståhl, 847–66. Amsterdam: Elsevier.

Parsons, C. 1995. "Quine and Gödel on Analyticity." In *On Quine: New Essays* (CUP), ed. Paolo Leonardi and Marco Santambrogio. Cambridge: Cambridge University Press.

Potter, M. 2000. *Reason's Nearest Kin: Philosophies of Arithmetic from Kant to Carnap.* Oxford: Oxford University Press.

Richardson, A. 1994. "The Limits of Tolerance: Carnap's Logico-Philosophical Project in *Logical Syntax of Language." Proceedings of the Aristotelian Society,* supplementary volume, 67–82.

Ricketts, T. 1994. "Carnap's Principle of Tolerance, Empiricism, and Conventionalism." In *Reading Putnam,* ed. P. Clark and B. Hale, 176–200. Oxford: Blackwell.

———. 1996. "Carnap: From Logical Syntax to Semantics." In *Origins of Logical Empiricism,* ed. R. Giere and A. Richardson, 231–50. Minneapolis: University of Minnesota Press.

Stein, H. 1992. "Was Carnap Entirely Wrong, After All?" *Synthese* 93:275–95.

I I

On the Origins and Development
of the Vienna Circle

On the Austrian Roots of Logical Empiricism

The Case of the First Vienna Circle

THOMAS UEBEL

University of Manchester, England

What are the determinants of logical empiricism as a distinctive school of philosophical thought? The answer to this question, once thought simple, has become complicated. Is logical empiricism to be regarded, for instance, as a school of thought in the analytic or the Continental tradition? Not that long ago, this question would have appeared just silly, but with the neo-Kantian influence upon Reichenbach, Carnap, and Schlick more widely recognized nowadays, it is no longer all that strange but hints at connections beyond the obvious. The remaining oddness of the question, of course, suggests that we sharpen this question still further.

Was logical empiricism just another version of neo-Kantian philosophy—and, if not, what may have been the other element that went into it? The answer to this question may differ depending on which wing of early logical empiricism we focus upon. Concentrating on its Viennese center here (and disregarding Berlin), I propose to investigate the thesis that there existed something like a "first Vienna Circle" already before World War I and that it was the work of this first Circle that provided the suitably "Austrian" counterweight to the neo-Kantian influence on the Vienna Circle proper around Schlick. In other words, it was the cooperation of members of this early circle of philosophically minded Austrian scientists—the mathematician Hans Hahn, the physicist Philipp Frank, and the economist Otto Neurath—with the scientifically trained German philosophers Moritz Schlick and Rudolf Carnap that accounted for the distinctive force of Viennese logical empiricism.

The thesis at issue, readers will no doubt have noted, hovers uneasily between banality and preposterousness. My task is to show that it is neither. Two seemingly obvious objections involve misunderstandings. Dealing with them right away will clarify and throw into relief the two parts of my thesis. Thereafter, I consider whether the claims made can be substantiated.

1.

The first objection holds that there is one distinction too many here, a distinction without a difference, namely that of *Austrian* as opposed to *German*, especially with regard to philosophy. The objection, in other words, disputes that part of the thesis under investigation that could be called the Neurath or Neurath-Haller thesis (to name it by its two most forceful advocates).[1]

Now as the term will be used here, to speak of an "Austrian philosophy" is to claim neither that it developed without any influence from German philosophy nor that all philosophy done in the Habsburg empire or the First Republic was Austrian in the relevant sense. Nevertheless, it is to claim that one can identify a distinct tradition that is broadly empiricist and anti-idealist in epistemological matters (as far as nonformal science is concerned), objectivist in ontological respects and attentive to matters of logic and language. Following Neurath himself, we may cast Bolzano as its fount, and count Zimmermann, Brentano, Meinong, Höfler, and Mach into its broad mainstream, developing into Husserl's phenomenology and Twardovski's school of Lvov from the Brentano school and the Vienna Circle more indebted to Mach.[2]

The most important aspect that makes this postulated line of development *non*-German—which is not yet to say that it was self-consciously anti-German—is that all the putative participants saw themselves in opposition to Kant. On this account, the philosophers mentioned provided legitimating reasons for what all along was a brute sociological fact, already stressed by Neurath: the suppression of Kant's philosophy as Protestant Enlightenment thought by the Austro-Hungarian state and the Catholic church. What makes this philosophy Austrian in our sense is certainly not the soil on which it arose or the nationality of its practitioners (Brentano was German), but the distinctive presence and/or absence of philosophical influences that came together in it. (I will come back to this.) And what makes it significant that this philosophy is Austrian in the intended sense is that it guards us against regarding logical empiricism, in the light of much recent work, as simply another form of neo-Kantian philosophy—against overcompensating, as it were, for the earlier neglect of Kantian elements in logical empiricism.

The second objection comes from the opposite side and states that stressing the importance of the "first Vienna Circle" is nothing new. Without quite using this name—but frequently employing phrases like "our original Viennese group" (1949a, 33–36)—Frank talked about the ideas explored by the group comprising Hahn, Neurath, and himself in the two historical accounts of the development of logical empiricism that serve as introductions to his two collections of essays in the 1940s. And Rudolf Haller specified—with reference to the combination of the outlook of Mach with the views of Poincaré, Duhem, and particularly Rey (whom Frank had already singled out for the incisiveness with which he characterized the crisis of his present-day philosophy of science)—distinctive aspects of the philosophies of the members of this "first Vienna Circle," as he in fact christened it. What then is there to be added to what could be called the Frank or Frank-Haller thesis?[3]

This question is a good one, but it must also be noted that Frank—presumably not wishing to upset the process of acculturation of logical empiricism in its American exile by what may even have been controversial among its survivors—was rather discreet in claiming original status for the first Vienna Circle. Having woven it into his narrative of the development of the movement as a basic biographical coordinate, but never emphasizing it as a thesis on its own, Frank allowed many readers not otherwise primed to miss this aspect of his histories. So the Frank thesis bears elaborating, which Haller did by characterizing the first Vienna Circle as the conduit for the radical conventionalist theses upon which the Circle's later antifoundationalism was to be built. My point here is to reconsider the first Vienna Circle in the light of the recent reappreciation of the neo-Kantian influence on logical empiricism and sharpen our appreciation of it in the light of some new results.

The issues concerning both of the two theses are threefold and concern their content, their significance, and their evidential standing. I will not add to the defense of the Neurath thesis undertaken by Haller except to comment upon the social mechanisms by which the Austrian philosophical tradition was transmitted and realized. But the question of evidence for the Frank-Haller thesis must be broached briefly.

2.

Just as once the Neurath-Haller thesis was met with a certain skepticism (perhaps still is), so a certain skepticism lingers (or could easily be awakened) concerning the Frank-Haller thesis. Yet just as the Neurath thesis does not just represent an imaginative construction by a notorious sloganeer and pamphleteer, but correctly charts a philosophical lineage, so the Frank thesis amounts to more than a retrospective reconstruction in the course of providing a legitimating disciplinary history for logical empiricism in its American exile. That the disciplinary histories for logical empiricism on occasion show considerable ingenuity in their legitimating narratives has been demonstrated by Alessandra D'Acconti's still unpublished Cambridge dissertation (1995). So, is talk of the first Vienna Circle but a case of uncritical literalism, of taking actor's categories as definitive?

In the "Historical Introduction" to his first collection of translated essays, Frank gave the first description of the early group "from which the 'Vienna Circle' evolved 20 years later" (1941a, 8).

> About 1910 there began in Vienna a movement which regarded Mach's positivist philosophy of science as having great importance for general intellectual life, but was clearly aware of [its] shortcomings . . . An attempt was made by a group of young men to retain the most essential points of Mach's positivism, especially his stand against the misuse of metaphysics in science. With regard to those points, however, where Mach stood in opposition to the present course of the development of science, they planned a reconstruction of his doctrines. To this group belonged the mathematician H. Hahn, the political economist Otto Neurath, and the author of this book, at the time an instructor in theoretical physics in Vienna. The attempts at this reconstruction were at first made rather gropingly; they were only preparations. We tried to supplement Mach's ideas by those of the French philosophy of science of Henri Poincaré and Pierre Duhem, and also to connect them with the investigations in logic of such authors as Couturat, Schröder, Hilbert, etc. The attitude toward the atomic theory was indicated to us by the ideas first of L. Boltzmann and then of A. Einstein. (6–7)

In his expanded "Historical Introduction" to the second collection of his essays, Frank dates these meetings in Viennese coffeehouses from 1907 onward and describes in greater detail their early thought with particular reference to Abel Rey's book on conventionalist philosophy of physics (Frank 1949a, 1–10).

But what independent evidence is there for the existence of the first Vienna Circle beyond Frank's reminiscences? Unfortunately, no minutes remain of its meetings—nor is it likely that any were taken—and both Hahn's and Frank's *Nachlass* is lost. But we can turn to what is left of the correspondence of Neurath.

When Neurath received word that Hempel was about to enter into the Circle's protocol sentence debate, he wrote:

> I am much looking forward to your paper. I hope I will not appear in it as someone in apostasy from Wittgenstein. In a certain sense I am glad that you now will have to read my older works in an, as it were, professional capacity. From these

you will see that everything which I stand for nowadays dates back to before the common period, the tremendous importance of which I certainly do not wish to discount. I just do not like it when you Germans (Reichsdeutsche) . . . think of Austria as some kind of appendage and fail to take account of the original French-English-Austrian development and only take cognizance of what happened since Wittgenstein. Thus people always underestimate Philipp Frank. (February 2, 1935, WKA)

Here Neurath clearly suggested that before the "common period" of philosophical activity—presumably the Circle with Schlick—there lay another one. And it is to this earlier period that he not only traced back his own radical opinions—a claim born out by his 1913 writings—but also dated an association with Frank.

Some three years later Neurath made the following remark, again to Hempel: "In the Vienna Circle we would not have made much advance without the Schlick Circle and where can one find [someone like] him. I believe that things would not have worked out well without him—especially his semi-passivity was important, apart from his qualities" (April 26, 1938, WKA). Like Frank in his later histories, Neurath here put this earlier group—the first Vienna Circle—in the driver's seat as far as the development of Viennese logical empiricism is concerned. Whether this is totally correct we need not decide for now, but we may recall that in his eulogy for Hahn, Frank called him "the real founder of the Vienna Circle" (1934). Hahn, in 1921 himself newly returned to the University of Vienna as *Ordinarius* in mathematics, was instrumental in engineering the call to Schlick as *Ordinarius* in philosophy. (In this connection Frank spoke of "new men" having been recruited after World War I to the task which the early group pursued [1949a, 26].)

Neurath and Hahn knew each other for a long time.[4] As is well known, Neurath married Olga Hahn, Hans Hahn's sister, in 1911, but his acquaintance with the Hahn family dates back still earlier. Otto and his first wife, Anna Schapire, organized a reading service for Olga, who had lost her sight due to an illness; she thus was enabled to complete her doctorate. In the summer of 1934 Neurath remarked to Hempel: "We are still devastated by Hahn's death. 35 years of related efforts in different fields. The shared youth with Poincaré, Duhem etc." (August 16, 1934, WKA). This dates their friendship to the turn of the century, with Hahn already studying at the university and Neurath still at school, the K. K. Staatsgymnasium im XIX, Bezirk, the same one Hahn had attended. "Hahn and I have been friends for many years—since our Gymnasium time," Neurath once wrote to Carnap: "He, the older, taught me a lot of things. We, Frank and other[s] read Spinoza in the 'Rahnhof' . . ." (June 16, 1945, RC 102-55-11 ASP). This observation is consistent with his earlier remark and with Frank's retrospectives.

Still, if we were inclined to doubt the veracity of Frank's histories, why should we believe Neurath's off-the-cuff retrospective remarks? Moreover, to put it bluntly, why should coffeehouse discussions and small talk with brothers-in-law constitute a (small) philosophical movement?

What is required is contemporaneous evidence for the first Circle. Luckily there exists something approaching this. In a letter to his early mentor, the German sociologist Ferdinand Tönnies, Neurath wrote:

> Next winter a local Privatdozent in mathematics plans to conduct a seminar on the foundations of math[ematics] and mech[anics] (in connection with the works of Poincaré). He has now asked me to co-lead the seminar with him. (Since I can't habilitate myself yet, this will be done unofficially, with him announcing it alone but me leading it together with him.) I gladly agreed because I hope to learn a lot from this. Already we are meeting twice a week and reading Russell's [P]rinc[iples] of [M]ath[ematics] also occasionally Machs's History of Mechanics. (No date, approx. February/March 1907, TNK)

Whether the seminar really took place with Neurath's co-leadership is not known, but this passage makes obvious references to the early group Frank identified. The "local Privatdozent" most likely was Hahn, who apparently involved himself and his colleagues in the investigation of logicism long before the publication of *Principia Mathematica*. Frank is not explicitly alluded to here, but given these physics topics it is unlikely that he stayed far away.

Needless to say, this contemporaneous report does not conclusively establish the truth of the relevant recollections of Neurath and relevant parts of Frank's histories of logical empiricism—but it does provide independent corroboration of their mutually supportive recollections. Thus I suggest that at least the first part of the Frank-Haller thesis be accepted, namely that something like a first Vienna Circle existed as the group Hahn-Frank-Neurath jointly pursuing interests in the philosophy of science.

3.

Yet what made this group so important, if at all? In the context of the thesis under investigation here this means: what made it into a conduit for influences countering the neo-Kantian ones on logical empiricism? We can begin by teasing out further details from underdescribed aspects of Frank's and Haller's accounts.

There is the question of the precise relation between the Neurath-Haller thesis and the Frank-Haller thesis. Frank himself presupposed the Neurath thesis when he outlined his own, but it is not obvious why there could exist no first Vienna

Circle if there existed no Austrian philosophy. In virtue of what then does the first Vienna Circle represent an instance of (or even depend on) Austrian philosophy? (We'd like to hear more than that it meets the criteria of empiricism, objectivism, and *Sprachkritik*. Couldn't this have been accidental?)

Haller and Frank stressed the influence of French conventionalism. Was this particularly Austrian? And was it even meant to be identified as such? It had better not be—for surely it was precisely this influence that was also shared by the neo-Kantian Ernst Cassirer and found ample expression in his groundbreaking *Substance and Function* (1910). The French influence thus hardly seems able to serve as a feature to distinguish the members of the first Circle. While this conventionalist legacy does provide an excellent counterweight to the absolutism of Wittgenstein's *Tractatus* and so helps to distinguish different factions within the Vienna Circle—alternatively called the epistemologically atomist and the epistemologically holist faction (Haller 1989) or Schlick's group and the left Vienna Circle (Uebel 1998a, 1999)—it does not distinguish the Circle from the neo-Kantian competition. What it does, rather, is stress that neo-Kantians like Cassirer and the first Vienna Circle shared the influence of Poincaré and his challenge to the traditional Kantian conception of the a priori. And what distinguished the first Circle here is that, unlike Cassirer, it did not view what was to replace that conception as a development in the spirit of Kant, but rather one as *contrary* to his philosophical ambitions.

Of course, there is one obvious answer to the question as to what's so Austrian about first Vienna Circle thought—namely, Mach and his positivism. But this is not the only answer that can be given. What's significant about Mach is that he too viewed himself, despite being impressed by Kant early on, as an anti-Kantian, indeed much more so than Poincaré himself (with whom, as with Duhem, Mach imagined wider agreement than actually obtained). It is just this opposition to Kant that makes for the Austrian feature of the conventionalism of the first Vienna Circle (in contrast to the conventionalism of Cassirer) and that places it in the philosophical tradition going back to Bolzano. (By no means were all of those belonging to the Austrian tradition positivists, after all!)

4.

There is another way of asking in virtue of what the first Vienna Circle represents an instance of (or even depends on) Austrian philosophy. Why should they, who were not professional philosophers—none held a philosophy Ph.D. or an academic position in this field—nevertheless come to express a typical Austrian prejudice in their philosophical explorations? Even if we grant that something like an Austrian philosophical tradition existed, it is not obvious that they had to be part of it.[5]

It is important to be clear here about what we have just been led to ask. We are asking both a specific question concerning the first Vienna Circle and a more general question about the Austrian tradition. Did there exist something like a popular intellectual substructure related to the superstructure of master texts marking the line Bolzano-Brentano-Meinong? Short of engaging explicitly with this philosophy—as none did, except for Hahn, who studied Bolzano's work extensively and added a commentary to the 1920 edition of his *Paradoxes of the Infinite*—the members of the first Vienna Circle can only be said to be engaged in Austrian philosophy if they at least shared in the intellectual substructure of this tradition.

The question asks not for criteria or even reasons but causes. This does not mean, of course, that no reasons could be given for adhering to an Austrian philosophy. But it means that we are looking here for something distinctive that the members of the first Vienna Circle could not help being or expressing. Short of attributing essentialist properties to them, this can only mean that we are looking for educational determinants for the thought style of the first Vienna Circle. Can such determinants be found? Yes: the elements in question are the logic textbooks used in the teaching of *Philosophische Propädeutik* in the upper two years of Gymnasium in Austria-Hungary. It is these textbooks and the related obligatory instruction in introductory (syllogistic) logic, epistemology, and philosophy of science, and in empirical psychology—it did not feature ethics or *Weltanschauung*—that provide the substructure for the tradition of Austrian philosophy. These textbooks convey ways of looking at conceptual and epistemological matters that for their readers soon became so familiar as no longer to merit mention unless explicitly challenged.

Can it be determined what the textbooks were that the future members of the first Vienna Circle studied? Yes: for the majority of them, Alois Höfler, student of Brentano and Meinong, was the author of the text they studied prior to their high-school graduation (Woksch 1897, 1901). This textbook, the *Philosophische Grundlehren. Teil 1: Grundlehren der Logik,* was a shortened version of the more scholarly *Logik,* which Höfler had published, as the cover indicates, "with contributions by Alexius Meinong," in 1890. (I include in "the majority of them" Hahn and Neurath, but not Frank; his difference in this regard can be shown not to matter since the propaedeutics textbook he used, while inferior, nevertheless agreed substantially with Höfler's.)[6] Now without saying so explicitly, Höfler's textbook, first published in 1890 with several editions up to 1906,[7] imbued its readers with the basic principles of Bolzano's logic, Brentano's descriptive psychology, and even the first object-theoretical distinctions (between content and object of representations), as well as with early conceptions of the hypothetico-deductive scientific method and the deductive-nomological character of scientific explanations (without discussion of their partially English provenance).[8]

Höfler's textbook, it is important to note, continued the undercover transmission of Bolzano's ideas begun by the textbook, which his own *Grundlehren* came gradually to replace, namely Zimmerman's *Philosophische Propädeutik*, published in two parts in 1852 and 1853.[9] Zimmermann's textbook was written, as is known now but generally was unknown at the time, after Bolzano's own blueprint in accordance with his express request to suppress his name in order to facilitate its wider use in the face of powerful clerical opposition (Winter 1975). The ruse worked brilliantly, as even Höfler's own textbook still shows. While Mach, who completed secondary school in 1855, may still have narrowly escaped Bolzano's underground influence, his Vienna Circle acolytes certainly did not. Quite independently of the role it played for the first Vienna Circle, Höfler's textbook is significant for its role in continuing to impart a broadly Bolzanean perspective on philosophy to generations of prospective university students.

Most likely, of course, most of these students promptly forgot at least the details of the conceptions they were taught, for without employment such things do fall into disrepair. Yet even though the members of the first Vienna Circle went on to study science, their propaedeutic instruction stayed with them longer, given their professed interest in philosophy of science. This is important, for it supports what we earlier determined to be the most significant Austrian feature of Austrian philosophy, its anti-Kantian stance. Given Bolzano's logic and epistemology (and Brentano's psychology), there simply remained no room for Kant's "synthetic a priori." In other words, while neither Zimmermann nor Höfler in their textbooks engaged in explicit anti-Kantian argument, the conception of logic and epistemology, which they do transmit effectively, renders the Kantian perspective unintelligible.[10]

Clearly, then, not only did there exist a mechanism for the transmission of Austrian philosophy apart from its master texts, but the mode of transmission of what was deemed the most basic philosophical knowledge even reinforced the typically Austrian characteristic of this tradition. Inasmuch as the thesis of the first Vienna Circle as a conduit of Austrian philosophy into the Circle around Schlick might be thought vulnerable by creating a suspiciously immaterial tradition, it may be considered cleared. If asked what relation obtains between the Neurath thesis and the Frank thesis, we can answer by pointing to their earliest training in philosophy.

5

It would be compatible with the story as told so far to assume that the members of the first Vienna Circle were utterly ignorant of Kant and of the challenge Kant raised for empiricism. While this used to be widely assumed to be the case for the Vienna Circle at large—another respect in which the recognition of its neo-Kantian influences has been salutary—it is, however, not even true for the members of the first Vienna Circle. And while they were all critical of Kant, the older Austrian philosophers at issue did not all teach that Kantian philosophy could be simply dismissed. Höfler in particular assumes a second important role here.

Already his textbook, despite its silence about the matter of the synthetic a priori elsewhere, carried in its appendix of excerpts from eminent philosophers a relevant passage from Kant's *Prolegomena*. More important still, however, is the fact that Höfler was the editor of the series "Publications of the Philosophical Society of the University of Vienna." In 1899 the society published an anthology, edited by Höfler, *Vorreden und Einleitungen zu klassischen Werken der Mechanik* (Prefaces and Introductions to Classical Works of Mechanics), which featured, besides texts of Galilei, Newton, D'Alambert, and Lagrange, also Kirchhoff, Hertz, and Helmholtz. This anthology provided the background to the discussions in the philosophy of physics of the time. More significant still for our present purposes is that in 1900 there followed Höfler's own *Studien zur gegenwärtigen Philosophie der Mechanik* (Studies in the Contemporary Philosophy of Mechanics) as a critical commentary on Kant's *Metaphysische Anfangsgründe der Naturwissenschaft* (Metaphysical Groundings of Natural Science), which also was published in a new edition by Höfler on behalf of the society. Höfler's commentary discussed Kant's topics from the perspective of the state of scientific and philosophical play in 1900. Importantly, Höfler expressly warned against the presumption that Kantian problems could simply be dismissed (1900, 8–9).

Neurath explicitly mentioned Höfler's editorial activity for the Philosophical Society as part of his rendition of the thesis of a distinctive Austrian tradition (but not specifically his Kant scholarship). While this remains evidentially neutral, it is surely suggestive that the members of the first Vienna Circle, from 1905 onward, took part in the society's activities. Many of the numerous, sometimes biweekly meetings of the Philosophical Society in that decade and still later dealt with issues related to the themes of Höfler's anthology and his Kant commentary.[11] (This focus on philosophy of science is reflected also in the *Yearbooks* the society published from 1902 until 1916, though it was by no means the only area of interest to members of the Philosophical Society as a whole.) Thus it is not at all outlandish to suggest that the members of the first Vienna Circle were cognizant of the post-Kantian nature of the problems being discussed.

As noted by Haller (1993), the Philosophical Society also provided a forum for the early philosophical work of Hahn, Frank, and Neurath. Their participation can be quantified, with a not uncharacteristic result.[12] Between 1906 and the end of World War I, Hahn and Frank lectured or led discussion evenings seventeen times; five of their lectures were published in the *Yearbooks* of the society. After World War I, Hahn and Neurath lectured or moderated discussions another ten times (Frank was then based in Prague). Their participation stopped, however, in 1927: that year the Philosophical Society became the local chapter of the Kant-Society. It is surely no accident that in the following year, Neurath and Hahn were active in the formation of the Ernst Mach Society, which consciously sought a broader audience for their by then maturing and distinctly anti-Kantian "scientific world-conception."

6.

The main phase of the participation of Hahn, Frank, and Neurath in the Philosophical Society took place roughly during the same period as the meetings of the first Circle. We may take it to complement their presumably more open-collar coffeehouse discussions. In this vein we may ask: What form did the work of the first Vienna Circle take? Of course, there is no manifesto they wrote and signed (though they can be expected to have known of the *Aufruf* issued by Petzold in 1912 for a society for scientific philosophy and to have been sympathetic to its aims). In what sense then, if any, is it justified to speak of the philosophical work of the first Vienna Circle? Do their individual publications gel, as it were, into an unspoken program?

It is philosophically relevant that the members of the first Circle have the following publications to offer in the period up to the end of World War I. Neurath offered (besides his efforts to reconceptualize economics) a wide-ranging review "On the Theory of Social Science" (1910), his contribution to the *Werturteilsstreit* of the Verein für Sozialpolitik (1913b), and three lectures in the Philosophical Society on the noncomputability of a social pleasure maximum (1912), on the relation of theoretical and practical reason and their common fallibility and foundationlessness (1913a), and on the methodology of history of science (1915/16). Frank gave three lectures in the Philosophical Society on the conventionalist dissolution of the principle of causality (1907), on the vitalism dispute (1908), and on the concept of absolute movement (1910), and, after a longish break, published his eulogy on Mach as an Enlightenment thinker (1917). Only Hahn seems to have had but one philosophically relevant publication to offer, a review of Pringsheim's *Vorlesungen über Zahlen- und Funktionslehre* (1919).

Can we discern a common theme in these works—beyond the fact that they all more or less belong to the philosophy of science? Certainly, both Frank and Neurath grappled with the lack of foundations of their respective disciplines, physics and economics. Because foundational crises are peculiarly philosophical, this encourages us to conclude that what unified the philosophical efforts of the members of the first Vienna Circle was precisely the attempt to comprehend the foundational crises that shook the science of their day. They were looking for a new philosophy of scientific knowledge.

This, of course, is precisely what Frank and Neurath claimed. To speak of the philosophical work of the first Vienna Circle accordingly is to speak of the shared effort to produce such a new theory of scientific knowledge. It would not detract from its merits to note that much of the philosophical work of the first Vienna Circle amounts to simply getting a measure of the magnitude of the challenge faced. In other words, it amounted to determining desiderata for the new philosophy of science.

Is it possible to say what those desiderata were? Consider Frank's narrative of 1949: "We admitted that the gap between the descriptions of facts and the general principles of science was not fully bridged by Mach, but we could not agree with Kant who built this bridge by forms or patterns of experience that could not change with the advance of science" (1949a, 8). Frank then developed the outlines of what he called the "new positivism," now known as conventionalism. Frank did not indicate what problems their thinking encountered in this direction, yet his narrative seems to imply that it was realized that assistance was needed if one wanted to succeed with the task of formulating the new philosophy of science. In fact, this realization was acted on by Hahn: as is well known, Hahn was instrumental in the appointment of Schlick for one of the open chairs in philosophy in 1922.[13]

The choice of Schlick we may presume to have been a reasoned one. These reasons in turn are bound to reflect the desiderata for the new philosophy of science that not only Hahn but also Frank wanted to see developed. Given that one of Schlick's areas of expertise was also one of Frank's, it is inconceivable that he would have been accepted by Hahn without having passed muster with Frank. Yet more to the point, it seems likely that he was selected precisely because of his work to date. Schlick, we begin to suspect, might as well have been "head-hunted" by Hahn and Frank. (Neurath was most likely not directly involved since at the time he was more than busy with the revolution of 1918–19 and its aftermath.)

Yet why Schlick? Two pieces of his tower over everything else: his *Allgemeine Erkenntnislehre* of 1918 and *Raum und Zeit in der gegenwärtigen Physik,* his monograph on the general theory of relativity, which was first published in 1917 and by 1920 was in its third revised edition. What distinguishes Schlick's work in these books is not only his intimate knowledge of physics, but also his deployment of what Thomas Ryckman has called the "semiotic conception of knowledge" (1991). An important aspect of that conception was the use of Hilbert's method of implicit definitions in accounting for the theoretical terms of the theory. In fact, the question of whether and, if so, how, something like the Hilbertization of the theoretical terms of scientific theories could succeed can lay good claim to being one of the central problematics of the Vienna Circle during its entire existence. Schlick early on formulated proposals, the sympathetic revision of which set a central theme for the Vienna Circle. (That Schlick also interpreted the conventionalist historicization of the a priori as an anti-Kantian move and not as an overdue repair only made for the icing on the cake.)

What follows from this for the first Vienna Circle? We can conclude that, at a minimum, Hahn and Frank were in overall agreement with Schlick's approach and proposals, in other words, with the theme his work set for the future Vienna Circle (or at least thought themselves in such agreement). It would be slightly bolder to claim that Schlick was selected because his approach and proposal were compatible with what Hahn and Frank had independently come to recognize as what was needed philosophically. The first Vienna Circle, in other words, had arrived at something like a program for a program like that of the later logical empiricism.

If this is right, then Hahn's agency bears witness to the fact that the first Vienna Circle was more than an occasional coffeehouse circle, despite the dispersion of its members after 1912 (Frank appointed in Prague, Hahn first in Czernowic then after his war service in Bonn, Neurath traveling extensively in the Balkans and after his war service moving between Heidelberg, Leipzig, and Munich). Rather, Hahn, Frank, and Neurath between them had formed something like the core conception of the later Vienna Circle for at least some of the epistemological problems of science. It was in the knowledge that the conception in question was shared and in need of professional promotion that Hahn worked for Schlick's appointment. (Call this the thesis of the philosophical agency of the first Vienna Circle.)

Clearly, this is not an altogether uncontroversial claim to make. Note, however, that this is neither to credit the first Vienna Circle with the solutions put forward and results gained in the Schlick Circle, nor even to claim that the first Vienna Circle had recognized as problematical all that was called into question in the Schlick Circle. Even so, the claim is substantial. The documents adduced earlier allowed us to defend the thesis of the first Vienna Circle as an intellectual fellow-ship against the charge of uncritical literalism. But the conclusion of philosophical agency goes beyond this, so the specter of the skeptical charge looms large again. It would be no use to point out that what was just deduced from our puzzle pieces is compatible with Frank's narrative—and not only because it goes beyond it in making more explicit and specific the claim of first Vienna Circle agency. What is needed is more evidence for Hahn's philosophical endeavors before 1919.

7.

Now it is true that according to the measure of sources available so far a certain paucity of such evidence for the thesis of the philosophical agency of the first Vienna Circle must be admitted. That Hahn welcomed Schlick's candidacy and promoted it is clear, but that does not yet positively legitimate the attribution of those beliefs to him which the thesis of agency would ascribe; it merely does not contradict it.

It can also be pointed out that Karl Menger, who completed his Ph.D. with Hahn as his advisor, accepted that Hahn, Frank, and Neurath are "the original planners of the Circle" in the sense of what we called the thesis of agency. "When Hahn accepted his first professorial position at the University of Czernovitz in the north-east of the Austrian empire, and the paths of the three friends parted, they decided to continue such informal discussions at some future time—perhaps in a somewhat larger group with the cooperation of a philosopher from the univer-sity" (Menger 1980, ix). Given that Menger's claim also is made only in a memoir, skepticism is not easily assuaged since no sources are given. Its independence from Frank's account can be called into question—even though Menger was a student of Hahn's and could have plausibly claimed to have heard the relevant anecdotes from his teacher. Similar comments hold for Schlick's student Herbert Feigl and his passing mention of the "'prehistoric' Vienna Circle evolv[ing] into the Vienna Circle of the logical positivist" ([1969] 1980, 59).[14]

Are we then reduced to trying to argue by elimination that the thesis of agency must be true since its denial is inconsistent with what is less controversial than the thesis itself? Since our thesis is, among other things, a causal claim, such a strategy seems a counsel of desperation. Fortunately, new and independent evidence is at hand. Current reprints of the Hahn bibliography by K. Mayrhofer of 1934 are pref-aced by the note that reviews are not included since "they generally had no philo-sophical content" (in Hahn 1980, 132). Similarly, Menger's preface to a collection of Hahn's papers contains a footnote that claims that, besides the remark that

Russell will come to be regarded as the most important philosopher of his period, Hahn's reviews in *Monatshefte für Mathematik und Physik* are "merely brief summaries of [the] contents" of the books reviewed (Menger 1980, xviii). Yet it turns out that these reviews—apparently long neglected—are well worth another look for more than one reason. Admittedly, Hahn's reviews in the *Monatshefte* do not offer lengthy disquisitions on matters philosophical, but they do offer revealing ways of characterizing the contents of the books reviewed and plenty of asides. For our purposes, therefore, Hahn's reviews provide just what is needed: independent and contemporary evidence of his philosophical interests—moreover, they provide this evidence not just for the postwar period Menger referred to, but, importantly, also for the period of the first Vienna Circle.

Yet not only do we get this independent check on the philosophical concerns of the young Hahn by virtue of this contemporaneous evidence, in the *Monatshefte* we also find additional evidence of the philosophical concerns of the young Frank (over and above his essays). Frank, too, published there several revealing reviews during the time of the first Vienna Circle that also are well worth reading. Thus Frank's earliest review concerns the aforementioned book by Abel Rey and bears out fully the assertions Frank made concerning this book and its role in 1949. (As in his post–World War II narrative of the development of logical empiricism, Frank gave ample space to an original rendition of Rey's theme of the crisis of mechanistic physics.) In addition to corroborating his retrospective, this review, along with others, provides valuable additional information.

8.

While it is tempting to do so, a comprehensive account of these reviews cannot be undertaken here (the more philosophical ones are listed in the appendix). I will consider them in just enough detail to support the contention made earlier: that Hahn invited Schlick to join causes in the knowledge that his preference for something like the conception of scientific knowledge found in Schlick's work was shared with his first Vienna Circle colleagues. Before turning to the relevant details, let me quickly throw into relief their intellectual biographical context.[15]

By the time Hahn and Frank began to review for *Monatshefte* they had proved themselves promising practitioners in their own fields. By 1905, Hahn was a freshly habilitated mathematician who had just spent part of the two previous years as a postdoctoral fellow in Göttingen with Klein, Hilbert, and Zermelo and had begun to make significant contributions to the calculus of variations and the theory of functions. His first review of a philosophical book in *Monatshefte* was published in 1907. In 1909, Frank, who had studied for his doctorate still under Boltzmann and also had spent time studying in Göttingen with Minkowski, habilitated with a study of "The Role of the Principle of Relativity in the Systems of Mechanics and Electrodynamics," which placed special emphasis on Minkowski's just-published four-dimensional formulation of the special theory of relativity. (Earlier Frank had coined the term "Galilean transformations," since then in use.) Frank's reviews for *Monatshefte* begin in 1910.

Hahn and Frank, in a word, were up to the demands of having informed opinions about the foundational disputes in their special sciences. Not surprisingly, their reviews do not lack the harshness of judgment characteristic of young professionals. This is particularly clear in their reviews of books on mathematical and physical topics. Their decisiveness is muted when it comes to endorsing specific theses of the philosophical books reviewed, but notably not when it comes to recommending overall philosophical strategies.

Hahn's first philosophical review (1907) is of the German edition of Poincaré's *La Valeur de la Science* (orig. 1905, German trans. 1907). Hahn shows his familiarity with the original by criticisms of the translation (and that of the previous *La Science et la Hypothèse*) and lauds the German commentary but otherwise mainly gives a brief annotated rendition of the content while indicating the origin of the separate parts. Declaring Poincaré's Paris address of 1900 on intuition and logic in mathematics "of interest only for mathematicians," Hahn calls "of deepest significance" the core of conventionalism developed (anew) in the parts that follow. He calls the third part "purely philosophical" and replicates the essence of Poincaré's response to Le Roy's "extreme nominalism," according to which

> a scientific fact is but the creation of the scientists; the so-called laws of nature are but rules of action, which represent reality as little as rules of a game of cards. Against this point of view Poincaré argues that these "rules of action" make it possible to predict future experiences whose occurrence or nonoccurrence does not depend on us, so that obviously they are based on some kind of reality, and he investigates further what this reality is. But I cannot here enter further into the extremely original and incisive reasoning of this great mathematician. (Hahn 1907)

What Hahn called "extremely original and incisive" is of course nothing but Poincaré's antirelativist development of "structural realism." Poincaré's structural realism is so called only recently (Worrall 1989), of course, and there are reasons to doubt whether his early Viennese readers appreciated that Poincaré sought to ascribe reality also to theoretical relations, not only to experimental ones. But these very passages also contain Poincaré's remarkable reformulation of the objectivity of scientific knowledge as the intersubjective controllability of its assertions.[16] It was, I suspect, this point in particular that Hahn focused on.

Two years later Hahn reviewed Dingler's *Grundlinien einer Kritik und exakten Theorie der Wissenschaften, insbesondere der mathematischen,* recommending its study but making his dissent crystal clear. Focusing on Dingler's "standpoint of presuppositionlessness" and his aim to establish, via conventional determinations, the "absolute validity" of all laws, Hahn wrote that "much could be said against the standpoint of the author; the definitions may be correct, but his law of causality is not ours, his 'a priori' laws are not our laws of nature, [and] the science for which he is laying the foundation and which is free of contradictions, to be sure, does not have much in common with the sciences which interest us" (1909). Hahn, it seems, accepted Poincaré's argument that the conventional aspects of scientific theories do not diminish their objectivity, once the latter is properly understood. Thus he could reject as mistaken the attempt to reach absolute validity by postulating away all the messy bits of science as practiced—such a project vainly sought to fulfill the now overcome criteria of objectivity of old.

Back in his 1907 paper on the law of causality Frank had produced his own venture into radical conventionalism (aspects of which he still defended in 1916 but much of which by 1932 he conceded was overdone). Thereby Frank ran into sharp opposition by the Göttingen (Friesian) mathematician Gerhard Hessenberg, whose similar opposition to Dingler was noted by Hahn—perhaps indicating some reservations about own his friend's proposals. In any case, Frank's relatively long review (two and a half pages) of Abel Rey's *La Théorie Physique chez les Physiciens Contemporains* lauded the author for placing the question of the objectivity in the center of investigations. Frank also not uncritically remarked that Rey managed to systematize Poincaré at the price of inspiring "doubt whether it is in full agreement with [Poincaré's] views." It is unclear, however, whether Frank noted the difference between them with regard to Rey's claim that "all criticism of physics left standing an objective core, namely the empirical relations between measurable quantities." Rey disregarded the theoretical status of the structural relations Poincaré claimed to have established as real—Frank merely bemoaned Rey's relative lack of clarity in the second, philosophical part of the book. But Frank also wrote: "I wish to draw the attention of mathematicians particularly to the chapter on Duhem, because it discusses his efforts to create an axiomatic system of physics, efforts which touch on those of Hilbert and his students." (1910a) Clearly, Frank appreciated that abstract physical theory employed a language whose terms were defined not by ostension (however mediately) but by implicit definition.[17] For Frank there seems to have been no doubt that this convergence spoke in favor of the conventionalist perspective.

In a review of Planck's *Die Einheit des physikalischen Weltbildes*—which, as is well known, independently prompted a counterpolemic by Mach and a rejoinder by Planck—Frank thus claimed that the following "two statements connected by 'or' do not exclude each other": In Planck's words, "Is the picture physics gives of the world merely a more or less arbitrary creation of our minds or do we find ourselves pushed to the opposite view that there are many natural processes which are wholly independent of each other?" (Frank 1910b). Given that the two principles do not contradict each other, Planck's attack on Mach is deemed to spring from a misunderstanding: Mach's theory of elements is concerned with the empirical basis, Planck with the theory of physics as such. Frank thus opposed Planck's apparent presumption that certain physical results or principles are forever firm and unrevisable with reference to Poincaré (such unrevisability is inevitably bought for the price of conventionalizing the hypothesis), and he opposed Planck's claim of superior heuristic force for a realistic conception of physics by reference to Duhem's *Aim and Structure of Physical Theory*.

By 1910 Hahn and Frank thus showed themselves well versed in conventionalist philosophy of science and able to employ such reasoning to a variety of contemporary issues and foundational projects. Notable is the critical attitude toward Poincaré's stance on the theory of relativity: both Frank (1912) and Hahn (1915) feel compelled to remark on the lack of any mention of Einstein in his later writings on the matter. But this did not hinder their appreciation of the general philosophical attitude of conventionalism which they regarded as "Enlightenment work against the pseudo-problems of a purely grammatically oriented pseudo-philosophy" (Frank 1914).

It is particularly the linkage of Duhem's observations concerning the language of mathematical physics and Hilbert's method for the axiomatization of geometry and physics that deserves attention. (Hahn's own review [1911] of Pasch's *Grundlagen der Analysis* makes clear his familiarity with his *Vorlesungen über neue Geometrie,* which had pioneered the idea of implicit definitions even before Hilbert.) By linking the two, Frank and Hahn in effect moved in precisely the direction of Schlick's thinking about the structure of scientific theories. The propositions of high-level physical theory are expressed in terms of concepts which are only implicitly defined. Of course, this conception raised the question of how to constrain conventionalist ingenuities, as the cases of Le Roy and Dingler show—*avant le lettre* precisely the *Problematik* of Schlick's *Allgemeine Erkenntnistheorie!*

With the usual fallibilist caveats in place it can therefore be said that Hahn's and Frank's reviews for *Monatshefte* support the claim that by 1921 Hahn appreciated exceedingly well—and in this felt in agreement with Frank (and Neurath)— Schlick's use of implicit definitions in his conception of scientific knowledge. (In fact, one of the caveats names the presumption that Hahn and Frank did not forget by 1921 what they had learned by the early 1910s!) That Hahn acted on it as documented constitutes what I called the philosophical agency of the first Vienna Circle.

9.

Both the Neurath and the Frank theses as well as Haller's support for them emerge strengthened from the investigation. There indeed existed a group that can justifiably be called the "first Vienna Circle" and of which it can be said that it served as the conduit for characteristically "Austrian" influences on the formation of logical empiricism. Before concluding let me briefly return to some residual skepticism concerning the Neurath thesis in the current context.

The self-consciously *anti*-Kantian attitude of the members of the first Vienna Circle (unlike that of most German philosophers thinking about Kantian problems in a fashion critical of Kant's own answers) does go some way toward explaining the distinctive rhetoric that distinguishes the later Vienna Circle from the Berlin group, to be sure, but it does not yet give specific content to their ideas. For that, we also need the thesis of the philosophical agency of the first Vienna Circle. Supposing it to hold as outlined above, it becomes clear that the intellectual self-image of the members of this early group also influenced the substance of what we have come to know as Vienna Circle philosophy, for the agency thesis extends to a particular choice of philosophical agenda and methodology, not just its rhetoric. Again, Hahn's and Frank's reviews in *Monatshefte* furnish support.

When Hahn reviewed a small volume of three of Mach's papers translated for the *Monist*, whose content had by then been incorporated into the German original of *Erkenntnis und Irrtum*, he highlighted a quotation on the search for a proof of the parallel axiom and gave expression to his own view on the importance of foundational investigations: "'With excitement we follow the steady expression of the ethical force of the will to know' etc. These are words, for which we are grateful to their author; we wish them the widest appeal. May the mathematicians who tend to reject investigations into the foundations of mathematics consider how a deep non-mathematical but natural scientific thinker views such investigations!" (Hahn 1908).

In fact, two things are here of note. The first is Hahn's pick-up on "the ethical force of the will to know." This is a highly significant marker of the concern— rarely surfacing concern in works of the Vienna Circle but there all the same!— with what might be called the will to rationality and the relation of knowledge to ethics.[18] Most relevant for present concerns, however, is the second point: even though as a practicing mathematician Hahn himself rarely worked and published on the foundations of arithmetic before the later 1920s, he recognized and defended their importance already early on against detractors in his own profession. In a similar spirit, but also focusing on the attitude that alone promised success in such investigations, Frank concluded concerning Rey: "His knowledgeable and thorough book on the theory of physics may help to overcome the disregard which philosophers and physicists often show for each other. It is only through common work in the border areas, by applying oneself to master unfamiliar terminologies, that something could be achieved here. It is surely desirable that a field, from which Descartes, Leibniz and Kant drew such inspiration, does not lie fallow because of conflicts of professional competencies" (Frank 1910).

Both Frank and Hahn recommended an interdisciplinary approach to foundational questions. While professional philosophers were clearly welcome to participate, Frank and Hahn were highly critical of the pretensions of a first philosophy. Frank's derogatory 1914 comment on "grammatically oriented pseudo-philosophy" suggests this, as does his 1915 complaint about his era as one "where a reawakened scholastics is concerned to advance what is but pseudo-precision." Relatedly, in 1916 Hahn dryly noted the "hyper-Kantian" standpoint of one author. Inasmuch as Kantianism of virtually all stripes has to assert the primacy of (its) philosophy in these foundational investigations, it is thus highly likely that already in the 1910s the members of the first Vienna Circle counted it amongst the "school philosophies" to be combated.

Together with their already noted allegiance to the Enlightenment and their early appreciation of efforts at the Hilbertization of empirical theories, we thus find *in nuce* in the publications of the members of the first Vienna Circle many of the elements that combined to undeniable effect in the Vienna Circle's semiofficial manifesto of 1929. Characteristically, this publication (Carnap, Hahn, and Neurath 1929) was authored by its left wing (in which the first Circle was *aufgehoben*).[19] That the manifesto did not please the Circle's public head, Schlick, to whom it was dedicated, had far less to do with its strident opposition to all "school philosophies" —which he tended to share (albeit in a more rarefied form)—than its political subtext. In any case, it seems undeniable that the fact that the members of the first Vienna Circle set the agenda for the later Vienna Circle in the way they did certainly reflects what we designated an Austrian feature of their intellectual socialization.[20]

10.

To conclude: beyond the fact that it points to a distinctive lineage of master texts, the thesis of an Austrian philosophical tradition is supported also by the existence of a series of textbooks in philosophical propaedeutics that give expression to the characteristic anti-Kantian approach of that tradition. It is due to being taught this type of propaedeutics that all the members of the first Vienna Circle (not only Hahn) can be counted into the Austrian tradition still besides their admiration for

Mach. As to the thesis of the first Vienna Circle itself, the claim supported here says that its members were not merely accidentally involved in the creation of the later Circle around Schlick, but they had independently produced something like a program for a program like that of logical empiricism and so promoted his appointment in Vienna. The important reception of French conventionalist thought by the first Vienna Circle was paralleled by the neo-Kantian Cassirer, of course. Unlike Cassirer, however, the first Vienna Circle did not see in this the seeds for a repair to the Kantian conception but for its final dismissal. In an all-too-brief slogan, the original Viennese program for a program might be put thus: with Hilbert and the French conventionalists for Mach against Kant! Whether it is of any consequence that the theorists who provided inspiration would not have endorsed this program must remain a topic for another occasion.

Appendix

Partial list of "philosophical" reviews in *Monatshefte für Mathematik und Physik* by Hahn:

1907. "H. Poincaré, Der Wert der Wissenschaft, 1906," vol. 18, 33–4.

1908a. "E. Mach, Space and Geometry in the Light of Physiological, Psychological and Physical Inquiry, 1906," vol. 19, 60–61.

1908b. "B. Weinstein, Die philosophischen Grundlagen der Wissenschaften, 1906," vol. 19.

1909. "H. Dingler, Grundlinien einer Kritik und exakten Theorie der Wissenschaften, insbesondere der mathematischen, 1907," vol. 20, 51.

1911. "G. Hessenberg, Grundbegriffe der Mengenlehre, 1906," vol. 22, 48–50.

1912a. "P. Duhem, Études sur Leonard de Vinci, 1909," vol. 23, 23–24.

1912b. "H. Poincaré, Biographie, bibliographie analytique des écrits (par E. Lebon), 1909," vol. 23, 24.

1912c. "H. Bergmann, Bolzanos Beiträge zur philosophischen Grundlegung der Mathematik, 1909," vol. 23, 24–25.

1915. "H. Poincaré, Sechs Vorträge über ausgewählte Gegenstände aus der reinen Mathematik und der mathematischen Physik, 1910," vol. 26, 23–24.

1916. "F. Enriques, Probleme der Wissenschaft, 1910," vol. 27, 22–24.

1917. "E. Zilsel, Das Anwendungsproblem, 1916," vol. 28, 37–38.

1928a. "B. Russell, Einführung in die mathematische Philosophie, 1923," vol. 35, 3–4.

1928b. "H. Weyl, Philosophie der Mathematik und Naturwissenschaft, 1927," vol. 35, 51–55.

1929. "B. Russell, Unser Wissen von der Aussenwelt, 1926," vol. 36, 7–8.

Partial list of "philosophical" reviews in *Monatshefte für Mathematik und Physik* by Frank:

1910a. "A. Rey, Die Theorie der Physik bei den modernen Physikern, 1909," vol. 21, 43–45.

1910b. "M. Planck, Die Einheit des physikalischen Weltbildes, 1909," vol. 21, 46–47.

1910c. "P. Gruner, Die Voraussetzungen und die Methode der exakten Naturforschung, 1909," vol. 21, 49.

1911. "H. Minkowski, Raum und Zeit, 1909," vol. 22, 7.

1912a. "J. B. Stallo, Die Begriffe und Theorien der modernen Physik, 2. Aufl. 1911," vol. 23, 31.

1912b. "H. Poincaré, Die neue Mechanik, 1911," vol. 23, 32.

1914. "H. Poincaré, Letzte Gedanken, 1913," vol. 25, 54–55.

1915. "H. Kleinpeter, Der Phänomenalismus, 1913," vol. 26, 46–47.

1916. "M. Planck, Das Prinzip der Erhaltung der Energie, 3. Aufl. 1913," vol. 27, 18.

1917. "E. Study, Die realistische Weltansicht und die Lehre vom Raume, 1914," vol. 28, 4–5.

Note: pagination of reviews in *Monatshefte* is separate from rest of journal.

Notes

1. The thesis is adumbrated in Carnap, Hahn, and Neurath (1929), fully developed in Neurath (1936), and further explored and defended by Rudolf Haller, for instance, in [1968] 1988 and 1986. The term "Neurath-Haller thesis" is taken from Smith (1994, 14).

2. "Let us note that in Austria the anti-Kantian and Leibnizian Herbartianism was well regarded . . . since Kantianism was viewed as an effluence of the French Revolution. It is not difficult to see why the state and the Church on the one hand favoured anti-Kantianism, but on the other hand feared the new scientism, even though they contributed to its progress. Bolzano, author of *The Theory of Science* and *The Paradoxes of Infinity* and other important works, whose importance is appreciated only today [1936, TU], was partly supported and partly persecuted. . . . The speed with which logical analysis came to predominate is remarkable. . . . There was a systematic effort to link the particularity of Austrian culture grounded on logic with certain pre-Kantian tendencies of a utilitarian and empiricist nature and thereby to avoid the Kantian interlude. One example: the *Ordinarius* in Vienna was the Herbartian Robert Zimmermann, who was taught exact and mathematical thinking still by Bolzano himself. . . . already Zimmermann had a discernible influence on our present period . . . Brentano worked in Vienna more in a logical and critical vein than in a metaphysical. . . . Notably Meinong too, whose theory of objects always proved stimulating for partisans of logical empiricism, continues Brentantan lines of inquiry. His student Mally works on logic. Also continuing Brentanean lines is Twardowski (Lvov), who has awakened interest in the problems of modern logic in Poland . . . In Vienna itself the logical efforts of the Brentano school were developed further by a man, who supported the beginnings of the Viennese school [*sic*] at the beginning of the 20th century by introducing discussions of the foundations of physics: Alois Höfler. . . . He was for a long time the editor of the "Publications of the Philosophical Society of the University of Vienna"; they show the enthusiasm with which at that time the problems were discussed which later on were to occupy the Vienna school. . . . the most decidedly empiricist standpoint was represented by Mach. . . . The anti-metaphysical outlook, which Mach promoted in Vienna, did not remain his alone. In contrast to the development in Germany, an entire generation went over to positivism, utilitarianism and empiricism" Neurath ([1936] 1981, 688–91). All translations from sources where no English translation is indicated in the bibliography, and of the correspondence marked WKA and TNK, are by the present author.

3. Frank (1941a, 1949a), Haller (1985, 1993, chap. 3).

4. Compare the biographical accounts collected in Neurath (1973), P. Neurath (1994), and Fleck (1996).

5. At least in our day, it cannot be said that philosophizing scientists can be trusted to be suitably informed about contemporary philosophy; cf. Salmon (1997).

6. This claim clearly needs to supported by a textual analysis, which, however, would lead away from the present topic; it is provided in Uebel (2000, chap. 4).

7. 2d ed. 1896; 3d rev. ed. 1902; 4th ed. 1906.

8. Particularly the latter two provide a ready example of the force of early teachings. Given their canonical formulation by Hempel only in the 1940s, the hypothetico-deductive conception of scientific theorizing, even the deductive-nomological conception of explanation, had nevertheless been presupposed in much of the relevant work of the Vienna Circle all along, without being explicitly remarked upon.

9. 2d rev. ed. 1860; 3d ed. 1867.

10. This claim also needs to be supported by extensive textual analysis; it is also provided in Uebel (2000, chap. 4).

11. Compare the list of talks, with titles and dates, given in the Philosophical Society in Reininger (1938), translated in the appendix of Blackmore (1995).

12. Again, see Reininger (1938) and Blackmore (1995).

13. The other two going to Bühler and Reininger; see Stadler (1979).

14. Note also that Feigl's historical sketch mentions Richard von Mises as another member of the "prehistoric" Vienna Circle, while Frank's reports do not. Given Frank's life-long acquaintance and occasional cooperation with Richard von Mises (and his introduction to his memorial volume), I follow Frank here and in my (2000).

15. Compare the biographical entries in Haller (1993) and Stadler (1997).

16. See Poincaré ([1905] 1946, 347–49); cf. Uebel (1998b).

17. Due to his own studies in Göttingen, Frank was aware of the then still unpublished efforts of the Hilbert school toward the axiomatization of physics (over and above that geometry)—like Hahn, but unlike Schlick and Carnap. For discussion of their varying receptions (or lack of it) of this program of Hilbert's in later years, see Majer (2002) and Stöltzner (2002).

18. In a similar fashion Frank expressed unreserved agreement with Poincaré's pronouncements on science and ethics in his review of *Dernières Pensées* of 1914. These markers of an important but so far mostly unexplored aspect of Vienna Circle thought must here remain so; but see Uebel (1998a).

19. This indicates another aspect of the relevance of the first Vienna Circle to the history of the Vienna Circle around Schlick: with the recruitment of Carnap, the former first Vienna Circle became the left Vienna Circle.

20. Some readers may still suspect, of course, that the Neurath thesis was not always used in the legitimate descriptive manner. It is undeniable, for instance, that we can discern in the later work of Hahn, Frank, and Neurath more than just anti-Kantianism, but a self-conscious attempt to avoid the philosophical *Sonderweg* oriented mainly to indigenous German thinkers (or ancient Greek ones with whom a peculiar affinity was claimed to obtain) and instead to seek the connection with "Western" empiricism and positivism. Nevertheless, the thesis of an Austrian philosophical tradition was not intended to reflect a national resistance that never was: in the straits of exile, in which Neurath found himself, the thesis of an Austrian tradition at best served to project an intellectual basis for a broad-based resistance still to be formed. That Meinong's student Ernst Mally—who as a logician garnered an honorable mention still in Neurath's historiographical sketch of 1936—already shortly after the *Anschluss* produced an explicitly National Socialist propaedeutics text (1938; for discussion, see Uebel [2000, chap. 8]) shows, of course, that even such a projective use is highly fanciful: philosophical lineages do not prescribe politics (however striking some allegiances no doubt are).

Abbreviations

ASP   Archive for Scientific Philosophy, Carnap Papers, Hillman Library, University of Pittsburgh, Pittsburgh, Pa.

TNK   Ferdinand Tönnies Nachlass, Schleswig-Holsteinische Landesbibliothek, Kiel, Germany.

WKA   Wiener Kreis Archiv (Nachlass Otto Neurath), Rijksarchief Noord-Holland, Haarlem, The Netherlands.

Bibliography

For the reviews by Hahn and Frank, see the appendix.

Blackmore, J. 1995. *Ludwig Boltzmann. His Later Life and Philosophy, 1900–1906.* Vol. 2. Dordrecht: Kluwer.

Bolzano, B. 1921. *Paradoxien des Unendlichen.* Ed. A. Höfler with annotations by H. Hahn. Leipzig: Meiner.

Carnap, R., H. Hahn, and O. Neurath. [1929] 1973. "Scientific World Conception: The Vienna Circle." In *Empiricism and Sociology,* trans. and ed. M. Neurath and Robert S. Cohen, 299–318. Dordrecht: Reidel, 1973. (Originally published as *Wissenschaftliche Weltauffassung—Der Wiener Kreis.* Vienna: Wolf.)

Cassirer, E. [1910] 1923. *Substance and Function.* Trans. W. and M. Swabey. Chicago: Open Court. (Originally published as *Substanzbegriff und Funktionsbegriff.* Berlin: Bruno Cassirer.)

D'Acconti, A. 1995. "The Genealogies of Logical Positivism." Ph.D. diss., Cambridge University.

Feigl, H. [1969] 1980. "The Wiener Kreis in America." In *Inquiries and Provocations. Selected Writings, 1929–74,* Dordrecht: Reidel, 57–94. (Originally published in *The Intellectual Migration, 1930–1960,* ed. D. Fleming and B. Baylin. Cambridge, Mass.: Harvard University Press.

Fleck, L. [1979] 1996. "Part 1: A Life between Science and Politics." In *Otto Neurath: Philosophy between Science and Politics,* N. Cartwright, J. Cat, L. Fleck, T. E. Uebel, 7–88. Cambridge: Cambridge University Press. (Original "Otto Neurath: Eine biographisch-systematische Untersuchung." Ph.D. diss., Karl-Franzens-Universität Graz.)

Frank, P. 1908. "Mechanismus oder Vitalismus. Versuch einer präzisen Formulierung der Fragestellung." *Annalen der Naturphilosophie* 7:393–409.

———. 1910. "Gibt es eine absolute Bewegung?" In *Wissenschaftliche Beilage zum 23. Jahresbericht der Philosophischen Gesellschaft an der Universität Wien,* 1–19. Leipzig: Barth.

———. 1934. "Nachruf: Hans Hahn." *Erkenntnis* 4:315–16.

———. 1941. "Introduction: Historical Background." *In Between Physics and Philosophy,* 3–16. Cambridge Mass.: Harvard University Press.

———. [1907] 1949. "The Law of Causality and Experience." In *Modern Science and Its Philosophy,* 53–60. Cambridge, Mass.: Harvard University Press. (Originally published as "Kausalgesetz und Erfahrung," *Annalen der Naturphilosophie* 6.)

——— [1917] 1949. "The Importance for Our Times of Ernst Mach's Philosophy of Science." In *Modern Science and Its Philosophy,* 61–79. Cambridge, Mass.: Harvard University Press. (Originally published as "Die Bedeutung der physikalischen Erkenntnistheorie Ernst Machs für das Geisteslebens unserer Zeit," *Die Naturwissenschaften* 5.)

———. [1938] 1949. "Determinism and Indeterminism in Modern Physics." In *Modern Science and Its Philosophy*, 172–85. Cambridge, Mass.: Harvard University Press. (Originally published as "Bemerkungen zu Ernst Cassirer: Determinismus und Indeterminismus in der modernen Physik," *Theoria* 4.)

——— 1949a. "Introduction: Historical Background." In *Modern Science and Its Philosophy*, 1–51. Cambridge, Mass.: Harvard University Press.

——— 1949b. *Modern Science and Its Philosophy*. Cambridge, Mass.: Harvard University Press.

———. [1932] 1998. *The Law of Causality and Its Limits*. Trans. R. S. Cohen. Dordrecht: Kluwer, 1998. (Originally published as *Das Kausalgesetz und seine Grenzen*. Wien: Springer.)

Hahn, H. [1919] 1980. "Review of Alfred Pringsheim: Vorlesungen über Zahlen-und Funktionentheorie." Trans. Hans Kaal. In *Empiricism, Logic, Mathematics*, ed. B. McGuinness. Dordrecht: Reidel. (Originally published as "Besprechung von Alfred Pringsheim: Vorlesungen über Zahlen- und Funktionentheorie," *Göttingsche gelehrte Anzeigen*, 1919, H. 9 u. 10.)

Haller, R. [1968] 1986. "Wittgenstein and Austrian Philosophy." Trans. B. Smith. In *Questions on Wittgenstein*, 1–26. Lincoln: University of Nebraska Press. (Originally published as "Wittgenstein und die österreichische Philosophie," *Wissenschaft und Weltbild* 21:77–87.)

———. [1985] 1991. "The First Vienna Circle." In *Rediscovering the Forgotten Vienna Circle*, 95–108. Dordrecht: Kluwer. (Originally published as "Der erste Wiener Kreis," *Erkenntnis* 22.)

———. [1986] 1991. "On the Historiography of Austrian Philosophy." In *Rediscovering the Forgotten Vienna Circle*, ed. T. E. Uebel, 41–50. Dordrecht: Kluwer. (Originally published as "Zur Historigraphie der österreichischen Philosophie," in *Von Bolzano zu Wittgenstein: Zur Tradition der österreichischen Philosophie*, ed. J. Nyiri. Vienna: Hölder-Pichler-Tempsky.)

———. 1991. "Atomism and Holism in the Vienna Circle." In *Advances in Scientific Philosophy*, ed. G. Schurz, G. Dorn, 265–79. Amsterdam: Rodopi.

———. 1993. *Neo-Positivismus: Eine historische Einführung in die Philosophie des Wiener Kreises*. Darmstatt: Wissenschaftliche Buchgesellschaft.

Höfler, A. 1890. *Grundlehren der Logik*. Vienna: Tempsky.

———. 1900. *Studien zur gegenwärtigen Philosophie der Mechanik. Als Nachwort zu: Kants Metaphysische Anfangsgründe der Wissenschaft*. Leipzig: Pfeffer.

———, ed. 1899. *Vorreden und Einleitungen zu klassischen Werken der Mechanik: Galilei, Newton, D'Alembert, Lagrange, Kirchhoff, Hertz, Helmholtz*. Leipzig: Pfeffer.

Mally, E. 1938. *Anfangsgründe der Philosophie. Leifaden für den Philosophischen Einführungsunterricht an höheren Schulen*. Vienna-Leipzig: Hölder-Pichler-Tempsky.

Majer, U. 2002. "Hilbert's Program to Axiomatize Physics (in Analogy to Geometry) and Its Impact on Schlick, Carnap and Other Members of the Vienna Circle." In *History of Philosophy and Science: New Trends and Perspectives*, ed. M. Heidelberger and F. Stadler, 213–24. Dordrecht: Kluwer, 2002.

Menger, K. 1980. "Introduction." In H. Hahn, Introduction to *Empiricism, Logic, Mathematics*, ed. B. McGuinness, ix–xviii. Dordrecht: Reidel, 1980.

Neurath, O. [1910] 1981. "Zur Theorie der Sozialwissenschaften." In *Gesammelte philosophische und methodologische Schriften*, ed. R. Haller and H. Rutte, 23–46. Vienna: Hölder-Pichler-Tempsky, 1981. (Originally published in *Jahrbuch für Gesetzgebung, Verwaltung und Volkswirtschaft im Deutschen Reich* 34.)

————. [1912] 1973. "The Problem of the Pleasure Maximum." In *Empiricism and Sociology*, trans. and ed. M. Neurath and R. S. Cohen, 113–22. Dordrecht: Reidel, 1973. (Originally published as "Das Problem des Lustmaximums," in *Jahrbuch der Philosophischen Gesellschaft an der Universität zu Wien 1912*. Vienna.)

————. [1913] 1983. "The Lost Wanderers of Descartes and the Auxiliary Motive." In *Philosophical Papers, 1913–1946*, trans. and ed. R. S. Cohen and M. Neurath, 1–12. Dordrecht: Reidel. (Originally published as "Die Verirrten des Cartesius und das Auxiliarmotiv," in *Jahrbuch der Philosophischen Gesellschaft an der Universität zu Wien 1913*.)

————. [1913] 1981. "Über die Stellung des sittlichen Werturteils in der wissenschaftlichen Nationalökonomie." In *Gesammelte philosophische und methodologische Schriften*, ed. R. Haller and H. Rutte, 69–70. Vienna: Hölder-Pichler-Tempsky. (Originally published in *Äusserungen zur Werturteilsdiskussion*, ed. Verein für Sozialpolitik.)

————. [1916] 1983. "On the Classification of Systems of Hypotheses." In Neurath *Philosophical Papers, 1913–1946*, trans. and ed. R. S. Cohen and M. Neurath, 13–31. Dordrecht: Reidel. (Originally published as "Zur Klassifikation von Hypothesensystemen," in *Jahrbuch der Philosophischen Gesellschaft an der Universität Wien 1914 und 1915*.)

————. [1936] 1981. "Die Entwicklung des Wiener Kreises und die Zukunft des Logischen Empirismus." In *Gesammelte philosophische und methodologische Schriften*, ed. R. Haller and H. Rutte, 673–703. Vienna: Hölder-Pichler-Tempsky. (Originally published as *Le developpement du Cercle de Vienne et l'avenir de l'Empiricisme logique*. Paris: Hermann & Cie.)

Neurath, P. 1994. "Otto Neurath (1882–1945). Leben und Werk." In *Otto Neurath oder Die Einheit von Wissenschaft und Gesellschaft*, ed. P. Neurath and E. Nemeth, 13–95. Vienna: Böhlau.

Petzold, J. [1912] 1993. "Aufruf." In *Science and Anti-Science*, by G. Holton, 13. Cambridge, Mass.: Harvard University Press.

Poincaré, H. [1905] 1946. *The Value of Science*. Trans. B. Halsted. In *Foundations of Science*, 199–356. Pittsburgh: Science Press. (Originally published as *La valeur de la science*. Paris.)

Reininger, R. 1938. *50 Jahre Philosophische Gesellschaft an der Universität Wien, 1888–1938*. Vienna: Verlag der Philosophischen Gesellschaft an der Universität Wien.

Ryckman, T. A. 1991. "Conditio Sine Qua Non? *Zuordnungen* in the Early Epistemologies of Cassirer and Schlick." *Synthese* 88:57–95.

Salmon, W. 1997. "The Philosophy of a Scientist: Critique of Weinberg." Presentation given at the Third International Fellows Meeting of the Pittsburgh Center for Philosophy of Science, Castaglioncello, Italy.

Schlick, M. [1917] 1979. "Space and Time in Contemporary Physics." Trans. H. Brose and P. Heath. In *Philosophical Papers*, vol. 1, ed. by H. Mulder and B. van de Velde-Schlick, 207–69. Dordrecht: Reidel. (Originally published as "Raum und Zeit in der gegenwärtigen Physik," *Die Naturwissenschaften* 5. Enlarged separate ed. Berlin: Springer, 1917; 4th rev. ed., 1922.)

————. [1918/25] 1985. *General Theory of Knowledge*. Trans. A. Blumberg. La Salle, Ill.: Open Court. (Originally published as *Allgemeine Erkenntnislehre*. Berlin: Springer, 1918; 2d rev. ed. 1925.)

Smith, B. 1994. *Austrian Philosophy: The Legacy of Franz Brentano*. La Salle, Ill.: Open Court.

Stadler, F. [1979] 1991. "Aspects of the Social Background and Position of the Vienna Circle at the University of Vienna." In *Rediscovering the Forgotten Vienna Circle,* ed. T. E. Uebel, 51–77. Dordrecht: Kluwer, 1991. (Originally published as "Aspekte des gesellschaftlichen Hintergrunds und Standorts des Wiener Kreises am Beispiel der Universität Wien," in *Wittgenstein, der Wiener Kreis und der kritische Rationalismus,* ed. H. Berghel, A. Hübner, E. Köhler. Wien: Hölder-Pichler-Tempsky.)

———. [1997] 2001. *Studies on the Vienna Circle.* Trans. C. Nielsen et al. Vienna: Springer. (Originally published as *Studien zum Wiener Kreis: Ursprung, Entwicklung und Wirkung des Logischen Empirismus im Kontext.* Frankfurt a. M.: Suhrkamp.)

Stöltzner, M. 2002. "How Metaphysical Is 'Deepening the Foundations'? Hahn and Frank on Hilbert's Axiomatic Method." In *History of Philosophy and Science: New Trends and Perspectives,* ed. M. Heidelberger and F. Stadler, 245–62. Dordrecht: Kluwer, 2000.

Uebel, T. E. 1998a. "Enlightenment and the Vienna Circle's Scientific World-Conception." In *Philosophers on Education,* ed. A. O. Rorty, 418–38. London: Routledge.

———. 1998b. "Fact, Hypothesis and Convention in Poincaré and Duhem." *Philosophia Scientiae* 3:75–95.

———. 1999. "Otto Neurath, the Vienna Circle, and the Austrian Tradition." In *German Philosophy Since Kant,* ed. A. O'Hear, 249–70. Cambridge: Cambridge University Press.

———. 2000. *Vernunftkritik und Wissenschaft: Otto Neurath und der erste Wiener Kreis.* Vienna: Springer.

Winter, E., ed. 1975. *Robert Zimmermanns Philosophische Propädeutik und die Vorlagen aus der Wissenschaftslehre Bernard Bolzanos.* Vienna: Verlag der Österreichischen Akademie der Wissenschaften.

Worrall, J. 1989. "Structural Realism: The Best of Both Worlds?" *Dialectica* 43:99–124.

Woksch, K., ed. 1897. *Jahresbericht 1896/97 des k. k. Staatsgymnasiums im XIX. Bezirk von Wien.* Vienna: Eigenverlag.

———. 1901. *Jahresbericht 1900/01 des k. k. Staatsgymnasiums im XIX. Bezirk von Wien.* Vienna: Eigenverlag.

Zimmermann, R. 1852. *Philosophische Propädeutik für Obergymnasien. Erste Abtheilung: Empirische Psychologie.* Vienna: Braunmüller.

———. 1853. *Philosophische Propädeutik für Obergymnasien. Zweite Abtheilung: Formale Logik.* Vienna: Braunmüller.

On the *International Encyclopedia,* the Neurath-Carnap Disputes, and the Second World War

GEORGE A. REISCH
Independent Scholar

> Of particular importance for me personally was his emphasis on the connection between our philosophical activity and the great historical processes going on in the world."
>
> —*Rudolf Carnap (1963, 23)*

Introduction

There is some irony in Carnap's recollection about Otto Neurath. However much Carnap was intrigued by this "connection" between philosophy and "great historical processes," Neurath would say that Carnap never understood it. And this would be a sore point for Neurath, because the last years of their collaboration, terminated by Neurath's death in late 1945, were marked by intellectual, professional, and personal discord—discord that concerns, from Neurath's point of view, this connection between logical empiricism and the world.

I will examine the collaboration between Neurath, Carnap, and Charles Morris as they organized and edited the *International Encyclopedia of Unified Science* and ran the other components of their Unity of Science movement. These included the annual International Congresses for the Unity of Science held in the 1930s; trying to rescue and re-create *Erkenntnis* as the *Journal of Unified Science;* creating a monograph series, the Library of Unified Science, and the "Unity of Science Forum" that appeared sporadically in *Synthese.*[1]

I will outline this collaboration and, on the basis of their personal correspon-

94

dence, examine several conflicts that arose among the editors—intellectual conflicts, professional conflicts, and the personal conflict between Neurath and Carnap. After introducing these issues, I will examine them from Neurath's point of view, specifically taking into account his expectations and hopes for postwar Europe. In the early 1940s, Neurath was not particularly easy to work with. But if we view his comments, decisions, and behaviors in light of his thoughts about the war, and his hopes for the Unity of Science movement in the postwar world, we can better appreciate how much Neurath saw at stake in these disputes.

Overview of the Collaboration

In 1934, Charles Morris, then at the University of Chicago, traveled to Europe for the Eighth International Congress of Philosophy in Prague. Morris met Neurath and Carnap and returned to Chicago intending to make his home university the American home of Neurath's Unity of Science movement. He gained support from the University of Chicago Press for Neurath's new encyclopedia of unified science, and he paved the way for Rudolf Carnap to join Chicago's philosophy department. By 1936, Carnap and the encyclopedia were in Chicago, while Neurath remained in Holland after fleeing Austria in 1934. The first monographs of the encyclopedia appeared in 1938, and, for about a year, they appeared every few months.

By 1940, however, there were problems at every turn. Neurath barely escaped Holland when it was invaded and spent six months interned in England. The encyclopedia project slowed down, causing the University of Chicago Press to worry about rising costs and subscriber dissatisfaction. In 1943, they decided to suspend the project. Neurath, by this time living in Oxford, deftly persuaded them to reconsider by appealing to the admirable "business as usual" spirit he saw around him in England.[2] And he promised that one or two monographs would soon be ready for printing. It turned out, however, that the only monograph that could be ready in time was Neurath's own, his "Foundations of the Social Sciences" (Neurath 1944).

Neurath's monograph, however, ignited a storm that had been smoldering. As we shall see, Carnap did not like the monograph, did not edit it, and therefore arranged for a disclaimer to appear in the pamphlet indicating that he had not acted as editor for that particular title. This made Neurath very angry, and they argued in increasingly personal terms until Neurath's death in late 1945. After that, except for a brief flourish of activity in the early 1950s under the Institute for the Unity of Science, founded in Boston by Philipp Frank, the Unity of Science movement declined and disappeared. The encyclopedia never grew beyond the first twenty monographs, the last of which appeared in 1970.[3]

Axes of Conflict

Neurath and Carnap were no strangers to arguing, at least because Neurath argued with almost everyone. His abhorrence of metaphysics is famous, and he confronted anyone whose ideas or, more commonly, choice of language he thought "metaphysical." Carnap was no exception. There were two main periods of discord—the protocol sentence debates of the late 1920s and early 1930s (Uebel 1992) and their arguments over semantics that began in the mid-1930s. During this second period, in early 1943, Carnap sent Neurath his book *Introduction to Semantics*. Neurath immediately filed his antimetaphysical charge: "I am really depressed to see here all the Aristotelian metaphysics in full glint and glamour, bewitching my dear friend Carnap through and through."[4] Neurath had several, overlapping objections to semantics. They cluster around the complaint that Carnap's semantics (and specifically the theory of truth and confirmation theory) either presupposed or implicated what Neurath called an "absolutistic" or nonpluralist understanding of scientific language.

Neurath emphasized his pluralism increasingly toward the end of his career, especially as he came to see that his colleagues did not share it. It meant several things. First, scientific language, like all language, is imprecise and multivalent. He saw meaning in a holistic way: a term or statement gains meaning through connections with other terms and statements that in turn are connected to others. Meanings are not, therefore, exhaustively specifiable since these chains of connection can be unbounded. Semantics, he believed, obscured that fact.

As for confirmation theory and truth theory, Neurath rejected Carnap's use of levels of language, one that refers to statements and another to the world—as in the formulation "the statement 'snow is white' is true if snow is white." He believed, as he once told Charles Morris, that the lower level was "an ontological one" that smuggled into semantics an Aristotelian metaphysics of real things and real properties that has no place in properly empiricist scientific language.[5]

Neurath's charges sometimes had a sociological aspect. If one protocolist says that snow is white and another says that snow is not white, any dispute about the truth of "snow is white" may falter in the sociological facts of the disagreement. We may analyze the situation further, he would agree, and possibly determine that one of the parties is lying or hallucinating; but the possibility remains that two scientists will bluntly disagree about the color of snow (or something else) and thus reveal a kind of singularity in semantical theory. It will be of no use for settling the dispute because it simply begs the questions in dispute. Neurath instead promoted a wider program, a "behavioristic" theory that would marshal the resources of unified science to study disagreements among actual scientists. Hopefully, a behavioristic program would yield a repertoire of techniques for resolving arguments and conflicts by "orchestrating" different beliefs and perceptions in order that science and its applications always proceed.[6]

Finally, Neurath's arguments during this period sometimes joined his thoughts about "unpredictability within empiricism." He felt that many empiricists mistakenly believed that predictability was an ordinary, if not essential, aspect of scientific phenomena. Even with a sophisticated unified science of the future, Neurath believed, we may *not* necessarily rely on science to make reliable predictions. In sociology, especially, he pointed out in his "Foundations of Social Science" (Neurath 1944), events in the long term may be affected by details (of geography, weather, or ecology) that are not practically knowable. Thus Neurath's "pluralism" also includes the view that we must regard the future as open, influenced by our choices, and not predictable or deducible on the basis of scientific, much less philosophical, theories.

PROFESSIONAL DISPUTES

Prior to the dispute over Neurath's monograph, Neurath, Carnap, and Morris collaborated fairly peacefully and effectively. Despite this argument over semantics as well as others, they shared hope that the Unity of Science movement would grow to affect intellectual life and education throughout the world. Neurath was the leader of the movement and editor in chief of the encyclopedia; Carnap and Morris nearly always accepted this.

One issue that strained their collaboration, however, was Neurath's distance from Chicago. Carnap and Morris often waited weeks for Neurath to receive their letters containing questions or problems to be solved, and they waited weeks for a reply. Sometimes Morris requested that Neurath leave small matters for him and Carnap to decide. But Neurath always retained control. Morris also wished that Neurath would relocate to the United States. But Neurath did not want the movement to become entirely American and believed it important to keep some of the operations in Europe.

In 1941, this issue came to the foreground. In the background lay the important fact that Neurath's movement was always cash-starved. The Press encouraged the editors to find external funding for their publications. Morris, especially, tried very hard, but none was found. The encyclopedia was funded by advanced subscriptions. If and when they succeeded in bringing *Erkenntnis* to America, the University of Chicago Press insisted the arrangement would have to be the same unless outside funding was found.

As the Press and Neurath negotiated with Felix Meiner to purchase *Erkenntnis,* hopes for funding suddenly appeared. The Rockefeller Foundation, Morris reported, might grant the movement $20,000 on the condition that all the movement's publications would be based in America. Morris urged Neurath to contact the foundation and propose "a concrete plan" for meeting this requirement (hinting as well that the plan include Neurath coming to America).[7] In the meantime, Neurath had also been talking with the English publisher Basil-Blackwell about purchasing *Erkenntnis.* Carnap and Morris objected to this possibility, for they believed the new *Journal of Unified Science* belonged in Chicago with the encyclopedia. Just as their hopes for this Rockefeller windfall were high, Neurath surprised them by announcing that he had reached an agreement with Blackwell. And, he told them firmly, the matter was closed. They were not happy, and the Rockefeller grant fell through.

PERSONAL DISPUTES

Besides these intellectual and professional disputes, the personal relationship between Neurath and Carnap soured in 1944 after Carnap placed his disclaimer in Neurath's monograph "Foundations of the Social Sciences" (Neurath 1944). Carnap was then living in Santa Fe, and his motivations were simple. Neurath and Morris rushed the publication in order to fulfill Neurath's promise to the University of Chicago Press that a monograph would soon appear. Per the usual procedures, Carnap expected to receive Neurath's manuscript for editing, but he received in the mail only page proofs. Changes, Morris told him, would prove time consuming and expensive. Even worse, Carnap found the monograph disorganized, unclear, and peppered with phrases and slogans he did not understand. Instead of holding up the monograph's publication and further annoying the Press, and possibly also Neurath, he decided simply not to edit the manuscript and have that fact indicated on the back of the title page.

When Neurath learned of this, he was deeply offended. He told Carnap that it was a public and personal slight.[8] To defend himself, Carnap sent Neurath a copy of the letter he had sent Morris in which he first explained his position and his decision. He assumed that once Neurath saw the grounds for his decision he would cease to be so upset about it. But Neurath only became more angry. In this letter, Carnap referred to Neurath as a "volcano" who did not take criticism well (without erupting), and said that the monograph in its final form would do the movement more harm than good.[9] Neurath then charged that Carnap was "grieving" him and not taking him seriously as an intellectual colleague. Instead of engaging Neurath as an author with an important contribution to make, Neurath reasoned, Carnap pulled his name and let the monograph go through—all to keep the infamous "volcano" from exploding.

Although Carnap emphasized that his decision was merely editorial and not meant as a personal judgment, Neurath dwelled on these personal implications. Carnap's disclaimer was just one more insult in a string of personal affronts stretching back to the days of the Vienna Circle. Neurath recalled that Schlick treated him rudely—once in print (Schlick 1937)—and that Carnap had earlier failed to give Neurath attribution when writing about protocol sentences in his *Die physikalische Sprache als Universalsprache der Wissenschaft* (Carnap 1934). As he reflected upon these and other slights, Neurath voiced fears for his reputation and influence in philosophy of science. He was perhaps not much respected as a contributor to logical empiricism—merely as its organizer or manager:

> I should say so: my friends and potential friends do not regard me as much important within their circle—except as far as I act as a kind of manager. . . . OK. Let them.[10]
>
> I think that I, to a certain extent, am one of the pillars of this movement, not only its "promoter" as people sometimes like to treat me.[11]

At times, both men were consumed and frustrated by their escalating argument. Unfortunately, it was unresolved when Neurath died in December 1945.

The Postwar Role of the Movement

When we take the ongoing war into account, the disputes I have been describing become more complex and interconnected. First, recall that the editors, in their own ways, hoped that the Unity of Science movement would be socially powerful. It would foster international cooperation and development of a universal language of science that could be used to better understand and manage modern life. As the movement languished during the war, Neurath hoped that it would grow and blossom during the subsequent peace. "As you know," he told Carnap and Morris in 1942, "I do not cease to manage things and I shall think that we have a lot to do for Europe after this war."[12] "This nazified Germany and Europe will need [some] good dishes, we shall present them."[13]

This helps explain Neurath's decision to remain living in England and also his decision to publish the *Journal of Unified Science* with Basil-Blackwell. He believed that Morris and Carnap shared the popular view that the war had ruined England and Europe—"I shall not share this 'point of view' no, no, no." He insisted on maintaining their "British-American branch" of the movement so that it may "play its role in the building up the intellectual atmosphere of future Europe."[14] The inconveniences caused by Neurath's location, he told Morris, were not very important given the world situation—"You know I feel we should work here, indirectly against Hitler by promoting an attitude which is really antinazi."[15]

<h2>THE IMPORTANCE OF ENGLAND</h2>

It was not just England's proximity to Europe that made this "British-American" branch so important. Besides finding governmental support for his ISOTYPE work, Neurath was happy in England and greatly admired aspects of English culture and sensibilities. While describing his new life to Carnap and Morris, he often described incidents and people that fed his lifelong interest in human happiness and its relations to social conditions. England impressed him: "We try to find out how people behave in various countries under certain circumstances. The aggregation in the Anglo-Saxon countries and in Holland seems to be preferable to us. And we think these aggregations allow the growing up of more happiness, even if more temperate. Less tensions as something loved and admired. Holidays, weekend, hobbies important, even when very restful and without excitement. Of course, there are deplorable things, too. But the whole aggregation compared with others is fine."[16] German education, he wrote, is more concerned with "efficiency" and productivity, whereas "the English education, from this point of view, is more human." There is a "sense for personal happiness" that is accepted as important to life.

He was also impressed with English people:

> It is impressive to listen to plain people here, how they avoid boasting and over-statements in daily matters. I collect "expressions", e.g. fire guard leaders explaining how people should get a feeling to be sheltered by the neighbours etc. and then explaining what is needed to act "quickly", to be "calm", and to have the "usual commonsense." I like this type of habit much more than the continental one, with "highest duty", "national community" "self sacrifice," "obedience", "subordination", etc. "eternal ideals" wherever you give a chance to open the mouth.
>
> The British way of living is nice, the compromise habit, the not believing in too many arguments, usual common sense, instead of skyhigh principles from which one tries to deduce concrete details—in vain, of course.[17]

These were people after Neurath's heart: they prized happiness as a leading goal in life, they distrusted arguments in favor of common sense, and they avoided abstract, metaphysical language. Had Neurath's career started in England, he might never have been moved to promote his infamous *Index Verborum Prohibitorum* (Reisch 1997).

Neurath came to believe that England could contribute to, if not actually lead, the kind of "world commonwealth" that he and others believed was necessary to secure long-term peace in Europe. He wrote Morris, "I hope we shall get some World Commonwealth, but I fear we shall get no such comprehensive body. I like the Anglo-Saxon world and would like to be protected by it some way or another. Federal Europe, I think so, leads to a leading Germany again. And that is dangerous in any case."[18] This hope lay behind Neurath's paper "International Planning for Freedom" (Neurath 1942). While advocating democratic planning to rebuild society and economy after the war, Neurath emphasized these themes that appear in his letters to Carnap and Morris. The task for planners and social engineers was to formulate plans for society that maximize human happiness (423). This can occur only in a cultural atmosphere where happiness is valued and where arguments and clashing points of view are at least tolerated, if not nourished (427). Planning, moreover, does not look to science to provide an "optimal" plan for the future (much less one deduced from "skyhigh principles"). Science can only provide a range of plans from which society chooses using "common sense arguments" (426).

According to his letters, Neurath believed that the Unity of Science movement could both contribute to, and thrive within, the sort of commonwealth he envisioned. The movement's internationalist, collaborative spirit, and the empirical language of science it would cultivate, would be useful: "Whatever the future COMMONWEALTH OF NATIONS WILL BE," he wrote, "to have a language in common for expressing comprehensive ideas will be of importance. How important will it be, to have expressions which enable a better understanding between thinkers from Europe, America, and the Far East, etc."[19]

As the end of the war came in sight, and Morris raised the question of resuming the International Congresses, Neurath spoke of the symbolic role the movement could play. The first postwar congress should be in Holland: "Imagine these people hungry for news, hungry for civilization. If only a few people from the USA came over, it could be a symbol of a new scientific world commonwealth."[20] Neurath pinned his hopes for the postwar world, the Unity of Science movement, and the rise of democratic planning in the same place—England.

Metaphysics and Persecution

Against this backdrop, we can also understand Neurath better in his disputes with Carnap. Before the issue about Neurath's 1944 monograph exploded, one of

Neurath's objections to semantics was that it was connected—if only in a loose, indirect way—to totalitarianism and persecution. Neurath's own experiences and the plight of others during the war prompted him to read and think about persecution in world history. He planned to write a book, *Brotherhood and Persecution*, and he told Carnap about his research as the two were hashing out their differences over semantics in the early 1940s.

Not surprisingly, some of his conclusions dovetail with his crusade against metaphysics. Persecutors, he believed, had strong metaphysical beliefs and styles of thought. Neurath reasoned like this: If human happiness is, or at least ought to be, a basic, guiding value, those who persecute others and deprive them of their happiness must be distracted from this goal. They are distracted and confused by metaphysical thinking in which abstract ideals eclipse concerns with human happiness.

Plato and a "platonic attitude" was one historical source of this problem, Neurath explained. Around the time that Neurath publicly engaged English classicists with his claim that Plato's *Republic* encouraged fascism and totalitarianism (Neurath 1945), he explained to Carnap that Plato's *Republic* promoted not only "pure and simple cruelties [and] pure and simple oppression as ideal" but also absolutistic, metaphysical tendencies that often assisted persecution:

> I found out that empiricists on an average are less prepared to become merciless persecutors . . . (for the higher glory of THE transcendent nation idea etc or something else) because they are not prepared to sacrifice their own and other people's happiness to something "idealist" and antihuman. The commonsense leads back to looking at human happiness.
> I think that this merciless habit in history very often is connected with absolutism in metaphysics and in faith. If one thinks [pluralistically,] that there are many possibilities in arguing then one cannot be very hard with argumentative conviction, only indirectly or by heart, but not in argument. . . . Pluralist arguing, which seems to be closely connected with empiricism, leads to a certain toleration.[21]

Empiricism and Pluralism, for Neurath, were allied against the axes of metaphysics, totalitarianism, and persecution.

Neurath told Carnap that many philosophers were guilty of sustaining these dangerous styles of thought. He criticized "the German leaders in philosophy and moral discussions" for their "over-personal, transcendent, anti-happiness" arguments as well as "our Austrian Philosophers, too"—"now you see how they formed an environment out of which grew Nazism." He also detected this dangerous absolutism in writings of relatively *scientific* philosophers such as Popper and Russell. Gingerly, he hinted that this is one reason he so strongly objected to Carnap's semantics: its absolutism and nonpluralism were correlated with these social dangers.

"Perhaps you will not think too strong of my 'intolerance' towards metaphysics when you think of this possible correlation."[22]

After Neurath learned that Carnap had called him a "volcano," however, Neurath appropriately erupted and added to his complaints about Carnap. Aside from this "possible correlation" between metaphysics in philosophy and the history of persecution, he now emphasized that Carnap himself had acted inhumanely by speaking about Neurath so disrespectfully:

Dear Carnap,
Thanks for the letter and the enclosed letter to Morris dealing with the FOUNDA-TION [of the Social Sciences] case. Let me tell you frankly and freely that this letter to Morris is even a more serious thing than your behavior [of removing your name].

Carnap's disclaimer and his remarks to Morris were a "denouncement" that had little to do with the manuscript in question. It was not related to editorial "criticism"—it was an "offence."[23] In this way, Neurath reformulated his critique and criticized not Carnap's positions or views but rather his basic *"attitude"* toward life and other people.

Two months later, Neurath remained extremely angry and frustrated over the situation. Sitting at his typewriter, Neurath described for Carnap two general attitudes or "ways of behavior" that he had observed in his life and in his research on the history of persecution:

One attitude tries to find HIGHEST ideals . . . in justice, duties, etc. [with] every-thing regarded as something within a systematic structure [that] we know al-ready or at least should try to know.
  Beliefs and actions are then judged against this structure as being "correct" or not. The other attitude places human happiness above all else. Here, "people liv-ing together are able to create a friendly and kind atmosphere, to think, how they may make one another as happy as possible." Of course, "very often unhappiness appears". But it is NEVER ACCEPTED AS UNAVOIDABLE WHEN CERTAIN "ENDS" should be reached. . . . Friendship and brotherhood are the basic attitude and NOTHING ELSE COUNTS, no conviction, no faith, no enthusiasm.

Carnap, Neurath gingerly explained, belonged to the first group:

Look, my dear and take it as an attempt to come into closer contact with one an-other when I tell you that the first group of people has certain signs which we find in Plato's REPUBLIC . . . which for me are the real danger, and that your habit is often to a certain degree as I see it not so far away from the description I have given above. . . . Now I think, that in spite of all your personal charm and kind-ness you have many serious signs of a Platonic attitude.[24]

Given Neurath's earlier diatribes against Plato, metaphysics, absolutism, and persecution (where he once used the phrase "Plato-Hitler"),[25] this letter was bound to escalate the ill will between them. Neurath knew this letter was risky: "I should prefer, believe me, to have with you a serious and kind correspondence about the Platonism in you, as should call it, the Puritanism in you, the Prussianism in you etc. I know this kind of description is almost a kind of 'giving names', a very risk . . . but I think you are too grand a personality to be afraid of such a discussion."[26]

Quite possibly, Neurath decided that the risk was too great. He did not send this letter.

Around this time, Neurath received a letter from Ina, Carnap's wife. She scolded them both for being stubborn and for letting their dispute get out of hand, but she concurred with Neurath's view that Carnap could be inconsiderate of others "when he thinks there is no good cause for the other's feelings to be hurt." "If he thinks what he does or says is right he will do or say it and it's the other fellow's job to overcome his soreness." She even echoed Neurath's image of persecutors whose attention is riveted on extramundane goals and ideals. Carnap is "the zealous Lutheran from Prussia," she wrote, who lacks "the saving grace of a light touch and of a felicitous formulation which might soften the blows which he is striking in the name of science, impartiality, and other suchlike gods."[27]

With words and phrases that seem to echo Ina's letter, Neurath rewrote his letter to Carnap and sent it: "As far as I can read in your arguments, you are thinking in terms of 'right', 'correct', 'duty', 'etc' which is far away from being kind and friendly, thinking of human happiness as a central item. You think . . . if you are in the right, other people's unhappiness is not counting much . . . the victim may bear his pain with courage."[28] Though he omitted his earlier discussion of the two types of attitudes, he still insisted that Carnap sacrificed Neurath's happiness on the alter of abstract ideals and duties. He wrote again only four days later and rehearsed the same charges. Carnap never responded and Neurath died about three months later.

The Big Picture and a Last Ditch Effort

Neurath berated Carnap for his metaphysical semantics and his decision to remove his name as editor for several reasons: (1) to save his personal relationship with Carnap; (2) to fight for his own reputation within logical empiricism; and (3) to prepare the public face of the encyclopedia and the Unity of Science movement—in the image of empiricism and pluralism—so that it might assume a role in the reconstruction of Europe. The character of the Unity of Science movement was at stake in a way that connected these intellectual, professional, and personal issues. Aware of Carnap's prestige and influence, Neurath feared that if he could not convince Carnap that semantics was philosophically unsavory (much less politically and historically dangerous), the movement itself could easily become increasingly absolutistic and nonpluralistic. To the same effect, he worried that his monograph —in which he tried to popularize pluralism—would look peripheral and marginal: this monograph was special, for it was published without Carnap's editorial imprimatur. Less than philosophically important, it merely contained some musings from logical empiricism's organizer and manager.

If so, Neurath reasoned, then the encyclopedia and the movement would be less suited for their potentially important role in the reconstruction of Europe. Instead of promoting the pluralistic, nonmetaphysical thinking that planning and reconstruction required (in part because pluralism reminds us that the future is open, malleable, and plannable), the movement would instead promote the absolutistic, nonpluralistic thinking that, thanks to Plato, helped create Europe's mess in the first place.

This cluster of concerns helps us understand one of Neurath's last maneuvers in his professional collaboration with Carnap and Morris. Late in 1944, in the midst of his argument with Carnap, Neurath proposed a change of course for the encyclopedia. They had decided by this time that the next volumes of monographs would address methods of science. Now, Neurath proposed that the next installments present *dialogues* about logical empiricism and unified science: "We should avoid to become a dull inner circle speaking of [its] problems only and we should avoid to pretend too much. A kind of dialogue, well assembled would PERHAPS be the right thing. Partly we ourselves appearing, partly invented persons." He had Hume's dialogues most in mind but also envisioned a modern answer to Plato's and Galileo's.

What was Neurath thinking? First, dialogues would give him a voice in the next installments of the encyclopedia and allow him to confront those views that he saw as unempirical and nonpluralistic—including those within logical empiricism: "Looking at the various papers written by members of our movement," he wrote, "there are certain differences even between us, which are sufficiently serious and therefore cannot be overlooked." Dialogues would let Neurath bring these differences to the foreground.

Second, dialogues would counter Neurath's worry that there was a growing dogmatic core of logical empiricism—especially one that excluded his pluralism. The new format, he explained, would introduce readers to "what we may call the CLIMATE OF LOGICAL EMPIRICISM, characterized by opinions which differ or are in the making."[29] As one of the interlocutors, Neurath would appear as a "pillar" in the movement, not the marginal figure, the mere organizer, he feared he was becoming.

Finally, publishing dialogues would allow the encyclopedia to play one of the roles Neurath hoped it would play in the postwar world.[30] The interlocutors would show by example how to avoid metaphysical language and grapple with contradictory views. If they cannot be reconciled and harmonized, the task nonetheless is to "orchestrate" them in ways that allow common action toward a common plan. In the encyclopedia, the dialogues would ideally produce ways to orchestrate and unify the sciences. But in the larger world, they would appear as paradigms for democratic planning. "You see," Neurath explained, "I think we should present ourselves as able to organize orchestration when asking people to support orchestration in social life."[31]

Morris and Carnap did not respond to Neurath's suggestion, and no dialogues were ever published. But the proposal points to this connection Carnap mentioned between "our philosophical activity and the great historical processes going on in the world." When, in the heat of their argument, Neurath reassured Carnap—"be sure of my intention to calm down ill feeling and to discover a pleasant way of compromise, if we cannot reach a more common basis of happiness and humanity"— the "common basis of happiness and humanity" Neurath hoped to achieve was for the Unity of Science movement and also for the larger, postwar world.[32]

Notes

1. For more on logical empiricism's Unity of Science movement, see Stadler (2001) and Reisch (1994, 1995, forthcoming). I would like to thank the curators of the following archival collections for permission to quote from their holdings: Archives of Scientific Philosophy, Hillman Library, University of Pittsburgh (ASP); Charles Morris Papers Project at the Peirce Edition Project, IUPUI (CMP); Otto Neurath Nachlass, Wiener Kreis Archive, Noord-Holland, Haarlem, The Netherlands (ONN); Presidents' Papers and Unity of Science Movement Papers (USM) at the Department of Special Collections, Regenstein Library, University of Chicago.

2. Neurath to McNeill, May 24, 1943, University of Chicago Press Papers, box 346, folder 3.

3. For a fuller account of the travails of the Unity of Science movement during the late 1940s and 1950s, see Reisch (forthcoming).

4. Neurath to Carnap, January 15, 1943, ASP 102-55-02.

5. Neurath to Morris and Carnap, November 18, 1944, CMP. The quotation occurs on page 8 of this long letter.

6. Neurath to Carnap, September 25, 1943, ASP 102-55-03. Neurath's ideas about "orchestration" stem from his debate with Horace Kallen over the political aspects of logical empiricism and the Unity of Science movement. See Reisch (forthcoming) and Neurath (1946).

7. Morris to Neurath, September 29, 1941, USM box 2, folder 1.

8. Neurath to Carnap, November 18, 1944, ASP 102-55-06.

9. Carnap to Neurath, October 7, 1944, ASP 102-55-07.

10. Neurath to Carnap, September 24, 1945, ONN folder 223.

11. Neurath to Carnap, September 22, 1945. ONN folder 223. Neurath did not send Carnap this letter. He rewrote it as his letter of September 24, 1945.

12. Neurath to Carnap, July 17, 1942, ASP 102-56-04.

13. Neurath to Morris, December 28, 1942, CMP.

14. Neurath to Morris, December 1, 1941, USM box 2, folder 14.

15. Neurath to Morris, January 7, 1942, USM box 2, folder 14.

16. Neurath to Carnap, September 25, 1943, ASP 102-55-03.

17. Neurath to Carnap, September 25, 1943, ASP 102-55-03.

18. Neurath to Morris, December 28, 1942, CMP.

19. Neurath to Morris, December 1, 1941, USM box 2, folder 14. See also Neurath to Morris, January 7, 1942, USM box 2, folder 14.

20. Neurath to Morris and Carnap, November 18, 1944, CMP.

21. Neurath to Carnap, September 25, 1943, ASP 102-55-03.

22. Ibid.

23. Neurath to Carnap, June 16, 1945, ASP 102-55-11.

24. Neurath to Carnap, September 22, 1945, ONN.

25. Neurath to Carnap, September 25, 1943, ASP 102-55-03.

26. Neurath to Carnap, September 22, 1945, ONN.

27. Ina Carnap to Neurath, August 24, 1945, ASP 102-55-10.

28. Neurath to Carnap, September 24, 1945, ONN, folder 223.

29. Neurath to Morris and Carnap, November 18, 1944, CMP.

30. "I hope very much that our type of arguing will find its place in this world." Neurath to Morris, January 7, 1942, USM box 2, folder 14.

31. Neurath to Morris and Carnap, November 18, 1944, CMP.

32. Neurath to Carnap, September 24, 1945, ONN folder 223.

References

Carnap, R. 1934. *The Unity of Science*. Trans. M. Black. London: Kegan Paul, Trench, Trubner & Co. (Originally published in 1932 as "Die physikalische Sprache als Universalsprache der Wissenschaft," *Erkenntnis* 2:432–65.)

———. 1963. "Intellectual Autobiography." In *The Philosophy of Rudolf Carnap*, ed. P. A. Schilpp, 3–84. LaSalle, Ill.: Open Court.

Neurath, O. 1942. "International Planning for Freedom." In *Empiricism and Sociology*, ed. M. Neurath and R. S. Cohen and trans. M. Neurath and P. Foulkes, 422–40. Boston: Reidel, 1973.

———. 1944. "Foundations of the Social Sciences." *International Encyclopedia of Unified Science* 2, no. 1:1–51. Chicago: University of Chicago Press.

———. 1945. "Germany's Education and Democracy." *Journal of Education* 77, no. 912.

———. 1946. "Orchestration of the Sciences by the Encyclopedism of Logical Empiricism." In *Philosophical Papers: 1913–1946*, ed. and trans. R. S. Cohen and M. Neurath, 230–42. Boston: Reidel, 1983.

———. 1973. *Empiricism and Sociology*. Ed. M. Neurath and R. S. Cohen, trans. M. Neurath and P. Foulkes. Boston: Reidel.

———. 1983. *Philosophical Papers: 1913–1946.* Ed. and trans. R. S. Cohen and M. Neurath. Boston: Reidel.

Reisch, G. 1994. "Planning Science: Otto Neurath and the *International Encyclopedia of Unified Science.*" *British Journal for the History of Science* 27:153–75.

———. 1995. "A History of the International Encyclopedia of Unified Science." Ph.D. diss., University of Chicago.

———. 1997. "Epistemologist, Economist . . . and Censor? On Otto Neurath's Infamous *Index Verborum Prohibitorum.*" *Perspectives on Science* 5:452–80.

———. Forthcoming. "From 'The Life of the Present' to the 'Icy Slopes of Logic': Logical Empiricism, the Unity of Science Movement, and the Cold War." In *Cambridge Companion to Logical Empiricism,* ed. A. Richardson and T. Uebel. Cambridge: Cambridge University Press.

Schlick, M. 1937. "L'école de Vienne et la philosophie traditionnelle." In *Travaux du IX Congrès International de Philosophie, IV L'Unité de la Science: La Méthode et les méthodes, Hermann et C*, 99–107. Paris: Hermann et Cie.

Stadler, F. 2001. *The Vienna Circle: Studies in the Origins, Development, and Influence of Logical Empiricism.* Vienna: Springer Verlag.

Uebel, T. 1992. *Overcoming Logical Positivism from Within: The Emergence of Neurath's Naturalism in the Vienna Circle's Protocol Sentence Debate.* Atlanta: Rodopi.

Carl Gustav Hempel

Pragmatic Empiricist

———

GEREON WOLTERS

University of Konstanz

I.

"Peter" Hempel, as he was widely known, passed away on November 9, 1997, at the age of ninety-two. In his later years he was indeed the last survivor of the heroic early days of logical empiricism. He studied in Vienna for a term in 1929, which proved most influential, and received his Ph.D. in Berlin in 1934, with Reichenbach as his supervisor.[1] Thus Hempel united in his person the two major schools of early logical empiricism—that is, the Vienna Circle and the Berlin School.

At this point I would like to digress and turn to two matters of a more political nature, *begriffs*-politics to be more exact. The first relates to the entry "Logical Positivism" in the *Cambridge Dictionary of Philosophy* that was published in 1995 under the editorship of Robert Audi.[2] Here a quotation from the first section follows: "Logical positivism, also called positivism, a philosophical movement inspired by empiricism and verificationism; it began in the 1920s and flourished for about twenty or thirty years. . . . In some ways logical positivism can be seen as a natural outgrowth of radical or British empiricism and logical atomism. The driving force of positivism may well have been adherence to the verifiability criterion for the meaningfulness of cognitive statements."

What are we supposed to learn from this? Well, we learn that logical positivism, or, as I prefer to call it, logical empiricism, has completely British roots. It is characterized as "a natural outgrowth" of radical or British empiricism and logical atomism. As everybody will agree, "radical or British empiricism" is somehow British, and for "logical atomism" the dictionary unsurprisingly refers us to Russell, who is described as a "British philosopher, logician, social reformer, and man of letters, one of the founders of analytical philosophy" (699). So I think it no exaggeration to conclude that logical empiricism in the dictionary is characterized as a philosophical movement with exclusively British or Anglo-Saxon roots. I should add that the rest of the article makes no mention of any proper names, or even geographical indications of a kind that could give a hint to the possibility that there might be historical influences not of a British origin.[3] An innocent reader would conclude that "logical positivism" is part of the British philosophical tradition, and of nothing else. Well, I think many readers of this volume know better, and know better in the greatest detail. Nonetheless, I am afraid that Fumerton's article might be an indicator of what "globalization" has in store for philosophy: not only an ever larger preponderance of Anglo-Saxon philosophers on the international scene, but Americans in particular. Given their achievements, that preponderance might well be deserved and desirable. However, what really concerns me is the possibility that the *history* of philosophy will be more and more read from a parochial perspective.

Now comes my second political point. Consider the title of the conference that inspired this volume: Analytical and Continental Aspects of Logical Empiricism. Again, this title suggests that analytic philosophy is Anglo-Saxon and Continental philosophy is what it says, and that these two philosophical enterprises are somehow opposed to each other. The latter we learn from the *Cambridge Dictionary of Philosophy* article "Continental Philosophy" by the well-known phenomenologist Joseph J. Kockelmans of Pennsylvania State University. Continental philosophy is "the gradually changing spectrum of philosophical views that in the twentieth century developed in Continental Europe and that are notably different from the various forms of analytic philosophy that during the same period flourished in the Anglo-American world." The former is also supported by the dictionary. John Heil, from Davidson College, writes in "Analytic Philosophy": "Analytic philosophers tend largely, though not exclusively to be English-speaking academics whose writings are directed, on the whole, to other English-speaking philosophers. They are the intellectual heirs of Russell, Moore, and Wittgenstein." All right, analytic philosophy is, at least, "not exclusively" restricted to English-speaking academics.[4] That is some improvement. I would like to suggest that we abandon altogether the analytic-continental distinction. By suggesting this I am also criticizing myself, because I was involved in the early stages of organizing the conference. I think now that the misleading connotations upon which this distinction is built far outweigh the convenience of using it.[5]

II.

Back to Hempel. Most of us are inclined to call Hempel a logical empiricist. Some of us have done so in writing.[6] In his later years, Hempel was often asked to give his view on the history of logical empiricism.[7] Surprisingly enough, in all these writings, Hempel does not seem to include himself in the movement. The only thing he apparently admits is that he was "closely associated" with both groups of logical empiricists—that is, the Vienna Circle and the Berlin group (1993, 2), or that he was "strongly influenced" by them (1997b, 20). In contrast, I have found only one occasion where Hempel does use the inclusive "we," and that is in an interview that he gave in 1982 to Richard Nollan, then the curator of the Pittsburgh Archives of Scientific Philosophy.[8] So the question arises: was Hempel—as most of us believe—a logical empiricist or not? The answer is a clear and unambiguous "yes and no."

In my view there are two reasons why Hempel distanced himself somewhat from the logical empiricists proper such as Carnap and Reichenbach. The first is of a more personal kind: I think that Hempel was too modest to step forward into the limelight when asked about the history of logical empiricism. On those occasions he talked about his *teachers* Reichenbach and Carnap. In Hempel's view, I take it, it was *they* who deserved the fame as founders of scientific philosophy. The second reason, however, for not calling himself a logical empiricist is philosophical. It is Hempel's later pragmatic extension of logical empiricism, Hempel's *pragmatic empiricism,* in short. It appears in two stages and two forms. In the first stage it is negative, or defensive, as it were, and in the second it is positive. Here it comes as an important and liberating turn in scientific philosophy in general and Hempel's in particular. Let me elaborate on these somewhat cryptic remarks.

III.

We may begin with these pragmatic considerations in logical empiricism by noting what has been aptly termed its "third dogma."[9] Wolfgang Stegmüller in the first monumental and authoritative volume—which is dedicated to Hempel—of his equally monumental and authoritative *Probleme und Resultate der Wissenschaftstheorie und analytischen Philosophie* calls the "third dogma of empiricism" the "conviction that the instruments of logic are sufficient for the explication of all basic concepts that are relevant for the philosophy of science" (2).[10] It is this conviction that under the heading of "explication" or "rational reconstruction" constitutes the cornerstone of logical empiricism's methodology. In his *Logische Syntax der Sprache* (1934), Carnap even believed that syntactic methods alone were sufficient for rational reconstruction. "Wissenschaftslogik," or logic of science, was nothing other than the syntactical analysis of the language of science. It was the influence of Tarski's semantical definition of truth with which Carnap became acquainted soon afterward that led him to admit semantical methods in addition to syntactical ones in the proper explication of methodological concepts. Carnap never gave up this position. It can be further characterized by its analogy with metamathematics. In Hempel's words, metamathematics "is not concerned with a descriptive account of mathematical research but rather with the formal characterization of correct proof and definition; it can be said to provide, in precise terms, certain objective standards for the validity of mathematical claims and procedures" (Hempel 1988a, 6). This holds—mutatis mutandis—also for explication or rational reconstruction.

In general, Carnap of course also acknowledged pragmatics—that is, he considered the speaker or the speakers of a language in his methodological exposition. Pragmatics for Carnap as well as for Hempel, as we will shortly see, "includes consideration of historical, sociological and psychological relations within the language community" in question (Carnap 1968, 79).[11] But he did not care about it, because pragmatic considerations do not enter into the process of rational reconstruction, which is purely logical, whereas pragmatics, for Carnap, is empirical.[12]

Carnap's program of rational reconstruction set the agenda for Hempel as well. Already his dissertation of 1934 was an attempt to give such an explication: in this case of "probability."[13] The next major concept is "confirmation." In a paper from 1943 Hempel attempts "A purely Syntactical Definition of Confirmation." Here we find for the first time what is probably Hempel's most important achievement: the explication of the concept of explanation where he aims to rationally reconstruct explanations as arguments. This conception was, then, fully developed in the classical Hempel-Oppenheim paper of 1948.

Interestingly enough, Hempel finally failed in all his attempts to arrive at purely syntactical and/or semantical reconstructions. And with respect to confirmation and explanation he failed for the same principal reason. In both cases the rational reconstruction arrives at a point where substantial pragmatic considerations in the sense outlined above cannot be avoided. But as soon as such considerations enter the picture, the idea of explication as a purely syntactical and/or semantical procedure—that is, the third dogma of empiricism—has to be abandoned.

Let us first have a look at confirmation. Hempel, and similarly but independently Carnap, tried to reconstruct confirmation syntactically "as a relation between a body of actual or potential evidence, formulated in an evidence sentence e, and a hypothesis, represented by sentence h." In other words, the question is how a given hypothesis can be justified on the basis of empirical data. It is clear that no evidence whatsoever can conclusively *prove* a hypothesis. In the case of hypotheses of the form

$$(*) \text{ for all } x\text{: if } Fx \text{ then } Gx$$

this results from the fact that all instances of the kind F that have been examined so far do not conclusively prove the hypothesis that *all F are G*. Even existential hypotheses (for example, "there are white ravens") can be proved only conditionally, "namely on the condition that the observation report serving as premise is true" (Hempel 1992, 123).[14]

The idea to regard successful instantiations of $(*)$—that is, cases in which Fs are, indeed, Gs—as confirmations fails because of the paradoxes it generates, namely the Hempel paradoxes on the one hand and the Goodman paradox on the other. With respect to the paradoxes found by Hempel himself, I do not think that they undermine his project of explicating "confirmation."[15] However, Goodman's paradox is a different matter; his solution to the paradox presupposes the possibility that a hypothesis of the form $(*)$ can be "projected" from examined cases to unexamined ones. This, in turn, presupposes that the constituent predicates are well "entrenched"—that is, that they have been successfully used in previously projected generalizations. Goodman's "grue," for example, is not of that sort.

In his "Postscript" (1964) to the republication of the classic "Studies in the Logic of Confirmation," Hempel says with resignation: "Thus the search for purely syntactical criteria of qualitative or quantitative confirmation presupposes that the hypotheses in question are formulated in terms that permit projection; and such terms cannot be singled out by syntactical means alone. Indeed, the notion of entrenchment that Goodman uses for this purpose is clearly *pragmatic* [my emphasis] in character" (Hempel 1965, 51). So, the project of a syntactical explication of the concept of confirmation has failed. But at this point Hempel does not seem to draw any consequences from this failure as to the principal adequacy of the method of rational reconstruction as devised by Carnap. The only thing he admits is that "confirmation" "cannot be adequately defined by syntactical means alone" (50).[16] Pragmatics still plays a role. And it is pragmatics of the sort characterized above by Carnap. In order to adequately reconstruct, that is, explicate, the concept of confirmation, one has to consider historical, sociological, and psychological aspects of its use.

IV

The same holds for Hempel's theory of explanation. First, there is the ambiguity of statistical explanation that leads inevitably to pragmatic presuppositions. And, second, Hempel's epistemic turn entails that it does not make sense in science to speak of "truth," for the timeless "ontological goal of truth" is unattainable for us mortals. Rather, "scientific theorizing is oriented . . . (at the epistemological goal) of optimal epistemic integration, or at epistemic optimality of the belief system we hold at any time," writes Hempel in one of his last papers, which as a manuscript was titled "The Irrelevance of the Concept of Truth for the Critical Appraisal of Scientific Claims" (Hempel 1992, 127). As Richard Jeffrey communicated to me, this title was toned down while being edited from "Irrelevance of the Concept of Truth" to "Significance of the Concept of Truth." The epistemic optimality of a belief system at any time, Hempel maintains, must refer to the historical state of our knowledge at such a time, and such reference, by definition, is pragmatic. This in turn implies that Hempel's adequacy condition, which states the truth of the *explanandum sentences* in his deductive-nomological (D-N) model of scientific explanation, has to be given up in favor of the pragmatic term "epistemic optimality."

Let us now turn to how pragmatic considerations enter inductive-statistical (I-S) explanations. As I suggested earlier, here we find the culprit in the ambiguity of statistical explanations. The ambiguity of the I-S explanation consists in the possibility that there can be two (or more) sets of explaining premises for one and the same *explanandum*. The sets of explaining premises are understood to be both true and compatible with each other. But in one case the explaining premises lend very high inductive support to the *explanandum*, whereas in the other the *explanandum* receives practically no support, or, which is logically equivalent, its negation is highly inductively supported.

Hempel's preferred example is John Jones, who suffers from a streptococcal infection (S_j). There is a high probability of recovering (G) from that infection after large doses of penicillin (P): $p\,(G|S \wedge P)$ close to 1. This gives us the following I-S explanation:

$$p\,(G \mid S \wedge P) \text{ close to 1}$$
$$\underline{S_j \wedge P_j \qquad \text{(makes practically certain)}}$$
$$G_j$$

On the other hand, we know that there are in rare cases streptococcus strains that are penicillin-resistant.[17] In those cases the nonrecovery, or mortality, is highly probable. Thus we have the following situation for the sad case that Jones had been infected with a penicillin-resistant variety of streptococcus (S^*):

$$p\ (\neg G|S^* \wedge P) \text{ close to } 1$$
$$\underline{S_j^* \wedge P_j} \qquad \text{(practically excludes)}$$
$$G_j$$

Similar to Carnap's concept of total evidence, Hempel here introduces his concept of maximum specificity. This epistemological recourse is a pragmatic move because the concept of maximum specificity refers to the actual historical state of scientific and other knowledge.[18] This *epistemological* way out has been severely criticized from an *ontological* point of view by Wesley Salmon. For my argument, however, this criticism need not be taken into consideration.[19]

Thus, as in the case of confirmation, Hempel also ends up having to more or less grudgingly admit that he cannot at this point manage explanation without reference to pragmatics.[20] But perhaps there might yet be logical solutions discovered.

So far we have only talked about Hempel's negative pragmatism, as it were. This negative pragmatism is a result of his failure to realize explications as devised by Carnap. In the next section I would like to turn to Hempel's pragmatic turn. That is, to the fact that pragmatics is now seen as a requirement for a working philosophy of science; in other words, pragmatic considerations are a necessary complement to rational reconstructions and, thus, constitute real progress in the philosophy of science.[21]

V.

If I see it correctly, the beginning of Hempel's pragmatic turn can be traced back to the mid-sixties. It is closely connected with the work and the person of Thomas Kuhn.[22] Hempel met Kuhn for the first time during his stay at the Stanford *Center for Advanced Study in the Behavioral Sciences* in 1963–64. Kuhn was at that time teaching at Berkeley. In a 1982 interview Hempel recalls: "I was very much struck by his ideas. At first I found them strange and I had very great resistance to these ideas, his historicist, pragmatist approach to problems in the methodology of science, but I have changed my mind considerably about this since then. In fact, a good deal of the thinking and writing I did subsequently was in one way or another influenced by the problems and issues that have been raised by Kuhn's writings" (Transcript, Philosophisches Archiv Universität Konstanz [PUK], 21).

How radical this influence was shows in Hempel's address to the Eleventh German Congress for Philosophy in Göttingen in 1975. The last sentence of this address draws the following conclusion that expresses a farewell to explication or rational reconstruction as we knew it: "The acceptance of a universal timelessly valid, rational methodology seems to me a philosophical chimera: it would certainly not be rational to hunt for chimeras" (Hempel 1977b, 33).

In his pragmatic turn, Hempel makes two major points. First, he shows that the projects of people like Popper and Carnap wherein methodological concepts and principles were thought to be purely analytic and logical are in reality flawed, as they contain essential descriptive, or "naturalistic," elements. These seem to be exactly those historical, sociological, and psychological aspects that—as mentioned earlier—Carnap called "pragmatic." And, second, Hempel is convinced that we need a methodology that starts with pragmatic considerations and arrives at normative conclusions only in a second step. This methodology he calls an "explanatory-normative methodology, or . . . *E-N-Methodology* for short" (Hempel 1979b, 50). It is "more comprehensive," compared to the standard explicative methodology of mainstream logical empiricism (Hempel 1988a, 22). Hempel, unfortunately, has given only a rough sketch of his E-N methodology.

Let us have a closer look at the first point.[23] Hempel shows that analytical conceptions of methodological principles like those of Popper and Carnap essentially contain empirical and explanatory elements. In Carnap's case, adequate explications must "conform to a reasonable extent to the pre-analytic, vague, use of the explicandum terms" (11). Thus they do not apply wherever the preanalytic concepts are *not* applied. On the one hand, this "similarity" requirement of explicative and preanalytic use imposes "empirical constraints" on explication. On the other hand, explication still remains to a large extent prescriptive and conventional, insofar as one has to choose one particular explication. This choice is motivated by what in one's view complies best with such methodological requirements as simplicity.[24]

Conformity to preanalytic use of terms, now, amounts to making "descriptive socio-psychological claims about shared intuitive judgments concerning the *explanandum* concept" (12). So Carnap's methodology, and, in fact, any methodology, turns out to be "Janus faced":[25] simultaneously prescriptive *and* descriptive, analytic *and* naturalistic.

Now to Hempel's second point. If one is obsessed with explication as the only method that has value in methodology, then one misses other very important issues. So if one believes, as did Carnap, that the scientific value of a hypothesis is solely determined by its degree of confirmation by empirical data, this does not do justice to the complex considerations that enter into the critical appraisal of a scientific hypothesis when a background of other hypotheses or theoretical principles is available" (14). Here, of course, for Hempel, theory holism enters. But a more important defect of the confirmation approach is that it misses important issues in methodology; first and foremost are the problems of theory dynamics and theory choice that have been studied by Kuhn and others. In Kuhn's conception it is the careful study of historic examples that provides us with methodological claims. But, according to both Kuhn and Hempel, these claims still have to be *evaluated*, which is a normative procedure.

I think that this result can be put nicely into a Kantian wording: Hempel explications without historical examples are empty, and historical examples without explications are blind.[26]

Not much remains of the old program of explication and rational reconstruction. Certainly there have to be "precise explications of consistent and valid logical and mathematical reasoning in examining the implications of scientific theories" (18). Also in the area of measurement or the testing of statistical hypotheses, logical means alone might well suffice. In other cases, particularly with respect to the *"desiderata"* of good theories, such as fruitfulness, simplicity, and scope, there might remain a broad intersubjective consensus about their applicability. This consensus, however, is dependant on the widely, though not completely, shared conviction that science has the principal goal of forming a general account of the world that is accurate. But consensus is not explication. None of those explications Hempel was after during his long career seem to have survived his own criticism: he does not mention "probability" or "confirmation" or "explanation" or "cognitive significance" as possible candidates for a more or less strict explication in the E-N methodology.

In his later years Hempel was ever more fond of what he called "the other Vienna Circle"—that is, Neurath. When on the occasion of the First Pittsburgh-Konstanz Colloquium in the Philosophy of Science in 1991 he received an honorary doctorate from the University of Konstanz, he gave a spontaneous emotional address. There he said:

> When people these days talk about logical positivism or the Vienna Circle and say that its ideas are *passé*, this is just wrong. This overlooks the fact that there were two quite different schools of logical empiricism, namely the one of Carnap and Schlick and so on and then the quite different one of Otto Neurath, who advocates a completely pragmatic conception of the philosophy of science. He says: we must look to how science proceeds in reality. (This he calls a "Gelehrtenbehavioristik" behavioristics of the researcher.) And this form of empiricism is in no way affected by any of the fundamental objections against logical positivism. It is flourishing and thriving. Well, now it is important that we look at Neurath and then say that it is his ideas that despite their somewhat primitive form give us hints to how the philosophy of science has to be developed further. This all is

wonderfully tenable, and all the traditional criticism is meaningless with respect to this form of empiricism. And we may look into the future with confidence and gather under the banner of pragmatic empiricism that may yet live long, flourish and thrive. (German transcript [PUK], 3)

Well, I think pragmatic empiricism, as envisaged by Hempel, is the completion of logical empiricism. It lives on, at least in spirit, in much of the work done in today's methodology and philosophy of science.[27]

Notes

I am grateful to Ringan Douglas Austin for linguistic support and helpful philosophical criticism. For the latter I thank also Wesley and Merrilee Salmon.

1. More biographical information in Wolters (2000).

2. The article itself is written by Richard A. Fumerton, University of Iowa, who is also responsible for "Ayer," "phenomenalism," and "protocol statement."

3. In the respective article the verifiability criterion of meaning is also traced back to British empiricism.

4. In addition, Wittgenstein might not be regarded by everybody as a real English-speaking academic.

5. Wesley Salmon pointed out to me that the rather idiosyncratic views on logical empiricism criticized here might come from the one-sided historical diet provided by Ayer's *Language, Truth, and Logic* that for some people in the Anglo-Saxon world is the only source of information about logical empiricism.

6. For example, Wesley Salmon in the opening sentence of his *Laudatio* on Hempel at the conferral of an honorary doctorate at Konstanz in 1991: "Carnap, Hempel, and Reichenbach, the three leading figures in the movement of logical empiricism" (Salmon 1993, 237).

7. Cf. Hempel (1973, 1975, 1977b, 1979a, 1981, 1982–83, 1991, 1993). There is a bibliography in Richard C. Jeffrey's edition of essays (Hempel 2000) as well as, in a supplemented version, in Wolters 2000.

8. A microfilm clone of the Pittsburgh archives is in the Philosophisches Archiv of the University of Konstanz. The "we" one finds on page 5 of the unpublished transcript that, to the best of my knowledge, has not been edited by Hempel.

9. Stegmüller's "third dogma" seems to be a generalization of what Wesley Salmon earlier called the "third dogma": "scientific explanations are arguments." Cf. Salmon (1998, chap. 6, "A Third Dogma of Empiricism"). This article was first published in 1977.

10. Translations from the German are mine, if not indicated otherwise.

11. Carnap takes French as an example here. Note that "pragmatics" in this sense is different from the "pragmatics" of American pragmatism.

12. This holds in any case for Carnap (1936–37, 454). Later he advocated a "pure" pragmatics. "But he did not make explicit what the nature of this work was to be" (Kalish 1967, 355). In Kalish's article one finds further information about the development of pure pragmatics.

13. A copy of this rare booklet (Hempel 1934) I owe to the generosity of Prof. Richard Jeffrey (Princeton). This is an abridged version of the thesis as it was submitted. An abridged version of the abridged version was published in *Erkenntnis* (Hempel 1935–36).

14. Hempel's considerations with respect to the paradoxes were "influenced by, though not identical in content with, the very illuminating discussion of the paradoxes by the Polish methodologist and logician Janina Hosiasson-Lindenbaum" (Hempel 1965, 20 n. 25.). Hosiasson-Lindenbaum (1899) was arrested by the *Gestapo*, the Nazi secret police corps, in September 1941 in Vilnius (now Lithuania) and killed in April 1942. Her husband, Adolf Lindenbaum (1904), an excellent logician, was killed by the Germans in the second half of 1941.

15. Cf. Hempel (1965, 48), where he maintains that also the apparently paradoxical cases *are* in reality confirmatory. Their apparently paradoxical character is only psychological and has to do with cognitive habits.

16. At this point Hempel did not think of a *semantic* approach, as proposed later by Salmon in "On Vindicating Induction" (1963).

17. Rare at least at the time when Hempel wrote this in 1965 (394).

18. Carnap on the other hand, in order to save the logical character of his concept of confirmation, declared his principle of total evidence as a methodological principle. According to this idea the inductive logic has to be supplied not by pragmatic, i.e., empirical, considerations but by methodological rules.

19. Hempel conceded in his "Nachwort 1976" to the German translation of "Aspects of Scientific Explanation" that he would prefer an "absolute," i.e., not epistemologically relativized type of statistical explanation, as intended by Jeffrey and Salmon, "if there would be a satisfactory solution of the problem of ontological ambiguity, be it through an adequate explication of the postulate of homogeneous reference classes for explanatory statistical laws or in some other way" (Hempel 1977a, 113). He finally arrives at a "modification" of his epistemic relativization: "The postulate of high probability has to be given up as erroneous; the representation [of explanations] as arguments I continue to regard as admissible, but not as necessary. I would generalize the postulate of maximum specificity as specified in section 3.7.5. I would retain the term 'explanation' in the correspondingly enlarged statistical sense.... I would like to once again emphasize that the problems that I have mentioned certainly do not show that a clear explication of an absolute concept [of statistical explanation] is impossible, and that I consider it very desirable" (123).

20. In his important paper on "provisos" Hempel (1988b) adds another point: inferring observational consequences from theories, and therefore giving explanations and predictions, makes use of the "proviso" that no unknown and disturbing forces are at work. Provisos, thus, are clearly relative to the actual state of science.

21. Surprisingly enough, Hempel's pragmatic turn seems to have passed practically unnoticed. For example, Jeffrey (1995) in his outline of Hempel's work does not mention it. Salmon (1999, 347f.), in his commemorative article, alludes to it only by stating the compatibility of views of Kuhn on the one hand and Carnap's and his own on the other, and reporting from a symposium (1983) on Hempel's philosophy "that our [i.e., Hempel's and Salmon's on the one hand, and Kuhn's on the other] profound agreements far outweighed our differences." Rutte (1991, 89) simply ignores the last twenty-five or so years of Hempel's philosophical activity by characterizing Neurath's influence on Hempel as "a passing one" and claiming: "They [Carnap and Hempel] remained empiricists who constructed the language of science by logical means as a *system* and never [*sic!*] truly accepted Neurath's conventionalism and holism." Added in 2002: In his edition of Hempel's essays, Richard Jeffrey finally acknowledges something like Hempel's pragmatic turn: "His [Hempel's] later essays record a growing sense of that [Carnapian] strategy as a dead end, at least for him—from which the way out might prove to be Neurath's *Gelehrtenbehavioristik*, empirical sociology of science, in something like its Kuhnian atavar" (Jeffrey, in Hempel 2000, 3). In James Fetzer's

volume on Hempel, Robert Nozick deals briefly with Hempel's later philosophy, though without taking account of its essentially pragmatic turn (Nozick 2000). Only Michael Friedman mentions it in his contribution (40).

22. This is just another case against the widespread belief that it was Kuhn who killed logical empiricism. For more criticism of the "Kuhn-killer-thesis," see Reisch (1991) and Irzik and Grünberg (1995). See also Irzik's contribution to this volume.

23. I am paraphrasing or quoting Hempel's presentation in 1988a.

24. As an example, Hempel adduces Carnap's reconstruction of inductive probability as rational credibility. A precise theory of rational credibility must refer to intuitive beliefs about types of bets it would be rational to engage in and the like (Hempel 1988a, 12).

25. Cf. the title of Hempel (1985).

26. Thus, Hempel seems to come back, in a sense, to his Kantian beginnings as a student.

27. This thesis is put forward in Salmon's commemorative article on Hempel (Salmon 1999, 346) even for the nonpragmatic conception of logical empiricism.

References

Carnap, R. 1936–37. "Testability and Meaning." *Philosophy of Science* 3 (1936): 419–71 and 4 (1937): 2–40.

———. 1968. *Introduction to Symbolic Logic.* New York: Dover Publications.

Fetzer, J. H., ed. 2000. *Science, Explanation, and Rationality: Aspects of the Philosophy of Carl G. Hempel.* Oxford: Oxford University Press.

Friedman, M. 2000. "Hempel and the Vienna Circle." In *Science, Explanation, and Rationality: Aspects of the Philosophy of Carl G. Hempel,* ed. J. H. Fetzer, 39–64. Oxford: Oxford University Press.

Fumerton, R. A. 1995. "Logical Positivism." In *The Cambridge Dictionary of Philosophy,* ed. R. Audi, 445–46. Cambridge: Cambridge University Press.

Heil, J. 1995. "Analytic Philosophy." In *The Cambridge Dictionary of Philosophy,* ed. R. Audi, 22–23. Cambridge: Cambridge University Press.

Hempel, C. G. 1934. *Beiträge zur logischen Analyse des Wahrscheinlichkeitsbegriffs.* Jena: Universitäts-Buchdruckerei G. Neuenhahn.

———. 1935–36. "Über den Gehalt von Wahrscheinlichkeitsaussagen." *Erkenntnis* 5:162–64. Translation in Hempel (2000, 89–123).

———. 1943. "A Purely Syntactical Definition of Confirmation." *Journal for Symbolic Logic* 8:122–43.

———. 1965. *Aspects of Scientific Explanation and Other Essays in the Philosophy of Science.* New York and London: Free Press and Collier-Macmillan.

———. 1967. "Confirmation: Qualitative Aspects." In *The Encyclopaedia of Philosophy.* Vol. 1, ed. P. Edwards, 185–87. London and New York: Collier-Macmillan and Macmillan.

———. 1973. "Rudolf Carnap, Logical Empiricist." *Synthese* 25:256–68. Reprint in Hempel 2000, 253–67.

———. 1975. "The Old and the New 'Erkenntnis.'" *Erkenntnis* 9:1–4.

———. 1977a. *Aspekte wissenschaftlicher Erklärung.* Berlin and New York: Walter de Gruyter.

———. 1977b. "Die Wissenschaftstheorie des analytischen Empirismus im Lichte zeitgenössischer Kritik." In *Logik—Ethik—Theorie der Geisteswissenschaften. XI. Deutscher Kongress für Philosophie, Göttingen 5.-9. Oktober 1975,* ed. G. Patzig, E. Scheibe, and W. Wieland, 20–34. Hamburg: Felix Meiner.

————. 1979a. "Der Wiener Kreis—eine persönliche Perspektive." In *Wittgenstein, the Vienna Circle, and Critical Rationalism: Proceedings of the Third International Wittgenstein Symposium, August 1978,* ed. H. Berghel, A. Hübner, and E. Köhler, 21–26. Vienna: Hölder-Pichler-Tempsky.

————. 1979b. "Scientific Rationality: Analytic vs. Pragmatic Perspectives." In *Rationality Today—La Rationalité Aujourd'hui,* ed. T. F. Geraets, 46–66. Ottawa: University of Ottawa Press.

————. 1982–83. "Schlick und Neurath: Fundierung *vs.* Kohärenz in der wissenschaftlichen Erkenntnis." *Grazer philosophische Studien* 16–17:1–18. Translation in Hempel (2000, 181–98).

————. 1985. "Der Januskopf der wissenschaftlichen Methodenlehre." In *Jahrbuch 1983/84 Wissenschaftskolleg—Institute for Advanced Study—zu Berlin,* ed. P. Wapnewski, 145–57. Berlin: Siedler.

————. 1988a. "On the Cognitive Status and the Rationale of Scientific Methodology." *Poetics Today* 9:5–27. Reprint in Hempel 2000, 199–228.

————. 1988b. "Provisos: A Problem Concerning the Inferential Function of Scientific Theories." *Erkenntnis* 28:147–64. Reprint in Hempel 2000, 229–49.

————. 1991. "Hans Reichenbach Remembered." *Erkenntnis* 35:5–10. Reprint in Hempel 2000, 288–94.

————. 1992. "The Significance of the Concept of Truth for the Critical Appraisal of Scientific Theories." In *Interpreting the World. Science and Society,* ed. W. R. Shea and A. Spadafora, 121–29. Canton Mass.: Science History Publications. Reprint under the title "The Irrelevance of the Concept of Truth for the Critical Appraisal of Scientific Theories." In Hempel 2000, 75–88.

————. 1993. "Empiricism in the Vienna Circle and in the Berlin Society for Scientific Philosophy. Recollections and Reflections." In *Scientific Philosophy—Origins and Developments,* ed. F. Stadler, 1–9. Dordrecht: Kluwer. Reprint in Hempel 2000, 295–304.

————. 2000. *Selected Philosophical Essays.* Ed. R. Jeffrey. Cambridge: Cambridge University Press.

Hempel, C. G., and P. Oppenheim. 1948. "Studies in the Logic of Explanation." *Philosophy of Science* 15:135–75. Reprint in Hempel 1965.

Irzik, G., and T. Grünberg. 1995. "Carnap and Kuhn: Arch Enemies or Close Allies?" *Brit. J. Philosophy of Science* 46:285–307. Jeffrey, R. 1995. "A Brief Guide to the Work of Carl Gustav Hempel." *Erkenntnis* 42:3–7.

Kalish, D. 1967. "Semantics." In *The Encyclopedia of Philosophy,* vol. 7, ed. P. Edwards, 348–59. London and New York: Collier-Macmillan and Macmillan.

Kockelmans, J. J. 1995. "Continental Philosophy." In *The Cambridge Dictionary of Philosophy,* ed. R. Audi, 157–58. Cambridge: Cambridge University Press.

Nozick, R. 2000. "The Objectivity and the Rationality of Science." In *Science, Explanation, and Rationality: Aspects of the Philosophy of Carl G. Hempel,* ed. J. H. Fetzer, 287–307. Oxford: Oxford University Press.

Philosophisches Archiv Universität Konstanz (PUK). Microfilm copy of the Carnap papers in the Pittsburgh Archives for Scientific Philosophy.

Reisch, G. A. 1991. "Did Kuhn Kill Logical Empiricism?" *Philosophy of Science* 58:264–77.

Rutte, H. 1991. "The Philosopher Otto Neurath." In *Rediscovering the Forgotten Vienna Circle: Austrian Studies on Otto Neurath and the Vienna Circle,* ed. T. E. Uebel, 81–94. Dordrecht: Kluwer.

Salmon, W. C. 1963. "On Vindicating Induction." *Philosophy of Science* 30:252–61.

———. 1993. "Carnap, Hempel, and Reichenbach on Scientific Realism." In *Logic, Language, and the Structure of Scientific Theories. Proceedings of the Carnap-Reichenbach Centennial, University of Konstanz, 21–24 May 1991*, ed. W. C. Salmon and G. Wolters, 237–54. Pittsburgh and Konstanz: University of Pittsburgh Press and Universitätsverlag.

———. 1998. *Causality and Explanation.* New York, and Oxford: Oxford University Press.

———. 1999. "The Spirit of Logical Empiricism: Carl G. Hempel's Role in Twentieth-Century Philosophy of Science." *Philosophy of Science* 66:333–50.

Stegmüller, W. 1983. *Probleme und Resultate der Wissenschaftstheorie und Analytischen Philosophie.* Vol. 1 (Erklärung, Begründung, Kausalität). 2d rev. ed. Berlin: Springer.

Wolters, G. 2000. "Carl Gustav Hempel: Pragmatischer Empirist." *Zeitschrift für allgemeine Wissenschaftstheorie* 31:205–42.

I I I

The Riddle of Wittgenstein

The Methods of the *Tractatus*

Beyond Positivism and Metaphysics?

———————

DAVID G. STERN

University of Iowa

The Difficulty of the *Tractatus*

The *Tractatus* may well be the most difficult philosophical book written in this century. Two facts conspired to produce this result: The thoughts in it are very hard to explain—we are told; and Wittgenstein was singularly uninterested in or incapable of explaining his views to others. Almost everything he wrote was in the nature of a diary, a record of his thoughts; a conversation with himself or with God—hence, he did not feel the need to meet a potential interlocutor even halfway. (Coffa 1991, 142)

Despite the legendary difficulty of Wittgenstein's *Tractatus Logico-Philosophicus*, it has attracted an extraordinary variety of interpretations in the eighty years since its publication. Among the leading alternatives are readings that make epistemology, ontology, logic, semantics, ethics, religion, or mysticism central; construals of the text as realist, idealist, solipsist, phenomenalist, physicalist, or neutral monist; and those that identify Kant, Schopenhauer, Kierkegaard, Frege, Russell, or Carnap as providing the correct approach with which to understand the book. Moreover, the list barely begins to indicate the variety of conflicting approaches that have been seriously canvassed. Far from deterring interpreters, the text's extreme brevity and difficulty and the lack of a single authoritative interpretation have made it easier for every kind of reader to make his or her philosophical projects the key to understanding the book.

The conviction that it must be possible to give a single coherent exposition of the book's doctrines or its method is, I believe, an illusion. In *Wittgenstein on Mind and Language* I examine the conflicting impulses that shaped the composition of the *Tractatus* and the different readings that Wittgenstein himself gave on a number of subsequent occasions (Stern 1995, chaps. 1–3, 6). While the preface affirms that he had found "on all essential points, the final solution of the problems" of philosophy, the solutions he provided were extremely schematic. A large part of what makes the *Tractatus* such a fascinating and elusive book is that it is the product of two opposed and unstable forces, both of which are present in the preface. On the one hand, its author had a metaphysical vision: the definitive solution to the leading problems of philosophy. On the other hand, he was gripped by an equally powerful antimetaphysical drive, the aim of drawing a limit to language and to philosophy. Wittgenstein's philosophy is the product of two tendencies, one of them antipositivistic and the other in a more subtle way positivistic. They are not diametrically opposed to one another. But there is great tension between them: "Each of the two forces without the other would have produced results of much less interest. . . . But together they produced something truly great" (Pears 1986, 197–98, on the later philosophy). Because of this unresolved conflict between Wittgenstein's metaphysical and antimetaphysical tendencies in the *Tractatus*, one can only give a unified and systematic interpretation of the book if one carefully selects and construes the appropriate passages.

Nevertheless, the history of *Tractatus* interpretation is for the most part a history of wishful thinking, each successive group of interpreters seizing on the passages they have found most interesting in order to reconstruct the doctrines they knew must be there. My principal concern in this chapter is the role of problems of philosophical method in the interpretation of Wittgenstein's philosophy. If we take a step back from the arguments about what Wittgenstein really meant by the *Tractatus*, we will see that there is much to be learned about the book from the very fact that it has given rise to such diametrically opposed interpretations.

A Brief History of *Tractatus* Reception

The development of orthodox views about the method and subject matter of the *Tractatus* over the past eighty years roughly parallels broader trends in analytic philosophy over the same period. We can distinguish five principal phases in the development of *Tractatus* interpretation. The following list provides a name for each approach, a concise summary, the date at which the approach enters the literature, and references to some of the influential early expositions of that interpretation.

1. The logical atomist reading: a book on logic in the tradition of British empiricism, 1920s onward (Russell 1922; Ramsey 1923).
2. The logical positivist reading: the book that inspired the antimetaphysical scientific worldview of the Vienna Circle, 1930s onward (Neurath, Carnap, and Hahn 1929; Schlick 1930).
3. The metaphysical reading: a book that argues for a metaphysical system, based on a theory of logic and meaning, 1960s onward (Anscombe 1959; Stenius 1960).
4. The irrationalist reading: a book that advocates an ethical or religious worldview, 1970s onward (Engelmann 1968; Janik and Toulmin 1973).
5. The therapeutic reading: a book that rejects all philosophical doctrine, 1980s onward (Winch 1987; Diamond 1991a).

Obviously, this rough-and-ready chronology of the leading developments in *Tractatus* interpretation, set out in more detail below, is an oversimplification. While most work does belong in one of these categories, there are exceptions that resist this classification or make use of more than one of these approaches. Furthermore, while each successive phase has ushered in new ways of reading the book, previous approaches still have considerable currency; in particular, the metaphysical reading remains the most widely accepted approach. For instance, Alberto Coffa's history of the semantic tradition, the source of the quotation at the beginning of section one, provides a metaphysical reading of the *Tractatus*, while explaining the logical positivist construal as the product of Wittgenstein's conversations with the Vienna Circle.

For the first ten years after the *Tractatus*'s publication, most commentators followed the lead given by Russell's introduction to the *Tractatus* and Ramsey's review of the book for *Mind*, approaching the book as a development of Russell's work on logical atomism and as a contribution to the philosophy of logic and mathematics. While this approach was eclipsed during the heyday of logical positivism, subsequent reappraisals of Russell's philosophy have led, in turn, to fresh ways of thinking about what Wittgenstein could have learned from Russell (Pears 1967; Marion 1998; Landini 2000).

During the second phase, which lasted from the early 1930s to the 1950s, the *Tractatus* was usually read as a work of neopositivism, a contribution to the verificationist and antimetaphysical program of the Vienna Circle. Following Schlick and Carnap's lead, the logical positivists found inspiration in the central role the book gave to the distinction between sense and nonsense, its reliance on Russellian and Fregean logic, and its dismissal of most previous philosophy as cognitively empty. The *Tractatus* provided them with an account of the nature of language that showed how traditional philosophy is the result of conceptual confusion. By demarcating meaningful from meaningless discourse, the book thereby exposes the nonsensicality of metaphysics. In particular, they took the logico-semantic doctrines they found there as a point of departure for their own work: a semantic account of a priori truth and a verificationist theory of empirical meaning (Coffa 1991; Friedman 1999). While the verification principle is never explicitly mentioned in the *Tractatus*, Wittgenstein did stress it in his discussions with members of the Vienna Circle and in his writings from the early 1930s. On the other hand, the logical positivists rejected the central role that the *Tractatus* gives to the distinction between showing and saying. Like Russell, Carnap found Wittgenstein's arguments for the impossibility of a semantic theory unconvincing, arguing that whatever could not be stated about a language using the resources available within that language could always be expressed within a suitable metalanguage.

In the 1960s the logical positivist reading fell out of favor with the next generation of readers, who inaugurated the third phase of *Tractatus* interpretation by rediscovering the Kantian, Fregean, and logico-metaphysical aspects of the book. The positivist reading was rarely given a close examination; instead, it was dismissed out of hand as obviously partisan and question begging. It was pointed out, as if it sufficed to refute the logical positivist reading, that there was almost no discussion of epistemology in the *Tractatus*, but a great deal of metaphysics. The Vienna Circle—and the Wittgenstein who met with them in the late 1920s and early 1930s—had read into it a positivistic epistemology and a verificationist semantics that is not to be found there. Furthermore, it is hard to maintain a consistent position on the emptiness of all philosophical statements: even if the positivist has no trouble with the idea that neither Hegel nor Heidegger succeeds in asserting anything, most positivists still maintain that positivism itself is not simply cognitively empty. Consequently, the metaphysical reading arose as a reaction to these perceived failings on the part of the positivistic approach. (For some provocative discussion of these issues, see Conant [2002] and Gunnarsson [2001]).

According to the metaphysical reading, the positivists' emphasis on the role of the distinction between sense and nonsense, and on semantic theorizing, had led them to overlook the full significance of the distinction between showing and saying, between the form and content of our language. Strictly speaking, what the book allows us to *say* is in line with the positivist tradition. Still, while metaphysical doctrines cannot, according to the *Tractatus*, be *stated*, they can be *shown*. This amounts to accepting the positivistic view that Wittgenstein draws limits to what can be said and relegates what cannot be said to silence. However, it adds the crucial qualification that the limits are drawn in order to show what cannot be said, that the unsayable is still graspable, in some sense, even if it is unsayable. There are certain insights expressed in his words that cannot be directly stated, and the attempt to do so inevitably leads to nonsense. Nevertheless, these insights can be expressed indirectly; the nonsense does succeed in showing what cannot be said. In

The Riddle of Wittgenstein

this way, the book distinguishes between two kinds of nonsense: empty nonsense, the misleading kind of nonsense, and deep nonsense, the illuminating kind. While the former belongs on the Humean bonfire, Tractarian nonsense is philosophically valuable. The book that Carnap and his colleagues had embraced as providing the blueprint for a scientific worldview was actually a thinly disguised metaphysical treatise. A letter Wittgenstein wrote to Russell in 1919, replying to Russell's first questions about the book, provides strong prima facie support for this criticism.

> Now I'm afraid you haven't really got hold of my main contention, to which the whole business of logical prop[osition]s is only a corollary. The main point is the theory of what can be expressed (gesagt) by prop[osition]s—i.e. by language— (and, which comes to the same, what can be *thought*) and what can not be expressed by prop[osition]s, but only shown (gezeigt); which, I believe, is the cardinal problem of philosophy. (Wittgenstein 1995, 124)

Thus, the distinction between showing and saying became the key to understanding the book, although there was very little agreement about the precise nature of the logical and metaphysical doctrines to be found there. Given that the *Tractatus* says so little about the nature of objects, and Wittgenstein himself entertained a variety of different views, it is hardly surprising that able interpreters were able to find arguments that supported almost any ontology there. The distinction between showing and saying permits the interpreter to hold that while all sorts of doctrines are not actually said in the text, they are nevertheless shown. Thus, everything that is explicitly excluded can be let in the back door, as implicitly shown. This creates enormous exegetical leeway: the question as to precisely what is supposed to be shown by the *Tractatus* has been the point of departure for a great deal of disagreement. In the wake of the publication of the *Philosophical Investigations*, the *Tractatus* was also sifted for evidence of the views that Wittgenstein must have been refuting in his later work. If the later Wittgenstein was a critic of metaphysical theories, and the *Tractatus* and *Investigations* were diametrically opposed, then it made sense to look for the theories Wittgenstein criticized in the *Investigations* in the text of the *Tractatus*. Leading expositions of a metaphysical reading include the work of Hacker (1972, 1996), Pears (1986, 1987), Hintikka and Hintikka (1986), McGuinness (1988), and Malcolm (1986).

Initially, the publication of the three pre-*Tractatus* notebooks from 1914 to 1916 provided grist for the metaphysical mill. But the extended discussion of ethical and religious themes in the last notebook helped to redirect readers' attention to the question of their place in what had been regarded as an austerely analytical text. Paul Engelmann, an old friend of Wittgenstein's who had known him during the First World War, argued that the book should ultimately be seen as a contribution to ethics. His case gained strong support from another letter of Wittgenstein's. This one was sent to Ludwig von Ficker, an editor who he had hoped might be persuaded to publish the *Tractatus:*

> The book's point is an ethical one. I once meant to include in the preface a sentence which is not in fact there now but which I will write out for you here, because it will perhaps be a key to the work for you. What I meant to write, then, was this: My work consists of two parts: the one presented here plus all that I have *not* written. And it is precisely this second part that is the important one. My book draws limits to the sphere of the ethical from the inside as it were, and I am convinced that this is the ONLY *rigorous* way of drawing these limits. In short, I believe that where *many* others today are just *babbling,* I have managed in my book to put everything firmly in place by being silent about it. And for that reason, unless I am very much mistaken, the book will say a great deal that you yourself want to say. Only perhaps you won't see that it is said in the book. For now, I would recommend to you to read the *preface* and the *conclusion,* because they contain the most direct expression of the point of the book. (Engelmann 1967, 143–44)

Engelmann's ethical interpretation of the *Tractatus* was taken up and popularized by Janik and Toulmin (1973). Their book sparked a fourth, irrationalist, phase of *Tractatus* interpretation, on which the book's argumentative aspects were subordinated to the broader goal of advancing an ethico-religious vision. Broadly speaking, such readings are still within the overall framework of the metaphysical interpretation: they look for a hidden doctrine that cannot be stated but is implicitly present. But they shifted the interpretive focus from explaining how certain ontological and semantical doctrines are *shown (zeigen)* by the logic of our language to the questions of how religious, ethical, or mystical insights *show themselves (sich zeigen).* Given that the book says even less about ethics, God, and the mystical than it does about the nature of objects, it is hardly surprising that these interpreters arrived at extremely diverse conclusions. Rather than supplanting the broad consensus among analytic philosophers that the book should be read as a metaphysical treatise, this approach instead had the effect of making the book attractive to a readership with little interest in logical analysis. (For a complementary history of these initial stages of *Tractatus* reception, see Frongia and McGuinness [1990, 1–38].)

In recent years, Cora Diamond, James Conant, Warren Goldfarb, Tom Ricketts, and others, inspired in part by earlier work by Rush Rhees and Peter Winch, have argued that the positivist, metaphysical, and irrationalist readings of the *Tractatus* are equally mistaken. Instead, they argue, the book belongs to the genre of anti-philosophical therapy: its aim is to liberate its readers from all substantive philosophical views. On this construal of the *Tractatus,* Wittgenstein first sets up the most plausible philosophical views that he can and then knocks them down. The passages that appear to set out the standard positivist, metaphysical, ethical, and religious views are there to draw us in, by setting out views that we will ultimately be led to recognize as nonsense. The real message of the *Tractatus*, they contend, is that *all* philosophy, including the philosophy ostensibly presented and endorsed within the *Tractatus* itself, is simply nonsense.

Beginning with the question of the relationship of Wittgenstein's *Tractatus* to the work of Frege, Diamond has instigated a rethinking of its overall relationship to the analytic tradition. Diamond's therapeutic reading is motivated by her conviction that we must take Wittgenstein at his word when he says that philosophy is nonsense. Throughout Wittgenstein's writings, she is struck by Wittgenstein's "insistence that he is not putting forward philosophical doctrines or theses; or by his suggestion that it cannot be done, that it is only through some confusion one is in about what one is doing that one could take oneself to be putting forward philosophical doctrines or theses at all" (1991, 179). The importance Diamond attaches to this point about Wittgenstein's way of doing philosophy can hardly be overstated. As she puts it, "there is almost nothing in Wittgenstein which is of value and which can be grasped if it is pulled away from that view of philosophy" (1991, 179). In sharp contrast with those irrationalist interpreters who see Wittgenstein's antipathy to philosophical argumentation as arising out of his cultural or ethical convictions, Diamond argues that Wittgenstein's conception of nonsense arises out of a far-reaching engagement with Frege's writing on the philosophy of language and logic. On the therapeutic reading, there is a much greater continuity between the early and the late Wittgenstein than is usually supposed, for their Wittgenstein was always deeply skeptical of the idea that philosophy could arrive at any kind of a positive doctrine. Instead, the principal change between his early and his later work was that he gave up the idea that language has an essence, a necessarily shared structure, which could be given a systematic and definite analysis.

Diamond's most detailed exposition of her reading of Wittgenstein and Frege is to be found in "Throwing Away the Ladder: How to Read the *Tractatus*" in *The Realistic Spirit: Wittgenstein, Philosophy, and the Mind* (1991). Goldfarb (1997) gives an excellent introduction to Diamond's account of the role of nonsense in philosophy, and Conant (2002) and Gunnarsson (2001) provide valuable and complementary explorations of Diamond's nonsense-oriented reading of the *Tractatus*. When this paper was written in 1999, the therapeutic approach had only begun to receive a critical response amongst Wittgenstein scholars (notably in Goldfarb [1997], Lovibond [1997], Reid [1998], and McGinn [1999]). However, it had not yet gained much attention from a wider audience, perhaps because this interpretation has mostly been advanced in specialized journal articles and book chapters. With the publication of Crary and Read's anthology, *The New Wittgenstein* (2000) and the extensive debate this fueled at the following year's Kirchberg Wittgenstein Symposium (Haller and Puhl 2001), the implications of a therapeutic reading of the *Tractatus* have since attracted a great deal of attention. (Owing to limited space, I have not incorporated references to much of this material here. For further discussion of recent developments in this controversy and related issues, see Stern [forthcoming b].)

Hilary Putnam (1994) and John McDowell have spoken of a "recoil" effect in the development of opposed philosophical positions: philosophical positions often emerge as the result of denying what is implausible about a preceding view. Each camp overreacts to the other, until the result is two equally untenable and extreme positions. Similarly, philosophers often take it for granted that anyone who rejects their own view must be adopting a diametrically opposed position. Materialists insist that everything is matter; idealists deny that anything is matter. Platonism insists that a realm of unchanging forms must exist; conventionalists deny that such a realm is possible. Putnam rightly characterizes the later Wittgenstein's distinctive approach to philosophy as a matter of identifying the presuppositions that all sides take for granted in these disputes. However, very few of the philosophers Wittgenstein criticizes understand his writing in this way. Accustomed to the recoil principle, they immediately apply it to Wittgenstein: he attacked mind-body dualism, so he must have been a behaviorist; he attacked platonism about mathematics, so he must have been a conventionalist. The recoil effect is also at work within the world of *Tractatus* interpretation. The positivist reading makes Wittgenstein's opposition to metaphysics by means of a logically ideal language a guiding principle, while the metaphysical reading attributes a systematic doctrine to him. Both the positivistic and metaphysical readings approach Wittgenstein's views as the product of a rational train of argument from first principles, while the irrationalist considers the argument of the book to be epiphenomenal, a way of attracting attention to its existential convictions. The therapeutic reading makes an even more radical break with its predecessors, denying that there is any positive philosophical view in the book at all. Defenders of the metaphysical reading respond that the therapeutic reading is a post-modernist travesty. Each successive interpretation roundly condemns the basic methodological tenets taken for granted by its predecessors, and its opponents respond in kind. In the next section, I take a step back from this debate and ask what we can learn from it.

The Methods of the *Tractatus*

The most basic issue that divides interpreters of the *Tractatus* is the question of the book's philosophical method: how does it go about doing what it does? The question is particularly prominent at the very beginning and end of the book, where the issue of method is in the foreground.

The foreword begins by saying that it "will perhaps only be understood by those who have themselves already thought the thoughts which are expressed in it." It goes on to claim that the problems of philosophy rest on a mistaken way of putting those problems, a misunderstanding of the logic of our language. The whole sense of the book, Wittgenstein says, could be summed up as follows: "What can be said at all can be said clearly; and whereof one cannot speak thereof one must be silent." Accordingly, the aim of the book is to "draw a limit to thought," or, more carefully put, to draw a limit "to the expression of thoughts" from within, for the notion of some*thing* that lies beyond that limit is incoherent. Talk of drawing a limit to thought presupposes that we can think both sides of the limit, which is precisely what Wittgenstein denies. "The limit can, therefore, only be drawn in language and what lies on the other side of the limit will be simply nonsense."

The final three remarks of the *Tractatus* reprise and expand upon these themes in the following words:

> The right method of philosophy would be this. To say nothing except what can be said, i.e. the propositions of natural science, i.e. something that has nothing to do with philosophy: and then always, when someone wished to say something metaphysical, to demonstrate to him that he had given no meaning to certain signs in his propositions. This method would be unsatisfying to the other—he would not have the feeling that we were teaching him philosophy—but it would be the only strictly correct method.
>
> My propositions are elucidatory in this way: he who understands me finally recognizes them as nonsense, when he has climbed out through them, on them, over them. (He must so to speak throw away the ladder, after he has climbed up on it.)
>
> He must get over these propositions; then he will see the world aright.
>
> Whereof one cannot speak, thereof one must be silent. (*Tractatus* 6.53, 6.54 and 7)

But how are we to understand a book that ends by saying it is nonsense, a ladder that must be climbed and then thrown away? On the one hand, much of the rest of the text of the *Tractatus* appears, at least to its logical atomist, logical positivist, metaphysical, and even irrationalist readers, to be advocating any number of distinctive and debatable philosophical doctrines. On the other hand, these "framing" passages, which begin and end the book, are quite insistent that all philosophical doctrines must be discarded if one is to see the point of the book. Most expositors have found themselves driven to agree with Max Black that we must discard the frame if we are to do justice to the rest of the book, although few are quite as forthright about it as he was. Black's commentary on *Tractatus* 6.53, 6.54, 7, entitled "How the *Tractatus* Is to Be Understood," begins with the following paragraph:

The book ends with celebrated and much-quoted statements that seem to accept utter defeat. The "right method of philosophy" would be to abstain from positive remarks, contenting oneself with demonstrating that every attempt to "say something metaphysical" results in nonsense. And the remarks of the *Tractatus* itself must be recognized as strictly nonsensical *(unsinnig)*, to be abandoned once their true character has been revealed. The rest is silence. It may be noticed that in the preface to the book, Wittgenstein has quoted the very last remark (7) as part of the "whole sense of the book": (What can be said can be said clearly and *poi basta*). The conclusion is profoundly unsatisfactory. That we understand the book and learn much from it is not to be seriously doubted. And the book's own doctrine of meaning and of the character of philosophical investigation must square with this. Wittgenstein may be too willing, at the very end, to equate communication exclusively with "saying". There is much on his own principles that can be shown, though not said. There is much in the book that he has shown: this ladder need not be thrown away. (Black 1964, 376–77)

Black denies that we really are supposed to throw away the ladder, because the ladder language of the *Tractatus* can lead to philosophical insight even as it commits us to denying that anything can be said on the topic. In much the same spirit of irritation and admiration, Gustav Bergmann wrote the following: "He did in fact propose an ideal language; only, he forbade *de jure* all discourse about it, that Moorean discourse, if I may so call it, which is the heart of the philosophical enterprise. *De facto*, as Russell observed, he managed to say a good deal about his ideal language" (Bergmann 1967, 49). Because the book does appear to be advocating and arguing for philosophical doctrines, most readers have charitably interpreted the opening and closing instructions as not being meant entirely seriously.

Diamond's reply is that we must take the preface and closing remarks of the *Tractatus*—the "frame" of the book—as the point of departure for interpreting the book as a whole. Her response to those who, like Black, refuse to let go of the Tractarian ladder, is that this amounts to a form of intellectual cowardice, "chickening out" (Diamond 1991a, 181). For if one says on the one hand that the book really is nonsense but on the other hand reads it as advocating any doctrine, explicit or implicit, then one lacks the courage of one's convictions. Diamond aptly observes that the problem of how one is to make sense of the method of the *Tractatus* is "particularly acute" in *Tractatus* 6.54:

> Let me illustrate the problem this way. One thing which according to the *Tractatus* shows itself but cannot be expressed in language is what Wittgenstein speaks of as *the logical form of reality*. . . . What exactly is supposed to be left of that, after we have thrown away the ladder? Are we going to keep the idea that there is something or other in reality that we gesture at, however badly, when we speak of "the logical form of reality", so that *it, what* we were gesturing at, is there but cannot be expressed in words?

That is what I want to call chickening out. What counts as not chickening out is then this, roughly: to throw the ladder away is, among other things, to throw away in the end the attempt to take seriously the language of "features of reality". . . . the notion of something true of reality but not sayably true is to be used only with the awareness that it itself belongs to what has to be thrown away. (Diamond 1991a, 181–82)

Warren Goldfarb (1997, 64) has proposed that we substitute the term "irresolute" for "chicken out." Like Diamond's expression, it captures the idea that a Black-style reading involves a certain kind of weakness—one wants to have one's cake and eat it too, wants to recognize that Wittgenstein says that philosophy is nonsense, but also make sense of the content of the *Tractatus*. Using the term "irresolute" is not only less tendentious; it also has a convenient opposite: Diamond's approach becomes a "resolute" reading. Diamond has endorsed this turn of phrase, noting that it captures the "failure-of-courage element" prominent in her talk of "chickening out" while also emphasizing another element, namely "a kind of dithering, which reflects not being clear what one really wants, a desire to make inconsistent demands" (1997, 98). This use of morally charged epithets makes finding the right philosophical position sound as if it were a matter of having enough moral fiber or a stiff enough upper lip. In this respect, it is uncannily akin to the role played by "resoluteness" *(Entschlossenheit)* in Heidegger's early philosophy. In *Being and Time*, Heidegger (1996, §60, 273ff) promotes resoluteness, the decisive taking of a stand, as the touchstone for an authentic life and the only way of getting beyond the meaningless alternatives offered by conventional conformity. Heidegger, confronted by the seeming impossibility of giving a rational justification of the right way to live, embraces resoluteness; Diamond and Goldfarb's Wittgenstein, facing an analogous nihilism about philosophy, likewise makes a virtue of necessity and encourages us to have the courage of his convictions.

However, the "frame" of the *Tractatus* cannot, taken by itself, resolve the question of the book's method. For, as we have already seen, it is very easy to read those opening and closing passages in ways that are compatible with the older interpretations. Logical atomist, logical positivist, and metaphysically minded readers all have ways of reinterpreting that frame in light of the other commitments they find in the remainder of the text. Indeed, the quickest way of doing justice to the antiphilosophical frame is an irrationalist reading on which the argumentative passages are a prelude to existential convictions that cannot be rationally defended.

If we are to read the frame in the way Diamond would have us do, then we will also have to find a way of reading the rest of the book that explains its role in the therapeutic project. First, there are those passages within the book that appear to be part of the frame rather than the ladder, such as those passages that set out the conception of sense and nonsense that Diamond considers its central contribution. One of these is the formulation of the Fregean context principle in *Tractatus* 3.3—"Only propositions have sense; only in the nexus of a proposition does a name have meaning"—and the remarks that follow it (Diamond 1991a, 112). Another such passage is the discussion of the nature of logic that begins with *Tractatus* 5.47, in which Wittgenstein argues that "there is no such thing as allowing a sign to figure in a *wrong* combination with other signs" (Diamond 1991a, 128; see also 196–97). Diamond does not explicitly include *Tractatus* 3.03—"Thought can never be of anything illogical, since, if it were, we should have to think illogically"—among those parts of the book that do state its method. However, Conant does so, and it certainly seems to be part of an expository discussion of one of the main claims of the preface. Conant (1993, 2002) also makes a strong case for including 4.112 among those parts of the text of the book that should be included under the rubric of "framing" passages: "Philosophy aims at the logical clarification of thoughts. Philosophy is not a body of doctrine but an activity. A philosophical work consists essentially of elucidations. Philosophy does not result in 'philosophical propositions', but rather in the clarification of propositions. Without philosophy thoughts are, as it were, cloudy and indistinct: its task is to make them clear and to give them sharp boundaries" (*Tractatus* 4.112).

The notion of "elucidation," identified in 4.112 as the essential component of genuine philosophical activity, is left almost entirely unexplained in the *Tractatus*. In 3.263, elucidations are glossed as propositions containing primitive signs, propositions that allow us to display their meaning once those signs are understood. The positivists took this as intimating an entry point for epistemology in the Tractarian system; the therapeutic reading, stressing 4.112 and 6.54, construes "elucidations" as akin to the *Investigations*'s "reminders" about the use of language. The role of elucidation in the therapeutic reading is oddly similar to the role of "objects" in metaphysical readings or "the mystical" in irrationalist readings: a notion that is barely specified in the text of the *Tractatus*, yet supposedly does an enormous amount of philosophical of work. We can already begin to see how the therapeutic reading's apparently clear-cut appeal to the method set out in the "frame" of the *Tractatus* is rapidly qualified by the need to introduce additional passages from the text of the book that also belong to the frame. Furthermore, this reading faces the problem that the preface includes the claim that the "*truth* of the thoughts communicated here seems to [Wittgenstein] unassailable and definitive," words which are more in the spirit of the transitional ladder language than the frame (*Tractatus* 29; compare Reid 1998, 100, and Hacker 2000, §3).

Much of what has been written in support of the therapeutic reading has had a one-sided methodological focus. Instead of giving a close reading of the philosophical arguments of the *Tractatus*, and elucidating how to see them as nonsense, their principal concern has been the "framing" passages. However, by far the most challenging task faced by the therapeutic reading is to give a detailed account of how the remainder of the *Tractatus* serves its supposedly therapeutic goals. As Hacker puts it, "they pay no attention to the other numerous passages in the *Tractatus* in which it is claimed that there are things that cannot be said but are shown by features of the symbolism" (Hacker 2000, §3).

Certainly, the text of the *Tractatus* has struck many of the best philosophers of the last century as a rich source of philosophical arguments for substantive views. In outline, the therapeutic response is quite clear: the arguments and views in question are present in the text of the *Tractatus*, but they were put there in order to help the reader see how to give them up. Conant (2002) provides an extended exposition of this interpretation of previous readings of the *Tractatus*. He compares the positivist and metaphysical readings with the two "voices" that Stanley Cavell (1979, 1996) finds in the *Investigations*'s dialogues. He connects the positivist reading with the voice of the plain, or the ordinary, on the one hand, and the metaphysical reading with voice of temptation, the voice that leads us into philosophical theorizing, on the other. The voice of the ordinary is content with what we say in everyday circumstances and is critical of the philosophical theories that the other voice tempts us with. Rather than reading Wittgenstein as an adamant defender of ordinary language, or as implicitly committed to a philosophical theory, Conant draws on Cavell's reading of the later Wittgenstein. He proposes that we read Wittgenstein as exploring the tensions between these views, helping us to climb "through them, on them, over them" (Wittgenstein 1922, 6.54), and so put them behind us. However, this is a large task, and one that has so far only been carried out in a piecemeal way (but see Diamond [1991a, 1991b, 1997] and Ricketts [1996] for some work in this direction). Perhaps the best way of approaching it is as an embryonic research program that will have to be judged by its results (compare Goldfarb 1997, 65–70). However, if one looks carefully at Wittgenstein's philosophical writing, one finds a number of incompatible methods. While I agree with Diamond and Conant that the text of the *Tractatus* can best be understood as the product of a conflict between the voice of temptation and the voice of the everyday, the metaphysician and the positivist, I part company with them when they assert that the author of the *Tractatus* was able to rise above that struggle.

Diamond's account helps us see just how Wittgenstein drew on Frege's work in the philosophy of logic in developing the conception of nonsense that plays such an important role in the *Tractatus*. But what is the status of this view of nonsense in the position she ultimately attributes to the *Tractatus*? One might think that this view, too, must be given up if we are to follow through on the purge of all doctrine the book supposedly advocates, but it is far from clear she is resolute about taking this final step. If we must throw away the ladder and not keep our feet planted on the top rung, where are we going to stand? Of course, Diamond is insistent on the importance of resolutely throwing away the ladder, but the very fact that she takes the philosophy of language she finds in Frege and the *Tractatus* so seriously makes it clear that she is also very much in the business of ladder-climbing. As a result, Diamond's own reading is closer to the logical positivists' than she might care to admit. For, like Carnap, she reads the *Tractatus* as setting out an antimetaphysical theory of meaning that provides the basis for a rejection of all previous philosophy as nonsensical. Like Carnap, she sees the distinction between showing and saying as a part of the book that must ultimately be discarded. The crucial question in assessing the relationship of her reading to the logical positivists' is the extent to which she ultimately endorses the philosophical views about sense and nonsense that she finds in the *Tractatus*. To the extent that she does so, she remains within the positivist tradition. For she does rely on the Fregean context principle (*Tractatus* 3.3) and the resultant views about the nature of meaning, sense and nonsense. These provide the basis on which to argue that the *Tractatus* aims to show that the philosophical doctrines contained in it are, strictly speaking, nonsense. These Wittgensteinian arguments undermining the ontological and metaphysical doctrines the book appears to set out—about such matters as the nature of objects and properties or possibility and necessity—rely, in turn, on our accepting that Fregean view about the nature of language, logic, and nonsense.

On this construal of Diamond's interpretation, "nonsense" is not just an expression used to emphatically dismiss a view, but also a term of art that depends on a theory of meaning derived from her reading of Frege. In that case, the *Tractatus* does contain a metaphysical core, even on the Diamond reading. Like the logical positivists, she rejects certain metaphysical doctrines but is left with a minimal semantic theory. We throw away certain Tractarian doctrines (simples, realism) but keep others in order to do so (logical form, elucidation). These doctrines provide the basis for the Tractarian conception of philosophy as nonsense. As Lynette Reid puts it, "It might be that Wittgenstein's intention to do away with metaphysics about the world depends on a great deal of unacknowledged metaphysics about language" (Reid 1998, 108).

On the other hand, it may be that Diamond is ultimately no more attached to the "framing" Tractarian views about meaning and nonsense than the ontology and theology that other readers have found in the "body" of the text. This would make the therapeutic reading a form of Pyrrhonian skepticism, which makes use of philosophical argument in order to do away with philosophical argument. Although most exponents of the therapeutic reading do not discuss the Pyrrhonist connection, Wittgenstein would have been familiar with the ladder image from his reading of Mauthner, one of the few authors cited in the *Tractatus* (4.0031): "Just as it is not impossible for the man who has ascended to a high place by a ladder to overturn the ladder with his foot after the ascent, so also it is not unlikely that the Sceptic after he has arrived at the demonstration of his thesis by means of an argument proving the non-existence of proof, as it were by a step-ladder, should then abolish this very argument" (Mauthner 1901, v. 1, 2, cited in Black 1964, 377).

Mauthner here summarizes the first use of the ladder simile in discussions of philosophical method at the end of the second book of Sextus Empiricus's *Against the Logicians*. In that passage, Sextus replies to the standard antiskeptical one-liner —that any proof that proof does not exist is self-refuting, or "banishes itself":

> Yes, say they, but the argument which deduces that proof does not exist, being probative itself, banishes itself. To which it must be replied that it does not entirely banish itself. For many things are said which imply an exception. . . . And even if it does banish itself, the existence of proof is not thereby confirmed. For there are many things which produce the same effect on themselves as they produce on other things. Just as, for example, fire after consuming the fuel destroys also itself, and like as purgatives after driving the fluid out of the bodies expel themselves as well, so too the argument against proof, after abolishing every proof, can cancel itself also. And again, just as it is not impossible for the man who has ascended to a high place by a ladder to overturn the ladder with his foot after his ascent, so also it is not unlikely that the Sceptic after he has arrived at the demonstration of his thesis by means of the argument proving the non-existence of proof, as it were, by a step-ladder, should then abolish this very argument. (Sextus Empiricus 1935, II.479–81)

If we read the *Tractatus* along these lines, then the therapeutic reading turns out to be a variant of irrationalism, with the Pyrrhonist twist that it gives a rational argument for irrationalism, an argument that can be discarded once it has succeeded. (For further discussion of this issue, see Reid [1998, 105–8].) Just like Sextus Empiricus, Diamond may regard the Tractarian theory of nonsense as just one more rung on a ladder that must ultimately be discarded. In "Ethics, Imagination, and the Method of Wittgenstein's *Tractatus*," she appears to align herself with this irrationalist approach:

> When I began to discuss Wittgenstein's remarks about ethics, I called them remarks about ethics, because the idea that there is no such thing as what they present themselves as, the idea that we are taken in by them in reading them as about ethics—that idea we cannot start with. So too the *Tractatus* itself. The reading it requires that it take us in at first, requires that it should allow itself to be read as sense, read as about logic and so on, despite not being so. What I have just said about the *Tractatus'* remarks "about ethics" goes then equally for its remarks about logic. (Diamond 1991b, 79)

Certainly there is no suggestion here that Wittgenstein's views about sense and nonsense are any more privileged than what he has to say about ethics or logic. Conant has expressed the same strongly resolute view: "the propositions of the *entire* work are to be thrown away as nonsense" (Conant 1989a, 274 n. 16). Warren Goldfarb raises a closely related issue when he asks whether there is not already some irresoluteness in Wittgenstein's use of "nonsense" as a term of criticism, for it presupposes certain "transitional" semantic views that are supposed to be discarded at the end of the *Tractatus*. Goldfarb's response is that "nonsense" should not be understood as a general term of criticism, but as a kind of shorthand for the particular ways of seeing how philosophical language falls apart when it is given a resolute push: "the general rubric is nothing but synoptic for what emerges in each case" (Goldfarb 1997, 71).

Diamond reads the image of throwing away the ladder as the key to understanding the *Tractatus* and uses it to attribute the philosophical methods of the *Investigations* to the author of the *Tractatus*. If this were so, one would expect the later Wittgenstein might have retained that image, or at least spoken positively of it. Instead, in a draft for a foreword, written in November 1930, in which he described himself as out of sympathy with what he called the "spirit of the main current of European and American civilization," and the values of the positivists, namely progress and constructive activity in science, industry, and architecture, he associated that spirit, and those values, with the image of climbing a ladder. Instead of aiming at progress, at trying to get somewhere else, Wittgenstein thought of himself as trying to understand where he was already:

> I might say: if the place I want to get to could only be reached by way of a ladder, I would give up trying to get there. For the place I really have to get to is a place I must already be at now.
> Anything that I might reach by climbing a ladder does not interest me.
> One movement links thoughts with another in a series, the other keeps aiming at the same spot.
> One is constructive and picks up one stone after another, the other keeps taking hold of the same thing. (Wittgenstein 1980, 7)

The *Tractatus* had been constructive, linking "thoughts with another in a series," and it was this aspect of his work that had attracted the Vienna Circle. If the author of the *Tractatus* had once thought of those constructions as a ladder language that would free his reader from the desire to climb ladders, he showed no sign of it in these remarks. Instead, he associated the image of a ladder with just that aspect of his earlier work that he now found most repugnant. Seven years later, shortly after writing the first version of the *Investigations*, Wittgenstein again rejected the notion of climbing a ladder in very similar terms: "You cannot write anything about yourself that is more truthful than you yourself are. That is the difference between writing about yourself and writing about external objects. You write about yourself from your own height. You don't stand on stilts or on a ladder but on your bare feet" (Wittgenstein 1980, 33 [December 1937]).

In recognizing the plurality of methods in the *Tractatus*, and emphasizing that the *Tractatus* is best understood as a self-conscious response to the problem of finding a consistent philosophical method, the therapeutic approach represents a valuable step forward in *Tractatus* interpretation. But resolutely rejecting both positivism and metaphysics leads Diamond to a form of irrationalism: if the favored arguments are self-immolating, and we resolutely discard the ladder, we are left not only without any doctrine but also without any positive philosophical view at all.

The easiest and perhaps the only way to motivate the therapeutic reading is to read back the later Wittgenstein's methods into the ostensibly argumentative and analytic framework of his earlier work. While it is an overstatement to speak of this reading as giving us a "new Wittgenstein" (Crary and Read, 2000), it did come about as the result of applying the ideas of a number of influential interpreters of the later Wittgenstein, and it is almost inconceivable that such an interpretation would ever have arisen had Wittgenstein died in the early 1930s. (I say "almost" because Gunnarsson [2001] does so by imagining a reader who only knows the preface and the concluding paragraphs of the *Tractatus*.)

Diamond and Conant make much of the distinction between what is said in the text of the *Tractatus* and what its author meant by that text. They emphasize that this contrast is particularly clear in *Tractatus* 6.54, where Wittgenstein does not say that one must understand the propositions of the *Tractatus*, but that one must understand *him*: "he who understands me finally recognizes [my propositions] as senseless" (see Diamond 1991a, 19; 1991b, 57; Conant 1989b, 344–46; 1991, 145; 1995, 270, 285–86). However, the authorial voice that they can hear so clearly at this point is hardly unequivocal. For if we step away from their preferred reading of their preferred passages, there is no evidence that the person who wrote the book actually read it in this way. As Hacker (2000, §4) has documented at length, what Wittgenstein actually said about the *Tractatus*, both in his own writings and when he discussed the book with Engelmann, Russell, Ramsey, the Vienna Circle, and his Cambridge students is quite incompatible with the therapeutic reading. Rather than explaining his aims in therapeutic terms, he always talked about doctrinal matters: repeatedly insisting on the importance of the show/say distinction, reaffirming certain views and modifying others, stressing the unrecognized inconsistency in the notion of elementary objects in his later discussions, and so forth.

Wittgenstein himself never represents the *Tractatus* as therapeutic in nature; that is, he consistently spoke of the book as having advanced arguments and views about the nature of language and world, not as trying to show that those arguments and views were nonsensical. In particular, his philosophical writings from 1929–30 are the work of a philosopher trying to develop views he had come to see as requiring further development (Stern 1995, chap. 5 and sect. 6.1). It is hard to avoid the conclusion that Ludwig Wittgenstein was the first reader of the *Tractatus* to fail to appreciate its therapeutic character.

Adopting a resolute stance, or attributing it to the *Tractatus*—a distinction that therapeutic readings often slur over—may ensure a certain consistency, but it has a corresponding hermeneutic drawback, which is that it guarantees that one will misread an inconsistent text. On Diamond's reading, the author of the *Tractatus* had given up all philosophical theories: every doctrine that appears to be endorsed in its numbered propositions has been discarded by the end. While this is certainly what Wittgenstein tells us at the beginning and end of his book, it is not so clear that he was able to carry out this intention.

The later Wittgenstein's writings are an extended critique of the idea that philosophy can provide us with an objective vantage point—a "view from nowhere" —a privileged perspective that shows us the world as it really is. He tries to show the incoherence of that goal, and of the related project of transcending our particular circumstances in order to provide an objective verdict on our everyday activities or commitments. Prior to the emergence of the therapeutic reading, almost all interpreters of the *Tractatus*'s skeletal arguments have taken them to lead to some such Archimedian standpoint, even though there has been almost no agreement about what Wittgenstein's arguments were and what kind of verdict he was arguing for. The therapeutic reading offers us a way of cutting through this Gordian knot of exegetical controversy about Wittgenstein's early philosophy. It proposes a diametrically opposed construal of the *Tractatus*, namely as a self-deconstructing antiphilosophical activity, a construal that is congruent with the later Wittgenstein's critique of traditional philosophical argument. But this flies in the face of Wittgenstein's own insistence that he was arguing for specific philosophical doctrines in the *Tractatus*, and his commitment to views about logic, language, and the foundations of mathematics that he had developed on the basis of his reading of Frege and Russell. Recoiling from the dogmatic extremes of most previous *Tractatus* interpretation, Diamond ends up with an equally extreme reading, one that fails to do justice to the very irresoluteness about the possibility of traditional philosophy that is one of the strongest characteristics of Wittgenstein's earlier work.

This tension between dogmatism and skepticism is also present in the discussion of philosophical method in the *Philosophical Investigations*, where Wittgenstein states that "the real discovery is the one that makes me capable of stopping doing philosophy when I want to. —The one that gives philosophy peace, so that it is no longer tormented by questions which bring *itself* into question. —Instead, a method is shown by examples; and the series of examples can be broken off" (Wittgenstein 1953, I §133; compare 1993, 194).

But his commitment to bringing philosophy under control was counterbalanced by the hold philosophy had over him. In a letter, Rush Rhees recalled a conversation with Wittgenstein that ended in the following way: "As he was leaving, this time, he said to me roughly this: 'In my book I say that I am able to leave off with a problem in philosophy when I want to. But that's a lie; I can't'" (Hallett 1977, 230). Wittgenstein aimed to end philosophy, yet in doing so, he was continually struggling with philosophical problems. (For further discussion of these passages, see Stern 1995 chap. 1 §3, esp. 19 ff.)

Oddly, there are striking parallels here with current arguments in the sociology of science over how to interpret Wittgenstein, with the twist that these are arguments about how to read the later Wittgenstein. Whereas most philosophers have wanted to turn the *Tractatus* into a systematic defense of a philosophical doctrine, sociologists of science have wanted to turn the later Wittgenstein into a systematic defender of their preferred form of empirical social science. In effect, they have seen the later Wittgenstein as providing us with access to an external standpoint, lying beyond our ordinary practices, which will justify their particular way of studying science.

Chicken Wars

Both David Bloor and Harry Collins wrote books on the sociology of knowledge in the early 1980s that made prominent use of Wittgenstein. Those books set out a reading of Wittgenstein that has become the received view, albeit a contested one, within the sociology of scientific knowledge, commonly known as SSK. Bloor's *Wittgenstein: A Social Theory of Knowledge* (1983) argued that a naturalized Wittgenstein proved the basis for his "Strong Programme" in SSK, a scientific theory of scientific and mathematical knowledge as social constructions. The Strong Programme had already been set out in earlier work that also drew on Bloor's reading of Wittgenstein, principally "Wittgenstein and Mannheim on the Foundations of Mathematics" (1973) and *Knowledge and Social Imagery* (1976). Collins's *Changing Order* (1985) begins by invoking a Wittgenstein-inspired skepticism about rule-following in setting out his "empirical relativist" program for studying scientific practice. While they are certainly not the only sociologists of science who have written on Wittgenstein, they have both emphasized the central place of Wittgenstein's treatment of rule-following in their respective arguments, and most of the subsequent discussion has been in response to their work. (However, see Phillips [1977] for an independent and contemporaneous Wittgensteinian sociology of knowledge.) Moreover, they are leading representatives of the two main styles of SSK that emerged in Britain in the 1970s: the macrosocial and more historical approach favored by Bloor and his colleagues at the University of Edinburgh, and Collins's and his coworkers' microsocial studies of scientific controversies, which were based until recently at Bath University.

Bloor's Strong Programme insisted that the sociology of science should not be confined to accounting for the institutional character of science, but should also account for the content and nature of scientific knowledge. Consequently, it proved to be a decisive moment in the formation of current approaches to the sociology of knowledge. Bloor's statement of the Strong Programme consists of the following four theses about the nature of an adequate sociology of knowledge:

1. It would be causal, that is, concerned with the conditions which bring about belief or states of knowledge. Naturally there will be other types of causes apart from social ones which will cooperate in bringing about belief.
2. It would be impartial with respect to truth and falsity, rationality or irrationality, success or failure. Both sides of these dichotomies will require explanation.
3. It would be symmetrical in its style of explanation. The same types of cause would explain, say, true and false beliefs.
4. It would be reflexive. In principle its patterns of explanation would have to be applicable to sociology itself. Like the requirement of symmetry this is a response to the need to seek for general explanations. (Bloor 1976, 7)

While philosophers usually pay little attention to the differences within SSK, and often equate the Strong Programme with SSK as a whole, the four theses are best understood as a point of departure rather than as a statement of a consensus. Most proponents of SSK reject the first thesis' commitment to a causal approach, conceiving of SSK either as an interpretive social science or as thick description that eludes systematic laws. Moreover, there has been much debate as to whether the fourth thesis' demand for reflexivity is appropriate and, if so, what it entails. However, what Collins has called the "central core" (Collins 1982, xi) of the Strong Programme, the theses of symmetry and impartiality, have become widely accepted within SSK. They have also been taken up within a number of other recent approaches to science, such as actor-network theory, feminist philosophy of science, and ethnomethodological studies of science that have distanced themselves from the specifically sociological approach characteristic of SSK.

By the early 1990s, Bloor and Collins had become canonical figures within SSK, as evidenced by their prominent role in anthologies such as *Science as Practice and Culture* (Pickering 1992a). Indeed, their work has led to a debate within science studies between those who agree with Bloor and Collins's advocacy of a thorough-going sociology of science and those who want to supplement or replace traditional sociological methods, a debate that is often framed as being about Wittgenstein interpretation. Exponents of both sociological and postmodern approaches to science studies have invoked Wittgenstein's writing on rule-following and practice. This is a leading theme in *Science as Practice and Culture*, a collection of essays on the turn to practice in science studies, in which Wittgenstein is the most frequently cited author. Nearly half the book is devoted to two interrelated arguments about methodology and epistemology, both of which turn on the significance of Wittgenstein's work for the study of science. The first exchange begins with David Bloor's defense of the Edinburgh approach to SSK and Michael Lynch's ethnomethodological reply. The Lynch-Bloor debate is over Wittgenstein's views on rule-following and their implications for our understanding of scientific knowledge. Bloor defends his interpretation of Wittgenstein as outlining a naturalistic and sociological theory, while Lynch replies that Wittgenstein denied that one could give such a theory of rule-following, arguing that Wittgenstein's conception of rule-following leads to an ethnomethodological approach to science. The second exchange begins with a piece by Harry Collins and Steven Yearly defending the Bath approach. There are two replies to Collins and Yearly: Steve Woolgar champions a reflexive approach, while Michel Callon and Bruno Latour write from the standpoint of actor-network theory.

Let us start by considering some passages that indicate Wittgenstein's role in SSK: two from introductory books on SSK and one from a leading work in that field. In *Science: The Very Idea*, Steve Woolgar explains the connection between Wittgenstein and SSK in the following terms:

> The call for a sociology of scientific knowledge attracted a lot of attention, not just because it proposed the sociological analysis of previously philosophical matters—the content and nature of scientific knowledge—but more significantly, because it emphasized the relativity of scientific truth.... In fact, this kind of relativity was no more than a particular case of a more widespread intellectual movement. In particular, SSK shows a marked affinity with a key notion in post-Wittgensteinian thought: skepticism about the view that practice (action, behavior) can be understood in terms of following rules (guidelines, principles). (Woolgar 1988, 45)

Scientific Knowledge, a recent textbook by three members of the Edinburgh Science Studies Program states that there are just two ways of "presenting the individual as an active agent in the context of the sociology of science." The first is

Bruno Latour's "actor-network" theory, which is briefly summarized before they turn to their preferred approach, which is "to characterize him or her as a participant in a form of life." They explain this expression in the following terms:

> The term is Wittgenstein's, and its use here is testimony to the relevance of Wittgenstein's work, directly or indirectly, to the work of many sociologists. Those who have taken up the work of Thomas Kuhn have thereby linked themselves to Wittgenstein; so have those who have extended ethnomethods into sociology of science. Harry Collins, who makes the most frequent explicit references to forms of life in science, has used the work of the philosopher Peter Winch as a line of access to Wittgenstein's ideas. [Bloor's] finitist account of the use of scientific knowledge in this book is another version of the same position. (Barnes, Bloor, and Henry 1996, 116)

Indeed, Wittgensteinian terms such as "form of life" and, to a lesser extent, talk of "language-games" and "world pictures" have become so widespread in SSK, as in many other areas of the social sciences, that they are commonly used without any citation or explicit reference to Wittgenstein. A good example of the wider use of "forms of life" in SSK can be found in Shapin and Schaffer's much-discussed *Leviathan and the Air Pump*, where the term is explicitly connected with Wittgenstein, but it is used to refer to social interests:

> We intend to display scientific method as crystallizing forms of social organization and as a means of regulating social interaction within the scientific community. To this end we will make liberal, but informal, use of Wittgenstein's notions of a "language-game" and a "form of life." We mean to approach scientific method as integrated into *patterns of activity.* . . . We shall suggest that solutions to the problems of knowledge are embedded within practical solutions to the problem of social order, and that differential practical solutions to the problem of social order encapsulate contrasting practical solutions to the problem of knowledge. (Shapin and Schaffer 1985, 14–15)

In each of these books, one would look in vain elsewhere for a more detailed interpretation of Wittgenstein's contribution; his role is that of a convenient canonical antecedent, rather than that of a live contributor to current debate. If one looks for a more detailed exposition of the grounds for this appeal to Wittgenstein in SSK, Bloor and Collins offer two of the most thorough bodies of work that address this topic.

Although nothing could have been further from Peter Winch's intentions, it was his interpretation of Wittgenstein in *The Idea of a Social Science and Its Relation to Philosophy* (1958) that proved to be the crucial link in the transformation of Wittgenstein's ideas about rule-following into a new sociology of knowledge.

Winch's reading of Wittgenstein on rule-following and skepticism provided the basis for Bloor's reading. (Bloor's indebtedness to Winch is most explicit in his earlier work, principally Bloor [1973]; for further discussion of the connections between Wittgenstein, Winch, and Bloor, see Stern [2000]). Both Bloor and Collins read Winch as showing that Wittgenstein appeals to the practices and conventions of particular social or cultural groups as an answer to a skeptical question about what it means to follow a rule. Winch's respect for the particularity of other forms of life, and the need to understand them from within, was enormously attractive to those who wished to approach scientific cultures along comparable lines, by combining Kuhn's notion of a paradigm with Winch's account of understanding another culture. The crucial move here was to conceive of the culture of a particular group of scientists—one of the senses of Kuhn's famously slippery term, "paradigm,"—along lines suggested, if not required, by Winch's discussion of forms of life. Bloor and Collins understand "forms of life" to refer to specific cultural or social groups, social entities comparable to the "primitive societies" discussed by Winch, or Kuhn's scientific research cultures. This way of understanding science as a culture led to a reading of Wittgenstein and Winch as providing the point of departure for a reinvigorated sociology of scientific knowledge.

In a section on "Wittgenstein and Rules," Collins sums up the sources of his reading of Wittgenstein as follows: "I take my interpretation of [Wittgenstein's] ideas from Winch (1958). For a very good introduction, see Bloor (1983)" (Collins 1985, 24 n. 9).

In particular, Bloor and Collins agree that to speak of sharing a Kuhnian paradigm or a Wittgensteinian form of life are two different ways of talking about the members of different cultural groups:

> To use Kuhn's (1962) idiom, the members of different cultures share different "paradigms", or in Wittgensteinian terms, they live within different forms of life. (Collins 1985, 15)
>
> What Wittgenstein called a "pattern of life" or a "form of life" can be thought of as a pattern of socially sustained boundaries. (Bloor 1983, 140)

While Bloor provided a more thorough and systematic interpretation of Wittgenstein as the principal philosophical antecedent of SSK, Collins's distinctive contribution was to provide detailed examples of how to do fieldwork on a practice-based approach. The principal methods he pioneered were sociological observation in the scientific laboratory and what became known as "controversy studies," which involved looking at every side of a disputed knowledge claim. Collins's principal case study of "normal science" in *Changing Order* concerns the role of skills and practices in the construction of TEA-lasers. The TEA-laser was chosen because it was a new item of scientific equipment, but one that could be constructed out of readily available and relatively inexpensive components. As it could not yet be bought off the shelf, the task of assembling such a laser and getting it to work properly provided a convenient example of an everyday, but not yet routine, scientific task. In Collins's retrospective summary of the motivations for his research on the building of TEA-lasers, the connections with the work of Wittgenstein, Winch, and Kuhn can be seen very clearly:

> In 1971, at the outset of these studies on lasers, the project was not envisaged as a study of replication but of knowledge transfer. The intention was to explore knowledge transfer in a manner informed by ideas drawn from the philosophy of social science and the history of science. The most important idea drawn from the philosophy of social science was that actors are to be understood as acting within a "form of life" (Winch 1958; Wittgenstein 1953). The idea was to have its counterpart, in the history of science, in the notion of "paradigm" (Kuhn 1962). With these ideas in mind the "communication network" of TEA-laser builders was explored. (Collins 1985, 171; compare n. 3)

Collins and Bloor both insist that their respective approaches of the sociology of knowledge are scientific, but their preferred conceptions of social science are different. Bloor, unlike Winch or Collins, has no qualms about conceiving of philosophy and social science as modeled on the natural sciences. Bloor's reading of Wittgenstein is behavioristic, while Collins's, drawing on Winch and Berger, is in the *Verstehen* tradition. However, Bloor and Collins do agree that forms of life amount to a sociological solution to skepticism about knowledge and thus provide the basis for a sociological solution to the philosophical problem of knowledge.

Bloor agrees with Collins that Winch's work leads to a strong form of relativism about logic and standards of appraisal, but he emphatically rejects the view that one should be a relativist about explanation. At the level of giving a causal explanation of belief, Bloor endorses a strongly naturalistic position: science, and a fortiori, the sociology of science, is committed to providing causal explanations. Thus, despite his relativism about standards of appraisal, Bloor is very far from being an "anything goes" epistemological anarchist: he maintains that there is a single scientific method that should be applied when determining why people believe what they do. Bloor's insistence on the importance of giving an objective and causal explanation of all belief is particularly clear in a paper coauthored with Barry Barnes, a colleague in the Edinburgh Science Studies Unit: "The position we shall defend is that the incidence of all beliefs without exception calls for empirical investigation and must be accounted for by finding the specific, local causes of this credibility. This means that regardless of whether the sociologist evaluates a belief as true or rational, or as false and irrational, he must search for the causes of its credibility" (Hollis and Lukes 1982, 23). While Bloor is a relativist about reasons, he is an objectivist about explanation, and the two views are closely connected: the relativism about reasons is a direct consequence of his objectivism about explanation.

Collins's sociological methodology of "participant comprehension" is explicitly modeled on Peter Berger's method of "alternation": learning to move between different forms of life, seeing one as natural, and then another (Collins and Yearly 1992a, 301–2, 323–24; Berger 1963). Crucially, Collins's "empirical program of relativism" does not go all the way down. While he is ready to apply it to the natural sciences, in order to unsettle an unreflective realism about their authority and objectivity, he insists that it would be inappropriate for him to apply it to his own program of research in the sociology of science. His solution is to offer the principle of "meta-alternation." Just as alternation is a matter of being able to move from one culture to another, meta-alternation is the ability to move between a skeptical, external perspective engendered by moving between different scientific forms of life and an involved, internal standpoint on which one takes the reality of one's surroundings for granted (Collins and Yearly 1992a, 302). Meta-alternation provides Collins with a principled basis for compartmentalizing the realism and the skepticism in his work that might otherwise come into conflict. This is the "crucial reflexive insight" that allows him to "reexamine the nature of science while at the same time doing science": "Science—the study of an apparently external world —is constituted by not doing the sort of thing that the sociology of scientific knowledge does to science; the point cannot be made too strongly. Sociologists of scientific knowledge who want to find (or help construct) new objects in the world must compartmentalise; they must adopt the 'natural attitude' of the scientist and not apply their methods to themselves" (Collins 1992, afterword to 2d ed., 188).

With this background in place, we can now turn to the debate between Collins and Yearly and their critics in *Science as Practice and Culture*. Both sets of criticisms can be seen as a radicalization of certain aspects of the approach to SSK pioneered by Bloor and Collins. Woolgar argues that the sociology of science should not only give a sociological account of the content of natural science, but must also (reflexively) apply those very methods to the claims made by those who work within SSK. Callon and Latour deny that there is a sharp distinction between the natural and the social, highlighting the role of that distinction in SSK's program of giving a sociological account of natural science. Instead, they want to replace a dualism of social and natural facts, and the assumption that the social can be used to explain the natural, by a monistic ontology of "actors," whose properties are entirely determined by the "networks" they belong to. Woolgar, and Callon and

Latour, see their respective approaches as a natural and more consistent develop-ment of the claims of the Strong Programme. Callon and Latour argue that their approach is truly symmetrical (thesis 3), in contrast to Bloor and Collins's privi-leging of the social; Woolgar argues that his approach is truly reflexive (thesis 4), in contrast to Bloor and Collins's privileging of the sociological.

Collins and Yearly's reply to their critics is that they have betrayed the guid-ing insights of the Strong Programme by making them overly skeptical and rela-tivistic. They characterize those who disagree with them as playing a dangerous game of "epistemological chicken": "The game of 'chicken' involves dashing across the road in front of speeding cars. The object of the game is to be the last person to cross. Only this person can avoid the charge of being cowardly. An early crosser is a 'chicken'" (Collins and Yearly 1992a, 301).

According to Collins and Yearly, their approach to SSK turns on using just enough skepticism to dislodge our ordinary acceptance of the everyday social world of "Mr. John Doe as he rides on the Clapham omnibus" (Collins and Yearly 1992a, 302) or the work world of scientific research communities. Because it unsettles the natural attitude of the everyday bus rider or laboratory scientist, their moderate relativism provides a "tenable . . . methodology for the study of science" (Collins and Yearly 1992a, 303). But, they continue, there is no need to be relativists about the account they offer of natural science, one on which its content is to be under-stood in terms of the causal role of certain social facts, namely the scientists' in-terests. Invoking the principle of meta-alternation, they reply that it is essential to compartmentalize the skeptical stance, so that it does not affect their own work. This is where their critics have cried "chicken," seeing an inconsistency between Collins and Yearly's use of relativism to disqualify the reasons that natural scien-tists offer for doing what they do and their refusal to be relativists about their own reasons.

Collins and Yearly's reply amounts to saying that sticking around when a speeding car is coming down the road is asking for trouble. When the only alter-native is to become roadkill, any sensible chicken will cross the road as quickly as possible: "In the absence of decisive epistemological arguments, how do we choose our epistemological stance? The answer is to ask not for the meaning, but for the use. Natural scientists, working at the bench, should be naïve realists—that is what will get the job done. Sociologists, historians, scientists away from the bench, and the rest of the general public should be social realists. Social realists must experience the social world in a naïve way, as the day-to-day foundation of reality" (Pickering 1992a, 308).

Unfortunately for Collins and Yearly, expediency is hardly a promising basis for a principled resolution. How are they and their colleagues to account for the privileged status they give to their preferred way of doing SSK? Just as Bloor and Collins turned Winch and Kuhn into the predecessors they needed, those who learned the most from them would repay them by turning their methods against them. In this case, the principle of symmetry was the primary locus of disagreement. Bloor and Collins are both committed to a sharp distinction between social facts and natural facts, social science and natural science. However, once one has turned a skeptical eye on the taken-for-granted procedures of the natural sciences, what is to prevent Woolgar, Callon, and Latour from turning an equally skeptical eye on the methodology of SSK itself? Do not the principles of symmetry and reflexivity require that we treat the sociology of science with the same skepticism as natural science? Can we take the procedures of Bloor's or Collins's versions of SSK for granted and at the same time treat the other sciences as social constructions that must ultimately be explained by an appeal to social interests? How can we rely on social facts in order to explain the construction of natural facts without applying the same procedures to the reliance on close description of "forms of life" and the history of scientific controversies that are SSK's preferred explanatory starting points? SSK casts doubt on the objectivity of reason-based theories of belief in order to replace those theories with a supposedly more objective causal account (Bloor) or a participant observer account (Collins). But since these notions are just as open to skeptical challenge as the explanatory strategies that they are supposed to replace, we are no better off than before. Paul Roth aptly sums up their predicament as follows: "Faced with a demand to substantiate the legitimacy of their approach, Collins and Yearly turn from bold challengers to reactionaries, much in the manner of positivists who could neither justify the verifiability principle nor bring themselves to give it up. In their case, they can neither legitimate the status given [their preferred solutions] nor give them up" (Roth 1994, 99; compare 1996, 56, 60.) Wittgenstein's philosophy held out the promise of a deus ex machina for Bloor and Collins, a universal skeptical solvent that would dissolve all competing programs, coupled with a theory of practice that would provide a firm objective basis for their own form of sociology of science. Like the sorcerer's apprentice, who tried to get his job done by borrowing a word of power from a venerable authority, it was much easier for the founders of SSK to get the skeptical argument started than to stop it where they wanted.

It is at this point that it is instructive to turn to the parallels between the philosophers' debates over the *Tractatus* and sociologists' debates over the *Investigations*. Both turn on issues of method and are pitched at an extremely abstract level, and both concern the correct understanding of Wittgenstein's way of doing philosophy. Generations of philosophers have turned the crystalline but obscure text of the *Tractatus* into congenial ontological and logical doctrines; contemporary sociologists of science have turned the conversational but elusive text of the *Investigations* into a validation of their preferred variety of social theory. In both cases, texts which are centrally concerned with problems of philosophical method and the question of whether it is possible to provide an objective, external validation for one's philosophical convictions are turned into one more argument in support of such a validation (compare Friedman [1998] for a more extensive discussion of this issue.)

In both cases, cries of "chicken" can be heard. As we have seen, in Diamond's hands it is an accusation of cowardice, of lacking the courage to stick to a single approach, and also indecisiveness, an inability to choose between one approach and the other. Diamond advocates that we be consistent and resolute, following through the approach outlined in the frame of the *Tractatus,* even as it leads to a thoroughgoing skepticism about any philosophical theory whatsoever. Her critics respond that the price of such foolhardy bravado—not stepping out of the way of the skeptical juggernaut—is to become the philosophical equivalent of roadkill. In the case of the disputes over SSK, Woolgar, Callon, and Latour occupy the position analogous to Diamond's. They claim they are consistently applying SSK's symmetry principle, while Bloor, Collins, and Yearly chicken out of applying its skeptical consequences to themselves in order to protect their own doctrines from dissolution.

We saw earlier that Diamond's critics can argue that the therapeutic approach leads to a dilemma: either it amounts to giving up doing philosophy altogether or it tacitly reinstates a positivistic philosophy of language. Collins and Yearly argue that a more consistent application of skepticism to SSK leads to a similar dilemma. One of the alternatives is to apply the skepticism to one's own methods, in which case one stays in the road and gets run over. The other is to jump out of the way of the speeding car—the skeptical refutation—at the last minute, in which case one is really no different from those who stepped out of the traffic a little sooner. What both sides of the "chicken wars" in the sociology of science fail to see is that Wittgenstein's principal target in his later philosophy is the very idea that they take for granted in making use of his work, namely the idea that we can take up a detached scientific stance on our normative commitments, whether it be to validate them or to undermine them.

Note

I would like to thank the Alexander von Humboldt Foundation, the University of Bielefeld, and the University of Iowa Faculty Scholar Program for supporting my work on this paper.

Earlier versions of this paper were presented at the University of Southern Florida, Tampa, and at a conference on "The Continental and Analytical Origins of Logical Empiricism: Historical and Contemporary Perspectives" at the University of Florence in Italy. Parts 1–3 of this paper form the basis for part of chapter 2 of Stern, forthcoming. The topics discussed in part 4 are developed further in Stern 2002 and 2003.

References

Anscombe, G. E. M. 1959. *An Introduction to Wittgenstein's "Tractatus."* London: Hutchinson.

Bergmann, G. 1967. *The Metaphysics of Logical Positivism.* Madison: University of Wisconsin Press.

Berger, P. 1966. *Invitation to Sociology: A Humanistic Perspective.* London: Penguin.

Black, M. 1964. *A Companion to Wittgenstein's Tractatus.* Cambridge: Cambridge University Press.

Bloor, D. 1973. "Wittgenstein and Mannheim on the Sociology of Mathematics." *Studies in the History and Philosophy of Science* 4 173–91. Reprinted in Collins 1982, 39–58.

———. 1976. *Knowledge and Social Imagery.* London: Routledge.

———. 1983. *Wittgenstein: A Social Theory of Knowledge.* New York: Columbia University Press.

———. 1992. "Left and Right Wittgensteinians." In *Science as Practice and Culture,* ed. A. Pickering, 266–82. Chicago: University of Chicago Press.

———. 1997. *Wittgenstein, Rules, and Institutions.* London: Routledge.

Cavell, S. 1979. *The Claim of Reason.* Oxford: Oxford University Press.

———. 1996. "Notes and Afterthoughts on the Opening of Wittgenstein's *Investigations.*" In *The Cambridge Companion to Wittgenstein,* ed. H. Sluga and D. Stern, 261–95. Cambridge: Cambridge University Press.

Coffa, J. A. 1991. *The Semantic Tradition from Kant to Carnap: To the Vienna Station,* ed. Linda Wessels. Cambridge: Cambridge University Press.

Collins, H. 1974. "The TEA Set: Tacit Knowledge and Scientific Networks." *Science Studies* 4:165–86.

———. 1983. "An Empirical Relativist Programme in the Sociology of Scientific Knowledge." In *Science Observed: Perspectives on the Social Study of Science,* ed. K. Knorr-Cetina and M. Mulkay, 85–114. London: Sage Publications.

———. 1992. *Changing Order: Replication and Order in Scientific Practice.* 2d ed., with new afterword. Beverly Hills, Calif.: Sage Publications.

———, ed. 1982. *Sociology of Scientific Knowledge: A Source Book.* Bath, England: Bath University Press.

Collins, H., and S. Yearly. 1992a. "Epistemological Chicken." In *Science as Practice and Culture,* ed. A. Pickering, 301–26. Chicago: University of Chicago Press.

———. 1992b. "Journey into Space." In *Science as Practice and Culture,* ed. A. Pickering, 369–89. Chicago: University of Chicago Press.

Conant, J. 1989a. "Must We Show What We Cannot Say?" In *The Senses of Stanley Cavell,* ed. R. Fleming and M. Payne, 242–83. London and Toronto: Associated University Presses.

———. 1989b. "Throwing Away the Top of the Ladder." *Yale Review* 79:328–64.

———. 1991. "The Search for Logically Alien Thought: Descartes, Kant, Frege and the *Tractatus.*" *Philosophical Topics* 20, no. 1:115–80.

———. 1993. "Kierkegaard, Wittgenstein, and Nonsense." In *Pursuits of Reason: Essays in Honor of Stanley Cavell,* ed. T. Cohen, P. Guyer, and H. Putnam, 195–224. Lubbock: Texas Tech University Press.

———. 1995. "On Putting Two and Two Together: Kierkegaard, Wittgenstein, and the Point of View for Their Work as Authors." In *Philosophy and the Grammar of Religious Belief,* ed. T. Tessin and M. von der Ruhr, 248–331. London: Macmillan.

———. 2002. "The Method of the *Tractatus.*" In *From Frege to Wittgenstein: Perspectives on Early Analytic Philosophy,* ed. E. H. Reck, 374–462. Oxford: Oxford University Press.

Crary, A., and R. Read, eds. 2000. *The New Wittgenstein.* Routledge: New York.

Diamond, C. 1991a. *The Realistic Spirit: Wittgenstein, Philosophy, and the Mind.* Cambridge, Mass.: MIT Press.

———. 1991b. "Ethics, Imagination, and the Method of Wittgenstein's *Tractatus.*" In *Bilder der Philosophie,* ed. R. Heinrich and H. Vetter, 149–73. Vienna and Munich: Oldenbourg, 55–90. Reprinted in Crary and Read 2000.

———. 1996. "Wittgenstein, Mathematics, and Ethics: Resisting the Attractions of Realism." In *The Cambridge Companion to Wittgenstein,* ed. H. Sluga and D. Stern, 226–60. Cambridge: Cambridge University Press.

———. 1997. "Realism and Resolution: Reply to Warren Goldfarb and Sabina Lovibond." *Journal of Philosophical Research* 22:75–86.

Engelmann, P. 1967. *Letters from Ludwig Wittgenstein with a Memoir.* Trans. L. Furtmüller and ed. B. F. McGuinness. Oxford: Blackwell.

Friedman, M. 1998. "On the Sociology of Scientific Knowledge and Its Philosophical Agenda." *Studies in the History and Philosophy of Science* 29:239–71.

———. 1999. *Reconsidering Logical Positivism.* Cambridge: Cambridge University Press.

Frongia, G., and B. McGuinness. 1990. *Wittgenstein: A Bibliographical Guide.* Oxford: Blackwell.

Goldfarb, W. 1997. "Metaphysics and Nonsense: On Cora Diamond's *The Realistic Spirit.*" *Journal of Philosophical Research* 22:57–74.

Gunnarsson, L. 2001. "Climbing Up the Ladder: Nonsense and Textual Strategy in Wittgenstein's *Tractatus.*" *Journal of Philosophical Research* 26:229–86.

Hacker, P. M. S. 1972. *Insight and Illusion: Wittgenstein on Philosophy and the Metaphysics of Experience.* Oxford: Clarendon Press.

———. 1996. *Wittgenstein's Place in Twentieth-Century Analytic Philosophy.* Oxford: Blackwell.

———. 2000. "Was He Trying to Whistle It?" In *The New Wittgenstein,* ed. A. Crary and R. Read, 353–94. Routledge: New York.

Haller, R., and K. Puhl, eds. 2001. *Wittgenstein and the Future of Philosophy: A Reassessment after Fifty Years. Papers of the Twenty-fourth International Wittgenstein Symposium.* 2 vols. Kirchberg am Wechsel: Austrian Ludwig Wittgenstein Society.

Hallett, G. 1977. *A Companion to Wittgenstein's "Philosophical Investigations."* Ithaca, N.Y.: Cornell University Press.

Heidegger, M. 1996. *Being and Time: A Translation of Sein und Zeit.* Trans. J. Stambaugh. Albany: State University of New York Press.

Hollis, M., and S. Lukes, eds. 1982. *Rationality and Relativism.* Cambridge, Mass.: MIT Press.

Hintikka, M., and J. Hintikka. 1986. *Investigating Wittgenstein.* Blackwell: Oxford.

Janik, A., and S. Toulmin. 1973. *Wittgenstein's Vienna.* London: Weidenfeld and Nicholson.

Kuhn, T. 1962. *The Structure of Scientific Revolutions.* Chicago: University of Chicago Press.

Landini, G. 2000. "The Russellian Nature of Wittgenstein's *Tractatus.*" Photocopied typescript, University of Iowa.

Lynch, M. 1992a. "Extending Wittgenstein: The Pivotal Move from Epistemology to the Sociology of Science." In *Science as Practice and Culture,* ed. A. Pickering, 215–65. Chicago: University of Chicago Press.

———. 1992b. "From the 'Will to Theory' to the Discursive Collage: A Reply to Bloor's 'Left and Right Wittgensteinians.'" In *Science as Practice and Culture,* ed. A. Pickering, 283–300. Chicago: University of Chicago Press.

———. 1993. *Scientific Practice and Ordinary Action: Ethnomethodology and Social Studies of Science.* Cambridge: Cambridge University Press.

Malcolm, N. 1986. *Nothing Is Hidden: Wittgenstein's Criticism of His Early Thought.* Oxford: Blackwell.

Marion, M. 1998. *Wittgenstein, Finitism, and the Foundations of Mathematics.* Oxford: Clarendon Press.

Mauthner, F. 1901. *Beiträge zu einer Kritik der Sprache.* Stuttgart.

McGinn, M. 1999. "Between Metaphysics and Nonsense: Elucidation in Wittgenstein's *Tractatus.*" *Philosophical Quarterly* 49:491–513.

McGuinness, B. 1988. *Wittgenstein: A Life. Young Ludwig (1889–1921).* London: Duckworth.

Neurath, O. 1973. *Empiricism and Sociology.* Ed. M. Neurath and R. Cohen. Dordrecht: Reidel.

Neurath, O., R. Carnap, and H. Hahn. 1929. *Wissenschaftliche Weltauffassung: Der Wiener Kreis.* Vienna: Wolf. Translated as "The Scientific Conception of the World: The Vienna Circle," in Neurath (1973, 299–318).

Ostrow, M. B. 2002. Wittgenstein's *Tractatus.* Cambridge: Cambridge University Press.

Pears, D. 1967. *Bertrand Russell and the British Tradition in Philosophy.* London: Collins.

———. 1986. *Ludwig Wittgenstein.* 2d ed., with a new preface by the author. Cambridge, Mass.: Harvard University Press.

———. 1987. *The False Prison.* Vol. 1. Oxford: Clarendon Press.

Phillips, D. L. 1977. *Wittgenstein and Scientific Knowledge: A Sociological Perspective.* Totowa, N.J.: Rowman and Littlefield.

Pickering, A., ed. 1992a. *Science as Practice and Culture.* Chicago: University of Chicago Press.

———. 1992b. "From Science as Knowledge to Science as Practice." In *Science as Practice and Culture,* ed. A. Pickering, 1–28. Chicago: University of Chicago Press.

Putnam, H. 1994. "Sense, Nonsense, and the Senses: An Inquiry into the Powers of the Human Mind." *Journal of Philosophy* 91:445–517.

Ramsey, F. 1923. "Critical Notice of the *Tractatus.*" *Mind* 32:465–78.

Reid, L. 1998. "Wittgenstein's Ladder: The *Tractatus* and Nonsense." *Philosophical Investigations* 21:97–151.

Ricketts, T. 1996. "Pictures, Logic, and the Limits of Sense in Wittgenstein's *Tractatus.*" In *The Cambridge Companion to Wittgenstein,* ed. H. Sluga and D. Stern, 59–99. Cambridge: Cambridge University Press.

Roth, P. 1996. "Will the Real Scientists Please Stand Up? Dead Ends and Live Issues in the Explanation of Scientific Knowledge." *Studies in History and Philosophy of Science* 27:43–68.

———. 1998. "What Does the Sociology of Scientific Knowledge Explain? Or, When Epistemological Chickens Come Home to Roost." *History of the Human Sciences* 7:95–108.

Russell, B. 1922. Introduction to *Tractatus Logico-Philosophicus,* ed. L. Wittgenstein and trans. (facing pages) C. K. Ogden. London: Routledge and Kegan Paul.

Schlick, M. 1930. "Die Wende der Philosophie." *Erkenntnis* 1:4–11. Translated in Schlick (1979, 154–60).

———. 1979. *Philosophical Papers.* Vol. 2. Dordrecht: Reidel.

Sextus Empiricus. 1935. *Against the Logicians.* Ed. and trans. R. G. Bury. 3 vols. Cambridge, Mass.: Harvard University Press.

Shapin, S., and S. Schaffer. 1985. *Leviathan and the Air Pump: Hobbes, Boyle, and the Experimental Life.* Princeton: Princeton University Press.

Sluga, H., and D. Stern, eds. 1996. *The Cambridge Companion to Wittgenstein.* Cambridge: Cambridge University Press.

Stenius, E. 1960. *Wittgenstein's Tractatus.* Oxford: Blackwell.

Stern, D. G. 1995. *Wittgenstein on Mind and Language.* Oxford: Oxford University Press.

———. 1996. "The Availability of Wittgenstein's Philosophy." In *Tractatus Logico-Philosophicus,* ed. L. Wittgenstein and trans. (facing pages) C. K. Ogden, 442–76. London: Routledge and Kegan Paul.

———. 2000. "Practices, Practical Holism, and Background Practices." In *Heidegger, Coping, and Cognitive Science: Essays in Honor of Hubert L. Dreyfus, Volume 2,* ed. M. Wrathall and J. Malpas, 53–69. Cambridge, Mass.: MIT Press.

———. 2002. "Sociology of Science, Rule Following and Forms of Life." In *Vienna Circle Institute Yearbook 9/2001: History of Philosophy of Science—New Trends and Perspectives,* ed. M. Heidelberger and F. Stadler, 347–67. Dordrecht: Kluwer.

———. 2003. "The Practical Turn." In *The Blackwell Guide to the Philosophy of the Social Sciences,* ed. S. Turner and P. Roth, 185–206. Oxford: Blackwell.

———. Forthcoming a. *Wittgenstein's* Philosophical Investigations: *An Introduction.* Cambridge: Cambridge University Press.

———. Forthcoming b. Critical Review of Ostrow 2002. *Inquiry.*

Winch, P. 1958. *The Idea of a Social Science and Its Relation to Philosophy.* London: Routledge and Kegan Paul.

———. 1987. "Language, Thought, and World in Wittgenstein's *Tractatus.*" In *Trying to Make Sense,* ed. P. Winch, 3–17. Oxford: Blackwell.

———. 1992. "Persuasion." In *The Wittgenstein Legacy, Midwest Studies in Philosophy 17,* ed. P. A. French, T. E. Uehling Jr., and H. Wettstein, 123–37. Notre Dame, Ind.: University of Notre Dame Press.

Wittgenstein, L. 1922. *Tractatus Logico-Philosophicus.* Trans. (facing pages) C. K. Ogden. London: Routledge and Kegan Paul.

———. 1953. *Philosophical Investigations.* Ed. G. E. M. Anscombe and R. Rhees and trans. G. E. M. Anscombe. Oxford: Blackwell.

———. 1980. *Culture and Value.* First published in 1977 as *Vermischte Bemerkungen,* German text only, edited by G. H. von Wright. Oxford: Blackwell.

———. 1989. *Logisch-philosophische Abhandlung: Kritische Edition.* Ed. B. McGuinness and J. Schulte. Frankfurt am Main: Suhrkamp.

———. 1993. *Philosophical Occasions, 1912–1951.* Ed. J. Klagge and A. Nordmann. Indianapolis: Hackett.

———. 1995. *Ludwig Wittgenstein, Cambridge Letters: Correspondence with Russell, Keynes, Moore, Ramsey and Sraffa.* Ed. B. F. McGuinness and G. H. von Wright. Oxford: Blackwell.

Woolgar, S. 1988. *Science: The Very Idea.* London: Routledge.

———. 1992. "Some Remarks about Positionism: A Reply to Collins and Yearly." In *Science as Practice and Culture,* ed. A. Pickering, 327–42. Chicago: University of Chicago Press.

I V

Philosophy of Physics

Two Roads from Kant

Cassirer, Reichenbach, and General Relativity

T. A. RYCKMAN

University of California, Berkeley

"O Kant, wer rettet dich vor den Kantianern?"
—*Reichenbach to A. Berliner, April 22, 1921*

Introduction

Kantian and neo-Kantian epistemological analyses of the theory of relativity comprised a not insignificant torrent in the flood of publications generated in the "relativity rumpus" following the announcement of the British Expedition's observational results in London on November 6, 1919. That was hardly surprising. For while many "schools" of philosophy felt obliged to pronounce upon the theory, often as a stunning confirmation of basic principles or outlooks, it was incontrovertible that the Einstein theory, with its demonstration of the non-Euclidean behavior of light rays in the solar gravitational field, minimally required modification or clarification of the Kantian account of geometry and so of the Transcendental Aesthetic and the doctrine of pure intuition. Most of this neo-Kantian literature, like that produced by the other "schools," poorly repays present perusal. But within a few months in 1920 and early 1921,[1] Ernst Cassirer and Hans Reichenbach each published a revisionist epistemological appraisal of the theory of relativity whose significance for the history of philosophy of science has not diminished. Though Cassirer's *Zur Einsteinschen Relativitätstheorie: Erkenntnistheoretische Betrachtungen* and Reichenbach's *Relativitätstheorie und Erkenntnis a Priori* are similar in various respects,[2] these works originate in neo-Kantian vantage points differing so fundamentally as to pose with considerable sharpness, at least in retrospect, two

159

paths for the future development of philosophy of science. It is a matter of record that philosophy of science followed the road taken by Reichenbach (and Schlick before him) rather than that proposed by Cassirer. But the distinct roads taken by Cassirer and Reichenbach are not peripheral tracks; they have a common origin in a difficulty *within* Kant, perhaps *the difficulty within Kantian epistemology* concerning the distinct mental faculties of "sensibility" and "understanding" as *independent sources of knowledge* and the nature of the relation between them.

In their respective epistemological reflections on the theory of general relativity, Cassirer and Reichenbach brought diametrically opposed notions of this relation to the central problem of "constitution of the object of knowledge." In turn, the diverging roads from Kant taken by Cassirer and Reichenbach imply importantly different conceptions of what has been called the "relativized a priori" within physical theories. The contention of this essay is that Cassirer, drawing upon his rich knowledge of the changing concept of the object of knowledge in the development of modern physical science, provided the more promising way to think about the role of a priori elements in the theoretical constructions of physics. To Cassirer, "reason" and "experience," a priori and a posteriori, are but dynamical "moments," separable only by analysis, whose changing interrelation neither requires, nor admits of, a clear-cut separation into opposing spheres. Yet this progressive "movement of thought" (*Denkbewegung*) can be tracked in the history of the modern physical sciences through the ever-growing replacement of "substance" or "thing" concepts, uncritical "anthropomorphic" modes of representation, by functional and relational concepts. With general relativity, this development has been still further "de-anthropomorphized": the conception of "physical objectivity" now incorporates a demand for general covariance: *invariance of the laws of nature* with respect to all frames of reference. The "concept of physical object" can no longer be mistaken for a "thing concept" but incorporates a principle of local symmetry.

Reichenbach, on the other hand, initiates his epistemological analysis of relativity theory under the presupposition of a fundamentally different construal of the relation of "reason" and "experience." He reformulated Kant's original distinction between sensibility and understanding, and the relation between them, in set theoretic language as a coordination between two sets: a process in which concepts—the fundamental equations of physics, inductively supported "axioms of connection" —are coordinated to sensible experience. According to Reichenbach, this "coordination" can be carried out only under the guidance of a necessary "subjective contribution of reason"—that is to say, through "axioms of constitution," principles first enabling this coordination to be defined. As constitutive of the object of knowledge, these "principles of coordination" retain the primary sense of the meaning of Kant's a priori; however, they do not retain the meaning of apodictic validity, for they are fallible, theory-specific, and therefore relative to a given stage of physical knowledge. According to the theory of general relativity, general covariance is such a principle, interpreted as having the meaning of "the relativity of the coordinates."

This difference marks a watershed for subsequent philosophy of science. For the Reichenbach conception of constitutive principles was quickly deemed to be too "thin" a role for the a priori to play. The result was that, with a bit of pressure from Schlick, it quickly collapsed into an empiricist and realist analysis of relativity theory wherein such principles become conventions, stipulated physical definitions of abstract theoretical concepts such as "metric." Thus the difference between Cassirer's and Reichenbach's epistemological analyses of relativity theory ultimately concerns the question of *whether there are nonanalytic a priori elements in physical theory*. Thanks in large measure to Schlick's rhetorical ability to pose the issue on his own terms, this was a very short debate that, in the eyes of "scientific philosophy," Cassirer lost (Schlick 1921).[3] Within a few years, the postulate that physical theories contain no *synthetic a priori* elements would become an ideological cornerstone of logical empiricism. Indeed, Reichenbach's later "classic" 1928 examination of physical geometry claimed to make "the strongest possible refutation of the philosophy of the *a priori*" ([1928] 1958, 67). Alone amidst the rubble of so many other failed doctrines of logical empiricism, the refutation of the synthetic a priori survived intact in the successor doctrines of Quinian holism, scientific realism, and Kuhnian-inspired relativism, deemed fundamental to the opposing projects of both modernity and postmodernity. Here lies a remarkable irony. Logical empiricism's success in enshrining this disciplinary article of faith, even from beyond the grave, has turned out to be but a Pyrrhic victory. Intended as a firewall to "metaphysical" obfuscations of science, the abjuration of synthetic a priori elements in physical theory has tacitly supported one perennial form of metaphysics, scientific realism, by casting its opponents in the impoverished guise of antirealism or relativism. Obviously, then, hindsight favors the road taken by Cassirer. But I do so also for reasons that are internal to the epistemological analysis of the theory of general relativity.

Cassirer's account concentrates on the transformation in the conception of physical objectivity attained in general relativity and in particular by the principle of general covariance. Thus, I think, together with Einstein and Cassirer, that this principle is not physically vacuous but "has considerable heuristic force" (to use Einstein's phrase) in the construction of physical theories. This is just to say that the principle has a "constitutive" significance. However, it has proven to be notoriously difficult to give precise meaning to this intended constitutive character of the principle of general covariance. Nonetheless, the fiber bundle geometrical approach of gauge field theories of fundamental interactions provides one reasonably clear illustration. Here the revolutionary transformation of physical objectivity in general relativity is made palpable by the analogy to gauge field theories: general relativity is a field theory with a *local symmetry;* indeed, it is the first such theory and a classical field theory to boot. "Revolutionary" because field theories with local symmetries have radically transformed conceptions of fundamental physical ontology or, better, of physical objectivity. After a brief sketch of how this goes, we shall turn back to the Kantian "faculties" of sensibility and understanding, the two roads taken by Cassirer and Reichenbach, and how they issue in different understandings of the significance of general covariance.

The Constitutive Character of General Covariance

In his first complete exposition of the general theory of relativity in 1916, Einstein posed the following requirement on physical theory, which he claimed led to an extension of the principle of relativity: *"The laws of physics are to be of such a kind that they apply to systems of reference in any kind of motion."* A few pages later, using the language of coordinates, he calls this condition the requirement of general covariance: *"The general laws of nature are to be expressed through equations which are valid for all coordinate systems, that is, are covariant with respect to arbitrary substitutions (generally covariant),"* and he claims that through this requirement "space and time have lost the last remnants of physical objectivity" (Einstein 1916, 287, 291). Einstein's repeated insistence that the principle of general covariance is not a mere requirement on how physical theories are to be formulated but also has nontrivial physical significance was and remains today a matter of controversy that has generated a sizable literature. The ensuing "eight decades of dispute," beginning with Kretschmann's objections to Einstein in 1917 and continuing into the present, have been extensively surveyed by Norton (1993). Here I shall try to outline a case for what I consider to be the best attempt to do justice to Einstein's insight that general covariance is a "limiting heuristic principle" that informs the search for the fundamental laws of physics.

Late in his life, Einstein made several attempts, to which John Stachel especially has called attention, to clarify his position in the decades-long controversy over the requirement of general covariance. Notably, the context of these last efforts is that of Einstein's unsuccessful unified field theory program. Here is perhaps the clearest example:

> On the basis of the general theory of relativity ... space as opposed to "what fills space" ... has no separate existence. ... If we imagine the gravitational field, i.e.,

the functions to be removed, there does not remain a space of the type [of Minkowski space-time], but absolutely *nothing*, and also not "topological space." For
the functions describe not only the field, but at the same time also the topological and metrical structural properties of the manifold. ... There is no such thing
as an empty space i.e., a space without field. Space-time does not claim existence
on its own, but only as a structural quality of the field. (Einstein 1952, 155)

It appears that the fundamental meaning of general covariance may be expressed thus: there can be no principled distinction between the structure of space-
time and its "contents"; in brief, "no metric, no space-time." Thus space and time
have lost "the last remnants of physical objectivity." Alternately, this can be phrased:
"The conception of position with respect to a background space-time is meaningless" (Rovelli 1997). That Einstein was actually not able to constrain his gravitational field equations to implement this Machian requirement should not blind
us to the guiding programmatic commitment that motivated its development and
was, for Einstein, the "most essential thing" of the theory: the *ambition* to remove
from physical theory, once and for all, the notion of a privileged frame of reference,
as background space-time structures that act (in the explanation of the inertial
motion of a body) but that in turn are not acted upon.[4]

Has there been any progress in illuminating this fundamental conception of
Einstein? Although a comprehensive discussion lies beyond the scope of this chapter, two relatively recent treatments may be briefly indicated; later on, I hope to
show that the second one rather strikingly vindicates Cassirer's idea concerning
the constitutive role of a priori elements in physical theory.

One notable attempt to clarify the situation regarding the physical significance
of general covariance is accounted in terms of a distinction between *covariance*
and *invariance* groups.[5] The covariance group of a theory simply specifies the admissible coordinate transformations (mappings from R^4 to R^4), which provide, as
it were, equivalent descriptions of the same physical state of affairs. In general relativity, the covariance group is necessarily the group of all admissible (one-one,
invertible, suitably continuous) coordinate transformations. This is, we shall see,
what Reichenbach in 1920 called "the relativity of the coordinates." But on the view
under examination, any nontrivial requirement of "general covariance" is in fact
misnamed. It is really the expression of a "principle of general invariance," for the
invariance (or symmetry) group of a theory is a transformation group that picks
out all the objects of the theory, if any, that are given once and for all—that is, are
never affected by physical processes and so do not belong to the set of state variables that distinguish different physical states of affairs. Such objects are the theory's
"absolute objects." For example, in special relativity the Minskowski metric η_{ik} is
an absolute object, for it is picked out by the invariance or symmetry group (the

Lorentz or the Poincaré group, see below) of special relativity. In general relativity, the metric g_{ik} is not an absolute object but is in principle fully determined by matter-energy distributions. Indeed, following Einstein's Machian-inspired injunction to banish all elements that act but are not acted upon, in general relativity there are supposed to be no "absolute objects" and so the invariance group is just the group $Diff\,(M^4)$ of diffeomorphisms (one-one, bicontinuous *point* transformations) of the space-time manifold. In virtue of thus picking out a theory's "absolute objects," the invariance group of the theory identifies the framework against which the dynamical laws may be formulated. Thus, in general relativity it is often said that M^4 is the only "absolute object," but a "trivial" one, since it has no non-topological properties. As a principle of "general invariance," it then may appear to be an extension of the principle of relativity from the Lorentz (or Poincaré) invariance of special relativity.

Yet this identification of general covariance with the diffeomorphism invariance of general relativistic models violates both the letter and the spirit of Einstein's dictum. Stachel, I think rightly, maintains that the presence of M^4 as an "absolute object" is not a minor difference with Einstein's conception. For, according to Stachel, in holding that space has existence "only as a structural quality of the field," Einstein is underscoring his heuristic postulate that "spatio-temporal individuation of the points of the manifold in a general-relativistic model is possible only after the specification of a particular metric field, i.e., only after the field equations of the theory (which constitutes its dynamical problem) have been solved" (Stachel 1989, 78). And Stachel has been insistent in stressing how the usual modern formulations of general relativity in terms of a class of diffeomorphically equivalent space-time models, $<M,g,T>$, $<M,h^*g,\ h^*T>$ where the latter are physically indistinguishable from the former under the diffeomorphism h^*, violates the spirit of this injunction by the initial posit of a differential manifold *(M) as* the space-time background for treating gravitational dynamics. In any case, there are unsolved internal difficulties with the proposal, for it has proved to be rather difficult to cleanly distinguish between "absolute" and "dynamical" objects by means of an invariance group, without in addition making controversial physical assumptions to guide the distinction (Maidens 1998).

But perhaps the most salient objection to this attempt to identify the physical significance of general covariance is this. On the assumption that physical significance resides only in the diffeomorphism group, then an uninteresting fact is asserted—that M^4 is—in some way—the spatiotemporal *positional* framework for the theory. Furthermore, this assertion takes no cognizance of the striking role played by the principle of equivalence within general relativity: that in those suitably small regions of space-time where gravitational effects can be considered to be negligible, it is possible to maintain the validity of all physical laws satisfying the special theory of relativity.[6] This means that in each of these various regions, the symmetry group of the special theory may act independently: it becomes a *local* symmetry group. Hence the distinction between *invariance* and *covariance* groups effaces the paramount distinction between *local* and *global* symmetries.

Accordingly, to unveil the physical significance of general covariance requires a distinction not between *invariance* and *covariance* groups, but between *local* and *global* symmetry groups. The cue here is taken from what field theorists have come to call the gauge principle or principle of local symmetries. From this perspective, general relativity is a *field theory of the gravitational interaction with a local symmetry*. Like all theories of fundamental interactions, it has *two* symmetry groups, a *spatiotemporal* group and a *local* group, where the latter is concerned with the interaction dynamics. This is distinct from special relativity, which possesses only one (spatiotemporal) global symmetry group, the six-parameter Lorentz group or the ten-parameter Poincaré group—the so-called inhomogeneous Lorentz group that includes translations. The spatiotemporal group of general relativity is *Diff* (M^4), but here, following Stachel, the "events" of space-time are not the points of the "base space" M^4. This is simply a manifold of points with the customary topology that enables it to be given an atlas of coordinate charts and so the smooth coordinate transformations of points lying in the intersection of overlapping charts —the necessary machinery for applying the concepts of the differential calculus to the manifold. Rather, the "point-events" of space-time "live above" M^4 as mappings from points of the cross-section of a bundle (E), which may be taken, following Prugovečki, to be the Lorentzian frame (or coframe) bundle over M^4 to each mathematical point $(P \in M^4)$. In this conception, the manifold-mapping *Diff* (M^4) is global and nonrestrictive, a mathematical expression of the physical equivalence of all general relativistic models. Thus it specifies an equivalence class of diffeomorphically equivalent Lorentz metrics, whereas any significant physical information is found only by the adoption of a specific Lorentzian metric from this class, corresponding to the reduction of the general linear frame bundle over M^4 to the Lorentzian frame bundle over M^4.[7]

Thus to see that the principle of general covariance pertains to a specifically local group, and not a global spatiotemporal invariance group, is to understand that, although it has the form of a global invariance principle, *it is really a principle about the invariance of laws of special relativistic physics in sufficiently small regions of space-time*. This recognition lies behind Cartan's affirmation, in 1931, that "the fundamental hypothesis of Einstein's theory is not, as many people have believed, that it is possible to formulate the laws of physics in any arbitrary system of coordinates, which is a simple tautology, but that in any sufficiently small region of space-time, the laws of classical physics, as they are comprehended within the special theory of relativity, are true to the first approximation" (Cartan 1931, 23).

Another way of making this point, according to Steven Weinberg, is that general covariance is "a statement about the effects of gravitation, and about nothing else" (Weinberg 1972, 92).[8] Through the principle of general covariance the Poincaré or Lorentz group is made into the *local* symmetry group of general relativity, whereas in special relativity it is the *global* spatiotemporal symmetry group. This is on account of the principle of equivalence, which minimally may be formulated so as to hold that the laws of special relativity obtain independently at each point of space-time.

The coordinate systems of the local group are defined in the frame bundle over each point: each one is an orthonormal frame (sometimes called a *vierbein* or a *tetrad*), which is free to rotate independently of the frames above the other points. The Poincaré or Lorentz group acts in this limited Minkowski space, a linear vector space. Since the frame bundles of the various points ($P \in M^4$) are disjoint, in this setting the Poincaré or Lorentz group is a local symmetry, a position-dependent or local transformation. As a result, the effects of gravity are "essentially expelled from the laws in the infinitesimal displacements around each point" (Auyang 1995, 40–41), the effects of gravity are "restored" by a connection (here, the gravitational potential, or "affine connection" [Γ_{ik}^{j}]) which reconciles the local laws at various points. This is the distinction pointed to by Friedman in the physical context of Einstein's theory as that between first-order and second-order laws (202 n. 13), or with Prugovečki (16, 52–53), between "infinitesimal" laws that obtain within the fibers of the various tensor bundles, and "local laws" that hold on the neighborhood of points in the base manifold of those bundles. The principle of *general covariance* then *"demands that the entire system of physical laws should be invariant under such position-dependent transformation"* (Auyang 1995, 42). This principle is indeed physically significant since its satisfaction requires that extra structure—the connection—be added to compensate for the rotations at various positions. The connection has *physical significance* as the gravitational potential, whereas in Newtonian space-time—the generally covariant formulation of Newtonian gravity—the potential is just a superfluous mathematical term that is forced to vanish by imposition of some artificial constraints (for example, Friedman 1983, chap. 3, sects. 3–4). Einstein's requirement of general covariance, that "the laws of physics be of such a kind that they apply to systems of reference in any kind of motion," is therefore an injunction that physical laws be formulated without a background space-time. This is then a requirement that physical laws have a local symmetry. To be sure, this clarification of the significance of general covariance introduces the concept of a frame above a manifold, rather than the usual setting of a chart within a manifold, a concept more familiar perhaps to gauge field theories than customary formulations of general relativity. But adopting this language makes it apparent that the same "logic" of local symmetries and connections is at work. However formulated, these field theories with local symmetries are philosophically revolutionary, for they have radically transformed the conception of the physical world portrayed by fundamental physical theory, or, in Cassirer's terms (as we shall see), the conception of physical objectivity.

The Problem of Dual Origin in Brief

The terminus a quo of the two roads is stated already in the first sentence of "Transcendental Logic" (part 2 of the "Transcendental Doctrine of Elements") and indeed in the single word *Vorstellung*, which is translated into English alternately as "representation" (by Kemp Smith) and as "presentation" or, indeed, even "conception" or "thought" (by Pluhar). "Our cognition arises from two basic sources of the mind. The first is [our power] to receive (re)presentations *(Vorstellungen)* (and is our receptivity for impressions); the second is our power to cognize an object through these (re)presentations (and is the spontaneity of concepts)" (A50/B74).

The operative distinction lies in the contrast between a faculty of passive receptivity ("through which objects are given to us") and one of active spontaneity ("through which they are thought"). Though he famously speculated that the "the two stems of human cognition . . . may arise from a common but to us unknown root" (A15/B29), Kant's view is that these two sources are independent faculties or powers of mind that play the distinct roles of "active" and "passive" in the synthesis that produces knowledge; the name "sensibility" was given to the faculty of receptivity, and the name "understanding" to that of spontaneity. The different roles are not only distinct but also necessary for knowledge. Kant stressed that the two faculties are *independent* sources of cognition, particularly in the section "On the Amphiboly of the Concepts of Reflection" (A271/B327), where the objection is lodged against Leibniz that he *intellectualized* appearances and against Locke that he *sensualized* all the concepts of the understanding. That only the synthesis of the two produces objectively valid judgments is perhaps a defining characteristic of Kant's epistemology. Kant has it that the concepts of the pure understanding must be related to sensible intuitions within a manifold of the pure intuitions of space and time; the concepts are accordingly given "schemata" (rules for application) according to the forms of intuition of time and space (A136/B175). On the other hand, is it really essential to Kantian epistemology to stress the independence of sensibility and the understanding, or does this only reflect the "two-front war" Kant simultaneously waged with Hume and with Leibniz?[9] Among current Kantian scholars, it has almost become orthodoxy that Kant's separation of the faculties is solely for purposes of analysis of the distinct contributions each makes to experience. At the same time, on the basis of certain Kantian passages, there is an almost irresistible tendency to reify what has thus been separated by analysis and to regard the faculties as separate per se. In any case, there arises the subsequent question regarding the relation between the two, a problem epitomized in the section of the "First Critique" regarding the schematism. Here, Kant, due to the "heterogeneous nature" of the two faculties notoriously posits a "third" to mediate between them that must nonetheless be "homogeneous" with both pure concepts and sensible intuitions (A138/B177). Oceans of ink have been spilt in expositing or critiquing this section, going back to Kant's earliest critics, Hamann and Maimon, and continuing through the Marburg School (see below) and Martin Heidegger (1929, 1977). Here is Frederick Beiser's paraphrase of J. G. Hamann's objection, voiced in his *Metakritik,* written in 1784 but not published until 1800: "Although Kant says that knowledge arises from the interaction between understanding and sensibility, he has so sharply divided these faculties that all interchange between them becomes inconceivable. The understanding is intelligible, nontemporal, and nonspatial; but sensibility is phenomenal, temporal and spatial. How, then, will they coordinate their operations?" (Beiser 1987, 41; see also 171, 292–93)

I do not know whether this is a typical example of such criticism, but I have chosen it precisely because it sets the stage on which the problem of the "constitution of the object of physical knowledge" must be treated, and from which the two roads of Cassirer and Reichenbach lead to quite distinct treatments of the problem of constitution.

Reichenbach's Relativitätstheorie und Erkenntnis a Priori

COGNITION AS COORDINATION

In his book, Reichenbach provided an elucidation of one sense of Kant's a priori as "constituting the concept of object."

> According to Kant, the object of knowledge, the phenomenal thing, is not immediately given. Perception does not give the object but only the material *(Stoff)* of which it is constructed through an act of judgment. In judgment a subordi-

nation *(Einordnung)* into a determinate schema is carried out, according to the choice of scheme a thing or a determinate type of relation develops. Intuition *(Anschauung)* is the form in which perception presents the material of knowledge; accordingly, intuition contains a synthetic moment. However, only the conceptual scheme, the categories, creates the object *(Objekt)*; the object *(Gegenstand)* of science is therefore not a "thing-in-itself" but rather an intuition-based reference structure *(Bezugsgebilde)*, constituted through categories. (1920, 46; 1965, 48–49)

This characteristic mediation between "conceptual scheme" and "perception" that produces the object of scientific knowledge is, according to Reichenbach, a permanent contribution to epistemology, to be retained even in the light of relativity theory. However, the language of critical epistemology , and in particular that concerning the relation of a representation to its object, needs updating. For one thing, relativity theory is not compatible with the Kantian theory of space and so all epistemological (as opposed to psychological) reference to intuition as the form in which perception necessarily presents the "material of knowledge" is inherently suspicious. Nor is the jargon of "synthesis" appealing. Reichenbach avoids both by following Schlick's lead in reformulating the Kantian account of the relation of concepts to perceptions by using the mathematically precise term of "coordination" *(Zuordnung)*. Hence, cognition in the empirical sciences is the result of a process of *coordination*. As in Schlick, a true empirical judgment is defined as univocal coordination; the criterion of univocality *(Eindeutigkeit)* is perception. But quite unlike Schlick, the element of reality, the real object, is determined only upon the coordination of mathematical concepts to what is given in perception. This reformulation of knowledge as resulting from a process of coordination is offered as a confirmation of the basic idea of Kantian epistemology: cognition requires both thought and intuition, reason and perception.

In motivating his reformulation, Reichenbach exploits the precise sense of coordination as a mapping between two sets; but he recognizes that, strictly speaking, it is a somewhat misleading analogy for the case of empirical knowledge, since it is a "notable fact" in such knowledge that "we carry out a coordination of two sets, of which one not only conserves its order through the coordination, but *whose elements are first defined through the coordination* (orig. *Sperrdruck* emphasis 1920, 38; 1965, 40). That is to say, in "the determination of knowledge through experience",

the defined side first determines the individual things of the undefined side, and conversely, the undefined side prescribes the order of the defined side. In this mutuality of coordination is expressed the existence of the real *(des Wirklichen)*. It is entirely indifferent whether one speaks of a thing-in-itself or whether one opposes doing so. That mutuality of coordination means that the real exists; this

is for us its conceptually graspable sense, and in this way we are able to formulate it. (1920, 40; 1965, 42–43)

What is expressly not intended is that concepts are coordinated to single perceptions. This is because the content of perception is much too complex. Nor does
perception contain a sufficient criterion for determining whether something belongs to the set of real things. This is shown by sensory illusions and hallucina-
tions. In such cases what seems to be required is a prior act of judgment, itself a
process of coordination. "Perception does not afford a definition of the real."[10]

Because of apparent similarities, it will be instructive to briefly contrast
Reichenbach's account of cognition as coordination with the earlier one of Schlick.
In his grandly conceived *Allgemeine Erkenntnislehre* (1918), Schlick famously adopts
the conception of knowledge as a coordination. But, unlike Reichenbach, Schlick
took a radically formalist view of all scientific concepts, both mathematical and
physical, as implicitly defined through their relations to one another in the deductive system developed from the axioms of the respective theory (1918; 1925, §7).
Concepts are designative signs, and "designation" is not constitution but presupposes, as Schlick made abundantly clear, a reality already formed. "Designation"
means nothing more than the coordination of a sign to an object of this reality,
which already exists, fully formed and individuated (1918, 20; 1985, 23). Accordingly, a judgment is defined to be true if its signs (concepts) univocally designate
objects within that part of reality under consideration; however, the criterion of
univocality *(Eindeutigkeit)* does not single out such a unique designation. This may
be done through the adoption of methodological principles (for example, greatest
overall simplicity) that have the standing of conventions (1918, 55–65; 1985, 59–72,
79). In intent, Schlick's account of knowledge as coordination is emphatically an
empirical realism wherein one may speak of knowledge of "things in themselves."[11]
There is neither a pure intuition nor pure forms of thought. The Marburg School's
claim that there is "an inconsistency in the Kantian assumption that thought already finds present in intuition a content independent of thought" and, hence, its
attempt to seek "forms of the real in general," merely conflates the real with its
conceptual wrapping (1918, 307; 1985, 361). Relations are "forms of the given." There
is no need for "thought to form reality," for "reality is already formed." In Schlick's
opinion, this *designative* account of knowledge "thoroughly disposes of" the Kantian or neo-Kantian conception of knowledge (1918, 306; 1985, 360). The "bridges are
down" between "concepts and intuition, thought and reality" (1918, 36; 1985, 38).

According to Reichenbach, the doctrine of implicit definitions expresses well
enough the character of the concepts of pure mathematics. But in physics, it must
be shown how the concepts, though formally definable through the fundamental
equations of a theory (its "axioms of connection"), apply to "reality." For unlike in

mathematics, where the relation of truth is purely immanent within the axiom system, the equations of physics are required to have validity for reality (1920, 34; 1965, 36–37). This is precisely where the idea of "constitutive principles," or "constitutive axioms," comes in. Reichenbach supposed an axiomatic conception of physics, in which the fundamental empirically attested laws or equations are posited as "axioms of connection." But the concepts occurring within these axioms are not considered *physically* defined until it can be shown how there can be a "univocal coordination" of these concepts to "reality." This is the purpose of the "constitutive principles," for it is through such principles that a coordination of concepts to reality is defined at all. They are required for the "logical constitution" of the concept of the object of physical knowledge.

Thus, unlike in Schlick, the system of implicitly defined mathematical concepts of the fundamental equations of physics do not "designate" but require a "mediating third" by which the coordination to the perceptually real *(Wirklichkeit)* is defined and thus the real may be univocally determined. Each fundamental physical theory presupposes a system of such principles in making the connection to experience. These principles are therefore a priori and are purely conceptual, not intuitive, in origin. But they are to be fallible, having none of the hallmarks of self-evidential knowledge ("immediately conscious knowledge a priori"). It will be apparent, I think, that the set theoretic language of "coordination" ill serves to elucidate the role of the constitutive principles a priori; as Schlick recognized, this language is inherently realist.[12] It is therefore not at all surprising that Schlick would strongly object to Reichenbach's attempt to doctor up Schlick's exact definition of cognition as a coordination with "constitutive principles" of coordination.[13]

REVISING THE KANTIAN DOCTRINE OF THE A PRIORI

This recasting of the dynamics of critical epistemology in the set-theoretic language of "coordination" is the setting for Reichenbach's central claim regarding how critical (that is, Kantian) epistemology must be revised in the light of the theory of relativity. Namely, that theory contradicts Kant's implicit assumption that there is a unique consistent system of such principles of coordination—and in two different ways. First, the theory of relativity demonstrates that an inconsistent system of such principles exists—that is, there is a system that does not yield a univocal coordination of concepts to reality. Second, it also demonstrates the existence of equivalent descriptions, each of which is a univocal coordination of concepts to reality. Each of these demonstrations mandates a necessary modification in Kant's doctrine of the a priori: elimination of the meaning of the a priori as "valid for all time" while retaining that of "constitutive of the object." It is of particular interest to us that Reichenbach proposes two distinct "methods" suggested by the theory of relativity to show how the required modifications may be effected while still remaining within the (now revised) framework of a priori constitutive principles. The two methods, that of "successive approximation" and of "analysis of science," are devised to illustrate how a single system of coordinating principles can be found that defines the univocal coordination of concepts to perceptual reality and so constitutes the object of physical knowledge. Such a system is inductively discoverable, not derived from the nature of reason. It is accordingly fallible and not absolute. But it is a priori, owing to its office in constituting the object of knowledge. On the basis of these two methods, essentially as complements to one another, Reichenbach made a further claim that belongs specifically to the epistemology of geometry rather than to the modification of epistemology per se. The theory of relativity shows that the nature of the metric of space(time) is not stipulated by a constitutive principle ("axiom"). Rather it is an "objective property" of nature that is determined empirically via an "axiom of connection" —the Einstein Field Equations.

This is, of course, a conclusion that is controverted in Reichenbach's "mature" neoconventionalist position of the "relativity of geometry," canonically exposited in his *Philosophie der Raum-Zeit-Lehre* (1928). Within a year or two, the problem of constitution of the object of physical knowledge completely withers away in Reichenbach's epistemological analysis of relativity physics. Why this is so has remained a matter of speculation. Various proposals have been made, especially concerning the influence of Schlick. But I think that there may well also have been reasons internal to Reichenbach's own conceptions that promoted the change. The two methods, crucially, are devices adopted to restore the univocality of the coordination of concepts to perceptions that constitutes knowledge in the physical sciences. But each of the aforementioned methods pulls in the direction of realism. The result is that Reichenbach is lead inexorably to a realism quite close to Schlick's, in which there is no longer any role for the notion of objects of physical science as constituted through concepts or conceptual structures. At the same time, they undermine Reichenbach's idea of a "relativized a priori," a fallible set of constitutive principles underlying physical knowledge (the agreement of mathematical physical theory with observation). To show this, we turn to the two methods in question.

THE "METHOD OF SUCCESSIVE APPROXIMATION"

A little-noticed aspect of Reichenbach's conception of principles of coordination is his assumption that systems of coordinative principles confront experiment and observation as a whole.[14] There is no logical compulsion to single out a particular one of the system of constitutive principles as invalidated by experience, for recalcitrant experience testifies only to the inconsistency of the totality of the principles. Since one of the relevant constitutive principles might well assert the Euclidean nature of physical space, it would seem Reichenbach will face here Duhemian problems of underdetermination that will lead to irreducibly subjective ("arbitrary") considerations in revising the inconsistent system of coordinating principles. That he does not is due to "the method of successive approximations." In fact, this method "represents the essential point in the refutation of Kant's doctrine of the *a priori*" since "it shows not only a way of refuting the old principles, but also a way of justifying new ones" (1920, 67–68; 1965, 70). The "method" itself is the metalevel application of a single coordinative principle to systems of coordinating principles, that of "normal induction." This is the injunction that, among all extrapolations and interpolations from experience, the "most probable" are to be chosen. The distinguished status of the principle of "normal induction" over the other constitutive principles has in fact already been presupposed. For only on its assumption is it possible to carry out a univocal coordination to experience, since "univocality of a cognitive coordination" is simply defined to mean that different empirical measurements may be taken to represent the same value of a given physical state variable (1920, 43; 1965, 45). The "method of successive approximations" has a similar normative directive. With its use it is both "logically admissible and technically possible to inductively discover new coordinating principles that represent a successive approximation of the principles used until now" (1920, 66; 1965, 69). That is to say, the old constitutive principle can be regarded as an approximation to the new one for certain simple cases. As a principle of correspondence, the method of successive approximations has the standing of an inductive maxim guiding the arrow of disconfirmation to a single coordinative principle that, in certain cases and within the limits of the inductive uncertainties of measurement and observation, can be seen to represent a too-limited case of a new, more general principle. That this process is not simply an accident of the history of physics is shown in the following methodological metatheorem stated by Reichenbach:

> For all imaginable principles of coordination the following statement is valid:
> For every principle, however it may be formulated, a more general one can be indicated that contains the first as a special case. (1920, 78; 1965, 81)

The essential ingredient in Reichenbach's refutation of Kant's "dogmatic" sense of a priori (as "valid for all time") is therefore a maxim governing scientific change in which advance occurs through successive monotonic generalizations of constitutive principles.

However, this metatheorem presents a consistency problem for Reichenbach's claim to have purged Kantian epistemology of its absolute elements. Reichenbach had maintained that "there are no most general (coordinating) principles" and "no most general concepts"; even the concept of "coordination" itself may prove to be too narrow a definition of cognition (1920, 83; 1965, 87). At the same time, the process of change of coordinating principles (hence, change in the object of scientific knowledge) has been stipulated as always proceeding in the determinate direction of successive approximation. The "relativization" of the a priori at the constitutive level of the coordinative process of cognition connecting individual theories to experience is compensated by being given a determinate direction according to an absolutist methodology of scientific change. For it does not appear that the "method of successive approximation" itself is fallible, but neither is it a conventional criterion for theory choice, as with Schlick's principle of "greatest overall simplicity." Nor can it be a Kantian "regulative ideal" regarding the task of constitution of the objects of science as an "infinite task," never to be completed (see further below). Rather it codifies and implements the realist intuition of a logic of approximate truth. In this way it has fiduciary standing, providing a nonarbitrary warrant for the decision about which coordinative principle to modify in order to reestablish a univocal coordination of concepts with recalcitrant experience.

It is again relativity theory that has provided an exemplary demonstration of the "method of successive approximations" to epistemology. For through its use, Reichenbach claims, the Einstein theory has demonstrated the falsity of Euclidean geometry as applying to the whole of physical space. At the same time this is a demonstration that the pseudo-Riemannian metric of general relativity is "an objective property of reality."

The argument for this conclusion is unfortunately garbled in Reichenbach's 1920 book; as he subsequently admitted, it is "not quite correct" (Reichenbach 1922, 44 n. 20). Furthermore, as presented there, it left Reichenbach open to the following objection, which had been forcefully made by Hugo Dingler (1920): If it is true that the coordination (which defines knowledge) is itself only defined through the coordinating principles, how can these principles themselves be contradicted in the facts of observation? The facts of observation involve physical measurements that can be made only with certain material appliances, whose construction presupposes the validity of Euclidean geometry and certain propositions of Newtonian mechanics.[15] But in a response to Hugo Dingler written in April 1921, the argument is clearly and ably presented. It is essentially this: The theory of relativity shows that the three following presuppositions are collectively inconsistent:

A. The (global) validity of Euclidean (actually, Minkowskian) geometry in "natural coordinates"

B. Equality of gravitational and inertial mass

C. Validity of the (laws of) special relativity in small domains where gravitational effects are negligible

Of these three, A may be readily recognized as a constitutive principle. But from all three presuppositions, relativity theory, regarding B and C as inductively warranted, draws the conclusion ¬ A. This is not a circular inference: A speaks of the global validity of Euclidean geometry; B and C require only that in physical measurements, despite the Euclidean presuppositions of the theory of measurement and instrumentation, the departures from the Euclidean being too small to be of consequence.

In his book, Reichenbach states that this analysis shows "that the Euclidean character of space must be given up if, by extrapolation according to Einstein's principle of equivalence (c.) [our C], a transition is made from the general theory to the special theory of relativity" (1920, 28; 1965, 29). The principle of equivalence, then, is Einstein's theory's approximative method for transiting between gravitational fields and gravity-free physics.[16] In the later paper, Reichenbach acknowledges that this inference is not logically compulsive. It is possible (as, for example, Dingler) to retain A and challenge, for example, C, in order to avoid the conclusion ¬ A. But this is indistinguishable from the postulate of a priori philosophy that A is necessarily valid as a condition of possible experience, a claim that here is precisely at issue.

Thus the "method of successive approximations" is an essential tool in the conclusion that physical space has been empirically determined not to have a global Euclidean structure. The positive case that the pseudo-Riemannian metric of general relativity is an "objective property of reality" is established through the application of the other method, that of "analysis of science."

WISSENSHAFTSANALYTISCHE METHODE

As seen above, Reichenbach held that Kant had been correct in arguing that the concept of the physical object "is determined through reason *(die Vernunft)* just as much as it is through the reality *(das Reale)* that it would conceptually formulate" (1920, 85; 1965, 88). Yet while right about the necessity of these two components of knowledge, Kant was wrong in thinking that the coordination of reason and experience produces only a single univocal coordination. Indeed, the theory of relativity shows that any such coordination contains arbitrary or subjective elements, and herein lies another lesson of the theory of relativity to epistemology (1920, 84–85; 1965, 88). The arbitrariness inherent in any univocal coordination is rooted in "the relativity of the coordinates." Every such coordination of equations to perceptual reality produces an "equivalent description" of that reality within some admissible coordinate system. At the same time, the theory of relativity has shown how to eliminate this arbitrariness of description through coordinate transformations. "The theory of relativity teaches that the four space-time coordinates can be chosen arbitrarily, but that the ten metric functions $g_{\mu\nu}$ must not be arbitrarily assumed; rather, they have entirely definite values for every choice of coordinates. Through this procedure, the subjective elements of knowledge are eliminated and its objective meaning *(Sinn)* is formulated independently of the special principles of coordination" (1920, 86; 1965, 89–90).

This teaching is generalized by Reichenbach into a new epistemological method for the sciences, "the method of analysis of science," whose goal it is to eliminate the subjective modes of description from the objective meaning of physical statements. The "wissenschaftslicheanalytische Methode" is "a sort of invariant theoretical method" to distinguish "that part of our scientific knowledge stemming from reason" from "the objective content of science, a content that, in the present form of science, is no longer clearly visible" (Reichenbach 1922, 39). The "arbitrariness of admissible systems" is the expression of the "structure of reason" *(die Struktur der Vernunft)* and, in fact, what Kant affirmed regarding "the ideality of space and time" has only now been exactly formulated through "the relativity of the coordinates." On the other hand, underlying metrical relations in nature are objective properties, but, due to "the relativity of the coordinates," there are determinate limits to our expressions regarding them. It is this invariance with respect to coordinate transformations that characterizes the "objective content of reality" *(objecktiven Gehalt der Wirklichkeit)* (1920, 86–87; 1965, 90–91).

Just as Kant's "analysis of reason" in the Transcendental Analytic is concerned to demonstrate that knowledge results only from a synthesis of different sources of cognition, so Reichenbach's "method of analysis of science" shows how this synthesis must now be analyzed into its "subjective" and "objective" constituents. It has therefore replaced Kant's analysis of reason: "The procedure of eliminating from the subjective form of description the objective meaning of a physical statement through transformation formulas, has, by indirectly characterizing this subjective form, taken the place of the Kantian analysis of reason . . . This is the sole way that affords us an insight into the cognitive function of our own Reason" (1920, 87–88; 1965, 91–92).

These passages make it apparent that the physical object of knowledge, conceptually structured through the coordination of mathematical equations to reality *(Wirklichkeit)*, is constituted as an object of experience only within a particular coordinate system. This indeed must be the case if "reality" is the perceptual reality of measurement and observation. The object of knowledge as thus constituted must then be refined through "the method of analysis of science" in order to determine what within it pertains to "the objective part of reality." According to the theory of relativity, that which belongs to the objective part of reality is just what is invariant under the admissible transformations of the coordinates. Once this is done it will be seen that the metric coefficients cannot in general have Euclidean values.

> If the metric were a purely subjective matter, then the Euclidean metric would have to be suitable for physics; as a consequence, all ten functions $g_{\mu\nu}$ could be selected arbitrarily. However, the theory of relativity teaches that the metric is subjective only insofar as it is dependent upon the arbitrariness of the choice of coordinates, and that independently of them it describes an objective property of the physical world. Whatever is subjective with respect to the metric is expressed in the relativity of the metric coefficients for the domain of points, and this relativity is the consequence of the empirically ascertained equivalence of inertial and gravitational mass. (1920, 86–87; 1965, 90–91)

The statement that "the relativity of the metric coefficients" is a consequence of the equivalence of inertial and gravitational *mass* (B, above) requires amendment since this equivalence, formulated as the "weak principle of equivalence" is also at the base of classical Newtonian gravity.[17] It must accordingly be understood as a reference to some stronger version of the principle of equivalence, such as that expressed in C, above. Exception may also be taken to the statement that the suitability of Euclidean geometry for physics has the consequence that "all ten functions $g_{\mu\nu}$ could be selected arbitrarily"; in fact, an Euclidean metric only results when, in a particular coordinate system, the ten functions $g_{\mu\nu}$ take the values 0 or 1 (±1 for Minkowski space-time). But the general point is clear enough. By "relativity of the metric coefficients for the domain of points" Reichenbach refers to the simple fact that the specific values of the components $g_{\mu\nu}$ at a point are functions of (are "relative to") the numerical values of the coordinates of that point in some coordinate system (chart). If the metric field is represented in one such chart χ^μ, then any representation of the metric field is admissible that is derived from the χ^μ chart through an arbitrary, continuous transformation of coordinates.

But what are we to make of Reichenbach's central contention that freedom (or arbitrariness) in the choice of admissible coordinate systems is the *subjective contribution of reason to physical knowledge?* The subjective part of the description of the physical world lies merely in the fact that, in order to make measurements it is necessary to refer to particular coordinate systems. This subjectivity is eliminated through the transformations linking all admissible systems that thus enable a physical statement to pertain only to the *objective properties of the physical world.* "It is obviously not inherent in the nature of reality that we describe it by means of coordinates; this is the subjective form that enables our reason to carry through the description" (1920, 86; 1965, 90). If the "contribution of reason" to coordination is merely now the selection of an admissible coordinate system, it becomes apparent that the systems of constitutive principles that, apparently, played a "thicker" role in earlier stages of science, have essentially been reduced, in general relativity, to a very "thin" one. Indeed, from the perspective of a modern formulation of the theory, this role is entirely trivial, since coordinate freedom is automatically ensured by the definition of the differential manifold M^4.[18] In Reichenbach's day, however, when the setting for space-time theories was not a modern differential manifold but a number manifold R^4, or one of its open subsets, this subtractive procedure appears more understandable. Here, as Norton (1992) has shown, there is a nontrivial problem that number manifolds have too much structure, structure that must be "transformed away" by enlarging the covariance group of the theory. This presumably is what Reichenbach had in mind. So there is something to what Reichenbach is apparently doing here.

But even so, parsing general covariance, or "the relativity of coordinates" in this way, as due to "the subjective contribution of Reason," is not really appropriate. To see this, recall that general covariance, understood as freedom to make arbitrary coordinate transformations, allows that once any solution to the field equations is found, it is possible to derive any number of other solutions merely by a different choice of coordinates. Of course, these solutions are only formally different; physically they are completely equivalent. But because of the freedom to choose coordinates, there are differential identities between the field equations; in fact, there are four identities for the ten independent field equations for the $g_{\mu\nu}$. Thus a constraint on the set of solutions is introduced. This constraint had already been precisely formulated by Hilbert in 1915: in a generally covariant field theory for n unknown functions, there can exist no more than n-4 independent equations.[19] Now Reichenbach provided no visible cognizance of this. But had he known, would he have still held that freedom to choose coordinates, as stipulated by a principle of coordination, is a "subjective contribution of Reason," not belonging to the "objective content of reality," since it actually plays a central role in linking the two sides of the Einstein Field Equations? Reichenbach's subtractive procedure appears to be guided by the idea that coordinate-free formulations of physical theories are inherently capable of describing the "objective content of Reality" without any "subjective contribution of Reason." And this would again seem to be an article of faith that only a realist might adopt.[20]

ASSESSMENT

In the event, Reichenbach soon dropped the language of coordinative principles or axioms and all talk of "constitution." It seems that he did so because of a growing realization that such principles actually could *not* play a significant constitutive role in anything like the Kantian sense. Moreover, in the very course of establishing his central claim that coordinating principles can have only an inductive warrant, Reichenbach erects a classically realist textbook account (in the Kuhnian sense) of scientific progress. Perhaps he also had been convinced by Schlick that his use of constitutive principles is a remnant of Marburg logical idealism that has no place in the essentially (critical) realist conception of cognition as a coordination. Still more, he probably also had been influenced by Einstein's lecture "Geometry and Physics," with its talk of coordination of metrical concepts to "practically rigid measuring rods" (Einstein 1921). By 1924, standing in place of constitutive principles and axioms are "coordinative definitions" that provide for purely extensional physical definitions of certain concepts, no longer as a constitutive process of coordination but in the much narrower role as *conventional* choices of material standards that realize the mathematical concept of "congruence." As "definitions," such stipulations can be neither true nor false. But only after they have been made does the problem of the geometry of physical space become an empirical matter.

Ironically, then, the price of surrendering his conception of the "relative a priori" is that Reichenbach had now to refrain from the conclusion, obtained from the "method of successive approximation," that global Euclidean geometry is not true of physical space. This, it seems clear to me, was not a step forward. Instead, he broadly follows Einstein's (1921) G(eometry) + P(hysics) formulation, allowing that Euclidean geometry can be preserved, if the laws of physics are altered through the addition of "universal forces." There is, however, a substantial difference in *how* this formulation is to be understood. Einstein's G + P "solution" to the problem of the geometry of physical space was put forward explicitly as provisional, given the absence of a theory of matter required for the closer analysis of the concept of "rigid body,"[22] whereas Reichenbach's solution in terms of "coordinative definitions" of metrical indicators and "universal forces" was accompanied by a new methodology whose express purpose is to analyze scientific theories into disjoint factual and definitional parts. The distinction between "descriptive" and "inductive simplicity" then follows, the former, governing coordinative definitions and producing "equivalent descriptions" having nothing to do with truth. The latter, implementing what remains of the constitutive principle of "normal induction," can have only a pragmatic justification. But the lure of "the method of successive approximation" will continue to instruct his analysis of physical geometry, which manifests a persistent striving to view the theory of general relativity through the lens of special relativity—that is, through the implicit "standard choice" of rigid rods and perfect clocks of special relativity as metrical indicators in gravitational fields.

Zur Einsteinschen Relativitätstheorie: Erkenntnistheoretische Betrachtungen

There are several points of contact between the Reichenbach and Cassirer monographs of 1920 that have tended to camouflage their fundamental difference. The similarities are readily apparent. Both are agreed that the fundamental question posed to epistemology is "How is knowledge possible?" and that its answer is to be sought in an analysis of knowledge in the exact sciences. They share the view that traditional empiricism has no plausible account of this knowledge. Both claim that, in at least one respect, the general theory of relativity has confirmed transcendental idealism's claim of the ideality of space and time in maintaining the arbitrariness or "relativity" of coordinates.[23] On the other hand, they concur that aspects of the Transcendental Aesthetic and Analytic must be revised or reinterpreted in the light of the theory of relativity. In particular, they are agreed that the traditional conception of the a priori as inviolable and eternally valid fixed categories must be abandoned. Furthermore, both books argue that the a priori elements in physical theories may change over time and that each such change represents a transformation of the concept of physical objectivity. So Cassirer and Reichenbach are both proponents of what has become known as "the relativized a priori." Although they will disagree concerning the functional role of the a priori in constituting physical objectivity, they nonetheless maintain some necessary role for a priori elements in the constitution of physical knowledge. Despite all these basic agreements, Cassirer was convinced of a fundamental difference with Reichenbach right from the beginning. When, in late June 1920, Reichenbach sent a copy of the typescript of his book to his former Berlin teacher Cassirer, now a professor in Hamburg, Cassirer's response is illuminating. While informing Reichenbach that he too had just sent a manuscript on Einstein's theory to press, and that their viewpoints were "almost interchangeable," Cassirer nonetheless wrote: "however, this does not precisely extend, so far as I can now see, with reference to the concept of *apriericity* and to the interpretation of Kantian doctrine which, in my opinion you take still too psychologically."[24]

These are not marginal differences for neo-Kantians. The charge of psychologizing Kantian doctrine is instructive. It is a familiar accusation made by Marburg neo-Kantians, and its basis lies exactly in the tendency mentioned above to reify as distinct the contributions to knowledge that Kant had only separated for purposes of analysis. Presumably, Cassirer was referring to Reichenbach's conception of cognition as a coordination *between* two separate realms of existence as a reformulation of the Kantian postulate of the dual origin of knowledge in intuition and thought. For from the same Hermann Cohen from whom Cassirer had acquired the conception that the sole method of epistemology was that of *Erkenntniskritik,*

Cassirer also inherited the interpretive view that sensibility and understanding are not to be understood as two independent capacities. For Cohen, and also Cassirer, the forms of sensibility are not passive but active attainments of the cognizing subject, ultimately having the same standing as the discursive structures of the understanding (Henrich 1994, 47). In Cassirer, "the problem of knowledge," in the Kantian sense of the necessary relation of "reason" and "experience," is ever shifting and so neither resolved nor, in any determinate sense, fully resolvable. But it can be tracked by the method of *Erkenntniskritik,* whose goal is the investigation and illumination of the "logical presuppositions" within the physical and mathematical science—that is, their transcendental conditions of possibility—to show that this relation is continually remade and transformed in a determinate direction within the development of science.

In his widely read earlier work of 1910, Cassirer had described "the problem of knowledge" as it appeared from the vantage point of the physical and mathematical sciences in the first decade of the century. Looking backward, Cassirer identified the central trend in the development leading up to the present as a transformation of the doctrine of the nature of concepts in the exact sciences. The primary change occurred along an axis from an abstractive or picture theory of concepts, characteristic of naive empiricism and realism, to an ever-growing greater reliance on purely functional, relational, and serial concepts. The nineteenth-century development of physical sciences had shown that empirical "content" is only determined through the serial relations of space and time, or magnitude and number, of dynamical interaction and interrelation of events. Here both the "form" and the "content" of cognition are reciprocally determining elements of the concept of physical object that itself remains irreducible. The highest point of this (prerelativistic) development was attained by Hertz and especially Duhem, who stressed that concepts are pure symbols for relations and functional connections within the real, and not in any sense copies or images of the real.

In Cassirer's genealogy of conceptual change, no particular principle of form or order characteristic even of the present state of science can be taken as immutable or as having apodictic validity. What remains unchanged through the successive changes in scientific knowledge is just the "objectifying function" itself, which Cassirer termed the "supreme law of objectification." Accordingly, the meaning of the a priori is delimited (as it is in some passages of Kant, cf. *Prologeomena* §19): a cognition is a priori not because it is prior to experience but because, and only insofar as, it is contained as a necessary premise of valid judgments concerning the "facts" of science (1910, 357; 1923, 269). The possibility is then open that those elements of physical theory viewed as "logical presuppositions" or as "free creations of the human mind" may be seen either as a priori elements or as conventions. For Cassirer, their nonconventional character rests upon two considerations. First, a recognition of the necessity of the objectifying function itself—that is, a recognition that physical objectivity is not given but constituted. Second, such elements are not conventional because they have emerged from within the train of development of earlier manifestations of this function as a later, and perhaps dialectically transformed, stage. The Hegelian influence on Cassirer's account of conceptual development is undeniable; at the same time, Cassirer's is but a weak historicism in which directed determination is found within history and not projected from it.

Toward the end of *Substanzbegriff und Funktionsbegriff*, Cassirer, in a footnote, greeted the recent appearance in print of Max Planck's December 1908 lecture "Die Einheit des physikalische Weltbildes" (1910, 408; 1923, 307). In this lecture, the opening broadside of what would be a vitriolic polemic with Mach and his followers over the nature of physical theory and the aims of physical science, Planck pointed to the "unity" of the "physical world picture" as what remained of constant value despite the comings and goings of particular physical theories. In Planck's striking expression, this is the ideal of "unity of all separate parts of the picture, unity of space and time, unity of all researchers, all nations, all cultures" (Planck 1910, 25).

Some ten years later, in *Zur Einsteinschen Relativitätstheorie*, Planck's thesis of unity through de-anthropomorphizing the "physical world picture" reappears as Cassirer's guiding theme for how the further development of physical theory, with the theory of relativity and especially the general theory, falls within the framework of the transition to purely "functional thinking" traced in *Substanzbegriff und Funktionsbegriff*.[25] The two further steps along this path taken by the theory of relativity in the interim are qualitatively quite different. First, in the transformation of the doctrine of measurement, relativity theory has shown that certain concepts ("length", "mass") are not properties of objects, but of relations of objects to frames of reference, which is an additional "de-anthropomorphic" step in the concept of the physical object. But it is the successive step taken by the general relativistic requirement of general covariance, of "the general invariance of laws of nature," that Cassirer sees as bringing a decisive new advance in the concept of physical objectivity. For this demand shows, as he later put it, that "the ultimate stratum of objectivity" lies in "the invariance of such relations and not in the existence of any particular entities" (Cassirer 1929, 473).

CASSIRER'S INTERPRETATION OF GENERAL COVARIANCE

The task undertaken in Cassirer's relativity monograph is quite specifically marked out: it is to determine the significance for epistemology of Einstein's claim that his theory has removed from space and time "the last remnants of physical objectivity." "What are we to understand by physical objectivity," Cassirer asks, "which is here denied to the concepts of space and time?" (1921/1957, 8; 1923, 356). It is not sufficient to merely observe that space in itself and time in itself do not satisfy Planck's often-invoked formulation of the criterion of physical objectivity—"What can be measured, exists." While this may be adequate for physics, measurement rests on presuppositions that require epistemological elucidation. Nor is it enough to understand Einstein's remark in the sense that space and time are forms of phenomena and not things, in the sense of naive realism. "That physical objectivity is denied to space and time by this theory must signify . . . something other and something deeper than the knowledge that the two are not things in the sense of 'naive realism'" (1921/1957, 10; 1923, 357). For none of the genuine concepts of physical objects—energy, mass, momentum, etc.—are such naive "thing-concepts." What is left still unaccounted is the "logically special position" *(logische Sonderstellung)* occupied by the *concepts* of space and time. Space and time are a further step of abstraction away from other physical concepts, "representing, as it were, concepts and forms of measurement of higher than the first order." Hence, any attempt to provide an answer to the question concerning the loss of "physical objectivity" by space and time is constrained to recognize the logically special, more fundamental, character of these *concepts.* As befits the method of *Erkenntniskritik,* which admits no superior epistemic authority outside of science itself, the answer must be sought in terms of the changing manifestations of the concept of physical object within physical science. So the specific task Cassirer has set himself is an examination of how "physical objectivity" is to be construed from within the physical perspective of the new theory such that it is denied to space and time.

In an earlier discussion, Cassirer already arrived at a preliminary result: the requirement of general covariance—namely, that laws be stated in a form valid for all frames of reference—represents a similar but different kind of advance in conceptual development *(Begriffsbildung)* as took place in the transition from classical mechanics to the special theory of relativity (1921/1957, 34; 1923, 79). In the latter instance, the validity of the general laws of nature was still restricted by reference to a class of determinate reference bodies; with general relativity this restriction is altogether removed. Although some determinate reference system is implied in testing these laws, "their meaning and value *(Sinn und Wert)* is independent of the particularity of these systems and remains identical with itself, whatever changes experience may bring to them." It is just this "independence from the arbitrary standpoint of the observer" *(Unabhängigkeit vom zufälligen Standort des Beobachters)* that is meant when we speak of an object of "nature" and of "laws of nature" as determinate in themselves. Measurement in one system, or in any of the unrestricted plurality of "justified" systems in the end yields only particularities *(Ein-*

zelheiten), but not the genuine "synthetic unity" of the object. With reference to Planck's Leiden lecture, these new requirements of physical objectivity mean that "the anthropomorphism of the natural sensuous world picture, the overcoming of which is the task of physical knowledge, has been compelled to take a further step back" (1921/1957, 37; 1923, 381).

The interpretation of general covariance as a further development of the methodological principle of "objectifying unity" is reiterated throughout the remainder of Cassirer's essay. Where experience had unexpectedly failed to find the preferred reference frame posited by Galilean-Newtonian mechanics for the motion of the solar system, or the motion of the earth in Michelson's experiment, the theory of general relativity made a virtue out of necessity by making it a requirement that there *cannot* and *must* not be such a preferred system. The general theory of relativity adopts the principle *(Prinzip)* "that for the physical description of the processes of nature *(Naturvörgange)* no particular reference body should be distinguished above all the others" (1921/1957, 38; 1923, 383). This is later designated as a "rule of the understanding" *(Regel des Verstandes)* hypothetically adopted within physics. The requirement of general covariance ("that all Gaussian coordinate systems are of equal value for the formulation of the general laws of nature")[26] is held to be a "principle that the understanding uses hypothetically as a norm of investigation in the interpretation of experience." The sole meaning and justification of this norm rests upon the fact that, through its application, it will be possible to attain the "synthetic unity of phenomena in their temporal relations" *(synthetische Einheit der Erscheinungen nach Zeitverhältnissen)*—that is, to lawfully explain the observed phenomena. The sole valid norm is not so conditioned: it is just the "idea of unity of nature, of univocal determination itself" (1921/1957, 76; 1923, 416).[27] With the requirement of general covariance, the general theory of relativity has given a new meaning to the "unity of nature" (1921/1957, 74–76; 1923, 415–16). In this regard, Cassirer's brief comments on the principle of equivalence are also telling. The equivalence between an inertial frame and one falling freely in a static gravitational field is not an empirical proposition, abstracted from particular observations. It is rather "a precept *(Vorschrift)* for the development of our physical concepts *(physikalische Begriffsbildung)*: a requirement made not of experience but only for our manner of intellectually representing it" (1921/1957, 89; 1923, 428).

Moreover, Cassirer recognizes, perhaps with an eye toward Kretschmann's "correction" of Einstein, that general covariance ("that the general laws of nature are not changed in form by arbitrary changes of the space-time variables"—called here the "principle of general relativity") may have the guise of a merely analytic requirement, nonetheless, the demand that there be such unities is, in principle, a *synthetic* requirement (1921/1957, 39; 1923, 384). As such, general covariance is "a universal *maxim* set up for the investigation of nature" (*eine allgemeine Maxime der Naturbetrachtung*): a formal restriction but also (here Cassirer quoted Einstein) "a heuristic guide in the search for the general laws of nature" (1921/1957, 33; 1923, 377). Similarly, Cassirer, citing the Kantian formulation of the object of knowledge as a "concept, with reference to which presentations have synthetic unity," recognized that the requirement that a physical theory be generally covariant (in "the form of equations and systems of equations covariant with respect to arbitrary substitutions") is a purely logical and mathematical relativization *(Relativierung)*. Yet it is through this relativization that the object of physics is determined as a "phenomenal object," though the phenomenal object is no longer connected with "subjective arbitrariness and subjective contingency." Unlike with Reichenbach, for whom "the relativity of the coordinates" indicated "a subjective contribution of Reason" as opposed to the "objective content" of physical knowledge, for Cassirer, it is deemed to be one of the "ideal forms and conditions of knowledge upon which physics rests as a science, which secures and at the same time grounds the empirical reality of all that physics regards as 'fact' and to which it accords the name objective validity" (1921/1957, 50–51; 1923, 393).

It is not until chapter 5 ("The Concepts of Space and Time of Critical Idealism and the Theory of Relativity") that Cassirer straightforwardly confronted what is usually posed as the primary obstacle to Kantian or Kantian-derived interpretations of the theory of relativity, the doctrine of pure intuition. Strikingly, Cassirer argued that the general theory of relativity, whose fundamental feature is characterized as having removed from space and time "the last remnant of physical objectivity," has improved on Kant in bringing about a clarification of the role of pure intuition in empirical cognition; in this regard, the general theory of relativity exhibited "the most determinate application and carrying through within empirical science of the standpoint of critical idealism" (1921/1957, 71; 1923, 413). While following the broadly critical idealist injunction that space and time are "forms of phenomena" and not "things," Cassirer also enjoined that they are conceptual "sources of knowledge," pure ideal *concepts* of the orders of "coexistence" and of "succession," as they were for Leibniz but not for Kant. In this regard, relativity theory, in robbing "pure intuition" of its sensible and chronometrical structure, has indeed *clarified* the Kantian meaning of the term, whose "most general sense . . . was certainly not always maintained by Kant equally sharply on account of his involuntary substitution of special meanings and applications" (1921/1957, 78; 1923/1953, 418). According to Cassirer, Kant's *intention*, with regard to the use of the term "pure intuition" was simply to express the "methodological presupposition" of the characteristic "thought-forms" *(Denkformen)* of connection and of ordering entering into all scientific knowledge; these are the concepts of number, of function, and of space and time. Such forms are not to be conceived as "rigid" but rather as "living and moving"; none is given to thought "at one stroke" but is only revealed through the process of "coming to be" in the concrete manifestations of scientific thought. But in the continual attempt of physics to bring these changing forms into a mutually determinative relation with the manifold of sensibility, the latter "progressively loses its 'fortuitous' ('*zufälligen*') anthropomorphic character and receives the impress of thought, the impress of systematic unity of form" (1921/1957, 80; 1923, 421). The loss of "physical objectivity" by space and time, triumphantly announced by Einstein, referred precisely to the appearance, in the general theory of relativity, of the concepts of space and time solely as functional forms of succession and of coexistence. Space and time designations (coordinates) possess no immediate physical significance but serve only the purpose of assigning individual phenomena their position in the space-time continuum.

ASSESSMENT

Cassirer identified the principle of general covariance as being, at the present time (that is, 1921), a local maximum, a culminating step in the continuous development of the conception of physical objectivity stretching back to the birth of modern science in the seventeenth century. As may be seen in the developments within the exact sciences themselves, the conception of "knowledge of reality" has traversed a long and laborious path from the "thing" concepts of perception through the "substance" concepts of classical mechanics before attaining the "invariance concept" of the theory of relativity. General covariance, the requirement that the "the laws of nature must be formulated so that they are equally valid for all reference frames, whatever their state of motion," is the completion of this last step (Cassirer 1999, 118).[28] It is the most recent refinement of the normative methodological principle of "unity of determination," of an ideal of physical objectivity that attempts to purge all traces of anthropomorphism from the worldview of physics. It is of some interest to note that Hilbert, apparently independently, came to the same view of the significance of general covariance: a "radical elimination" from the conception of physical objectivity of the "anthropomorphic slag" contributed by the senses and intuition.[29] So understood, general covariance is, for both Cassirer and Hilbert, *the* epistemologically salient aspect of the general theory of relativity. At the same time, however, Cassirer repeatedly underscored Einstein's emphasis that this new ideal of physical objectivity is but a norm, a "methodological maxim" or "regulative principle" for the intellectual treatment of nature.[30] This has the implication, as Michael Friedman recently observed, that Cassirer's conception of the a priori as a regulative ideal, consistent with the Marburg rejection of Kant's original distinction concerning sensibility and understanding as independent sources of cognition, is no longer a constitutive a priori in Kant's sense (see Friedman 2000, 117). That *regulative* ideals can play a *constitutive* role in cognition is then not of a piece with Kantian orthodoxy concerning the distinct roles of sensibility and understanding in cognition and their mutual relation. But it is agreed on all hands that general relativity needs occasion some modification in Kant's theory of cognition, while, it might well be argued, Kant's unnecessary retention of outmoded notions from eighteenth-century faculty psychology seems an obvious place to begin. Still more, an explicit recognition that a *completely* de-anthropomorphized conception of the processes of nature *is* a regulative ideal, an ideal limit, is a useful antidote to the pretensions of scientific realism that such representations are already at least approximately attained in our best scientific theories.

Concluding Remarks

In this chapter, I have argued that, stemming from a fundamental difference in the interpretation of the Kantian account of cognition, the 1920 monographs of Cassirer and Reichenbach on the theory of relativity point in diametrically different directions for subsequent philosophy of science. Each was certainly a product of its time. Yet I think that the "historical" philosopher Cassirer provided the more promising, if incompletely carried out, understanding of the role and justification of synthetic a priori principles in modern physics. He also produced a significantly more insightful appreciation of the epistemological innovation of the theory of general relativity. While lacking the language and mathematical tools of symmetry that are readily available today, Cassirer nevertheless succeeded in communicating that the revolutionary idea lies in the idea of local symmetry, of invariance of laws with respect to arbitrarily chosen reference frames, and that this indeed represents a transformation in the physicist's conception of objectivity. Of course, this is one of those epistemological ideas that was ahead of its time. But in the interim, suitably generalized to the "internal" spaces of gauge field theories, it has become the very template of theoretical construction. I doubt that this is just an accident of history. But whether it is or not, this occasion seems a particularly auspicious time to rethink the present alternatives of realism and relativism by considering roads not taken.

Notes

The epigraph is quoted with permission of the Archive for Scientific Philosophy, Hillman Library, University of Pittsburgh (HR 015-49-26). Translation (by author): "O Kant, who saves you against the Kantians?"

1. Though Cassirer's book went to press in the summer of 1920, before Reichenbach's, the latter was published in November 1920; Cassirer's, in January 1921. According to Cassirer's *Vorwort*, dated August 9, 1920, his book was read in manuscript by Einstein. We know that Einstein, to whom the book was dedicated, read at least parts of Reichenbach's book from marginalia in his copy. (I am grateful to Thomas Oberdan for providing me with photocopies of Einstein's marginalia.)

2. In the following, I have generally made my own translations from the original works; citations to the translations are made for ease of reference and purposes of comparison. I have used a reprinted edition of Cassirer's Einstein monograph, *Zur Einsteinschen Relativitätstheorie: Erkenntnistheoretische Betrachtungen*, whose pagination differs from the original edition. It is contained in E. Cassirer, *Zur modernen Physik* (Darmstadt: Wissenschaftliche Buchgesellschaft, 1957); page numbers to the German text are indicated "1920/1957."

3. By ignoring Cassirer's emphasis on the immanent development of the concept of objectivity within physical science since the seventeenth century, Schlick succeeded here in portraying Cassirer's conception of a priori principles as being too vague to be evaluated, hence lacking determinate content.

4. "I see the most essential thing in the overcoming of the inertial system, a thing that acts upon all processes, but undergoes no reaction. This concept is in principle no better than that of the center of the universe in Aristotelian physics" (Einstein to Georg Jaffe, January 19, 1954; cited in Stachel 1986; see note 8).

5. See Anderson (1967); Trautman, (1966, 319–36); and Friedman (1983) for a general discussion.

6. Giving a precise characterization of this condition is notoriously difficult; for a review, see Budden and Ghins (2001). Prugovečki (1995, 52) makes the case that unnecessary obstacles are created by coordinate formulations rather than concepts deriving from Cartan's idea of "moving frames." He formulates the strong (Einstein) equivalence principle thus: *"In any inertial Lorentz moving frame along a timelike geodesic g, all the nongravitational laws of physics, expressed in tensor coordinates with respect to that moving frame, should at each point along g coincide with their special relativistic counterparts expressed in tensor coordinates with respect to global Lorentz frames in Minkowski spacetime."*

7. Prugovečki (1995, chap. 2, "Classical Frame Bundles in General Relativity"). The selection of a specific Lorentzian metric enables a reduction of the general linear frame bundle over M (each of whose fibers consist of all linear frames above a given point $P \in M$) to the Lorentzian frame bundle over M. The *general linear frame bundle over M* is then the type of *principle bundle* encountered in classical general relativity; its *structure group* is the *general linear group* of all 4 x 4 real (and invertible) matrices.

8. For an illuminating discussion, see Auyang (1995, 40–42); also Ohanian and Ruffini (1994, 371–79).

9. For the issues involved, see Allison (1973, 73).

10. "Aber eine Definition des Wirklichen leistet die Wahrnehmung nicht" (1920, 39).

11. "Empirical realism" was Herbert Feigl's designation of Schlick's epistemology.

12. Coffa's comment (1991, 203) is instructive, for it reveals that the "dual origin" thesis comprises, for Coffa, "a Kantian framework": "Reichenbach shared Schlick's realist instincts, but he did not find the expression of these instincts within a Kantian framework any easier than Schlick had years earlier."

13. In Schlick's view, Reichenbach's account of knowledge as coordination comes too close to the fatal delusion entertained by Marburg neo-Kantians: "I believe that only the

undefined side determines—through the mediation of perception—the conceptual side, and not vice-versa. Your theory seems to me to emerge from the fact that it is so easy to confuse the *concept* of reality *(Begriff der Wirklichkeit)* with reality itself . . . an illusion to which the Marburg Neo-Kantians have fallen prey" (Schlick to Reichenbach, November 26, 1920 [HR 015-63-22], as translated in Coffa [1991], 204).

14. But see Torretti (1984, 342).

15. As given in his book Reichenbach's response to the obvious objection to this sort of (what Clark Glymour called) "bootstrapping" method of testing is patently dogmatic and question-begging (1920, 62; 1965, 64): "one formulates a theory by means of which the empirical data are interpreted and then checks for univocality *(Eindeutigkeit)*. If univocality is not obtained, the theory is abandoned. The same procedure can be used for coordinating principles. It does not matter that the principle to be tested is already presupposed in the totality of experiences used for the inductive inferences. It is not inconsistent to assert a contradiction between the system of coordination and experience."

16. In fact, it is not really possible to view the principle of equivalence as an instance of an approximative method, in the desired sense that the general case is to contain the special case as a limit. For in the special case, the Riemann curvature tensor necessarily vanishes, while it is necessarily nonvanishing in the general case. The two cases are mathematically inconsistent.

17. The "weak equivalence principle", sometimes called the Galilean equivalence principle, states, in one formulation, that *"the motion of any freely falling test particle is independent of its composition and structure."* See, e.g., Ciufolini and Wheeler (1995, 13–14). They also explicitly note that it "is at the base of most *viable theories of gravitation*" (91).

18. Such definitions invoke equivalence classes of coordinate atlases, and so, automatically, only chart-independent statements are held to be meaningful.

19. This theorem, a special case of Noether's second theorem, is stated without proof in Hilbert (1915); for discussion, see K. Brading and T. A. Ryckman (forthcoming). Cf. Pauli (1958, §56) or Ohanian and Ruffini (1994, 378). The modern view is this: Because of the twice-contracted Bianchi identities (i.e., the vanishing of the covariant divergence of the Einstein tensor, which is vital for linking the left-hand side of the Einstein Field Equations to the right-hand side, i.e., coupling the F to its matter sources) there are only six independent field equations for the particular values of the $g_{\mu\nu}$. There are, however, ten independent $g_{\mu\nu}$. So there are four additional independent conditions that must be specified in addition to the field equations to fully determine the ten independent $g_{\mu\nu}$. More explicitly, "In any generally covariant field theory in spacetime, by definition, one must have the freedom to make four independent general transformations on the coordinates. Once the coordinate frame has been fixed (e.g., four noncovariant conditions specified) the six field equations are sufficient to determine the form of the ten $g_{\mu\nu}$ with the components of the stress-energy tensor [i.e., the RHS of the Einstein Field Equations for matter sources-TAR] given" (Davis 1970, 266).

20. Michael Friedman has characterized Reichenbach's conception of the "relative a priori" somewhat differently. In accord with the difference noted in §2 above between covariance and invariance groups, Friedman maintains that, for Reichenbach, what is "relatively a priori" for a theory is determined by the invariance group of the theory. Hence, the distinction between the constitutive and empirical is, in current parlance, that between the *absolute* and the *dynamical objects* of the theory. According to Friedman (1994), this is to say that it is the function of the "relative a priori" to describe the spatiotemporal framework

Ryckman / Cassirer, Reichenbach, General Relativity

within which the particular dynamical laws operate. E.g., "The idea is as follows. Each of the theories in question (Newtonian physics, special relativity, general relativity) is associated with an *invariance group of transformations* that presents us with a range of possible descriptions of nature—a range of admissible reference frames or coordinate systems—that are equivalent according to the theory. The choice of one such system over another is therefore arbitrary, and Reichenbach's thought is that those elements left invariant by the transformations in question . . . are precisely the constitutive elements of the theory. . . . in special relativity the relevant group of transformations is the Lorentz group, and so . . . the underlying structure of space-time of Minkowski space-time is constitutively a priori; particular fields defined within this structure . . . do not count as constitutive. Finally, in general relativity the relevant group includes all one-one bidifferentiable transformations (diffeomorphisms), and so only the underlying topology and manifold structure remain constitutively a priori" (66). I do not contest that one could (contrary to Einstein, I think, for the reasons given in §2) explicate the "relative a priori" in terms of the difference between "absolute" and "dynamical" objects, but I do think doing so involves substantial reconstruction of Reichenbach's 1920 text. For example, Friedman's reconstruction assumes that Reichenbach had and employed an operative distinction between coordinate and point transformations. But I find any evidence for this assumption lacking in Reichenbach's text. This is not surprising, given the usual practice of the time among physicists (as noted above) to choose the number manifold R⁴ as the setting of the theory. Hermann Weyl was a notable exception to this common practice. I do, however, concur with Friedman's conclusion that "Schlick's conception (of scientific theories) does not, in fact, yield a *distinction* between the constitutive and the empirical, between the conventional and factual parts of science at all—even relative to a particular given theory" (67).

21. For discussion of the influence of this essay on Schlick and Reichenbach, see Ryckman (forthcoming), chapters 2 and 3.

22. See Ryckman (1996) for further discussion.

23. Reichenbach's term for general covariance (1920, 8–9, 86; 1965, 9, 90); Cassirer (1921, 71–72; 1923, 412–13).

24. Cassirer to Reichenbach (postcard), July 7, 1920, Hans Reichenbach papers in the Archive for Scientific Philosophy, University of Pittsburgh, Hillman Library, HR 015-50-09. Quoted by permission. I am grateful for Marlene Bagdikian's assistance in deciphering Cassirer's handwriting.

25. "In general epistemological consideration, [the theory of relativity] is precisely distinguished in that it carries out, more consciously and clearly than ever before, the progress from the copy theory *(Abbildtheorie)* of knowledge to the functional theory *(Funktionstheorie)*" (1921/1957, 49; 1923, 392).

26. It seems unlikely that Cassirer is here using "Gaussian coordinates" in the technical sense of geodetic normal coordinates (as adopted, for example, by Hilbert in 1917 in order to have a well-posed Cauchy problem in general relativity) but rather in the sense of "arbitrary coordinates." Hilbert (1917) is not cited by Cassirer.

27. It is this formulation to which Schlick (1921, 326) takes a vacuous.

28. "Die allgemeine Relativitätstheorie vollzieht dann den letzten entscheidenden Schritt; sie postuliert [,] daß für die Formulierung der Naturgesetze alle Bezugskörper K K' gleichwertig sein sollen, selches auch deren Bewegungszustand sein mag."

29. Hilbert, lecturing in 1921, endorsed Cassirer's thesis tying general covariance to physical objectivity *à la lettre:* "Hitherto, the objectification of our view of the processes in

nature took place by emancipation from the subjectivity of human sensations. But a more far reaching objectification is necessary, to be obtained by emancipating ourselves from the *subjective* moments of human *intuition* with respect to space and time. This emancipation, which is at the same time the high-point of scientific objectification, is achieved in Einstein's theory, it means a radical elimination of *anthropomorphic* slag *(Schlacke)*, and leads us to that kind of description of nature which is *independent* of our senses and intuition and is directed purely to the goals of objectivity and systematic unity" (summer semester 1921 lectures, entitled Grundgedanken der Relativitätstheorie, cited by and translated in U. Majer [1995, 283]).

30. "Daß es sich hierbei nicht um die einfache Feststellung eines Faktums, sondern um eine methodische Maxime, um ein 'regulatives Prinzip' für die Naturbetrachtung handelt, hat Einstein selbst betont" (Cassirer 1999, 118).

References

Allison, H. E. 1973. *The Kant-Eberhard Controversy.* Baltimore: Johns Hopkins University Press.

Anderson, J. 1967. *Principles of Relativity Physics.* New York: Academic Press.

Auyang, S. 1995. *How Is Quantum Field Theory Possible?* New York: Oxford University Press.

Beiser, F. 1987. *The Fate of Reason.* Cambridge, Mass.: Harvard University Press.

Brading, K., and T. A. Ryckman. Forthcoming. "Causality as a Condition of Possible Experience: Hilbert's Second Note on 'The Foundations of Physics.'"

Budden T., and M. Ghins. 2001. "The Principle of Equivalence." *Studies in History and Philosophy of Modern Physics* 32B:33–51.

Cartan, É. 1931. *Notice sur les travaux scientifiques.* Paris. Reprinted by Gauthier-Villars, 1974.

Cassirer, E. 1923. *Substanzbegriff und Funktionsbegriff: Untersuchungen über die Grundfragen der Erkenntniskritik.* Berlin: Verlag Bruno Cassirer. Trans. W. Swabey and M. Swabey, in *Substance and Function and Einstein's Theory of Relativity.* Reprint, New York: Dover, 1958.

———. 1921. *Zur Einsteinschen Relativitätstheorie.* Berlin: Bruno Cassirer. Reprinted in Cassirer (1957, 1–125). Trans. W. Swabey and M. Swabey, in *Substance and Function and Einstein's Theory of Relativity* (1923). Reprint, New York: Dover, 1958.

———. 1929. *Die Philosophie der symbolische Formen, Bd. III.* Berlin: Bruno Cassirer. Trans. R. Manheim as *The Philosophy of Symbolic Forms: The Phenomenology of Knowledge.* New Haven: Yale University Press 1957.

———. 1957. *Zur Moderne Physik.* Darmstadt: Wissenschaftliche Buchgesellschaft.

———. 1999. "Ziele und Wege der Wirklichkeitserkenntnis." (manuscript dated 1936–37) in *Ernst Cassirer: Nachgelassene Manuskripte und Texte, Bd.2,* ed. K. C. Köhnke and J. M. Krois Hamburg: Felix Meiner.

Ciufolini, I., and J. A. Wheeler. 1995. *Gravitation and Inertia.* Princeton: Princeton University Press.

Coffa, J. A. 1991. *The Semantic Tradition for Kant to Carnap: To the Vienna Station.* New York: Cambridge University Press.

Davis, W. R. 1970. *Classical Fields, Particles, and the Theory of Relativity.* New York and London: Gordon and Breach.

Dingler, H. 1920. "Kritische Bemerkungen zu den Grundlagen der Relativitätstheorie." *Physikalische Zeitschrift* 21:668–75.

Einstein, A. 1916. "Die Grundlage der allgemeinen Relativitätstheorie." *Annalen der Physik, Vierte Folge* 49 (1916): 769–822; reprinted in *The Collected Papers of Albert Einstein*, vol. 6., ed. A. J. Kox, M. J. Klein, and R. Schulmann, 283–339. Princeton: Princeton University Press 1996.

———. 1921. "Geometrie und Erfarhung." *Königlich Preussische Akademie der Wissenschaften. Physikalich-mathematische Klasse. Sitzungsberichte.* 123–30.

———. 1952. "Relativity and the Problem of Space." Appendix 5 in Einstein (1961).

———. 1961. *Relativity: The Special and the General Theory.* 15th ed. New York: Crown Publishers.

Friedman, M. 1983. *The Foundations of Space-Time Theories.* Princeton: Princeton University Press.

———. 1994. "Geometry, Convention, and the Relativized a Priori: Reichenbach, Schlick, and Carnap." Reprinted in Friedman (1999).

———. 1999. *Reconsidering Logical Positivism.* New York: Cambridge University Press.

———. 2000. *A Parting of the Ways: Carnap, Cassirer, and Heidegger.* La Salle, Ill.: Open Court.

Heidegger, M. 1929. *Kant und das Problem der Metaphysik.* Bonn: Friedrich Cohen. Trans. R. Taft as *Kant and the Problem of Metaphysics*, 4th ed. Bloomington: University of Indiana Press, 1990.

———. 1977. *Phanomenologische interpretation von Kants Kritik der reinen Vernunft (Marburger Vorlesung Wintersemster 1927–28).* Frankfurt am Main: Klostermann.

Henrich, D. 1994. "On the Unity of Subjectivity." Trans. R. Velkley as *The Unity of Reason*, ed. R. Velkley, 17–54. Cambridge, Mass.: Harvard University Press.

Hilbert, D. 1915. "Die Grundlagen der Physik; (Erste Mitteilung)." *Nachrichten der Königl. Gesell. der Wissenschaften zu Göttingen, Math.-Physik. Kl.* 395–407.

———. 1917. "Die Grundlagen der Physik; (Zweite Mitteilung)." *Nachrichten der Königl. Gesell. der Wissenschaften zu Göttingen, Math.-Physik. Kl.*, 53–76.

Kant, I. 1997. *The Critique of Pure Reason.* Translation of first (A) and second (B) German editions (1781, 1787) by P. Guyer and A. Wood. Cambridge: Cambridge University Press.

Maidens, A. 1998. "Symmetry Groups, Absolute Objects, and Action Principles in General Relativity." *Studies in History and Philosophy of Modern Physics* 29: 245–75.

Majer, U. 1995. "Geometry, Intuition, and Experience: From Kant to Husserl." *Erkenntnis* 42:261–85.

Norton, J. D. 1992. "The Physical Content of General Covariance." In *Studies in the History of General Relativity*, ed. J. Eisenstaedt and A. J. Kox, 281–315. Birkhäuser: Bosteon-Basel-Berlin.

———. 1993. "General Covariance and Foundations of General Relativity: Eight Decades of Dispute." *Reports on Progress in Physics* 56:791–858.

Ohanian, H., and R. Ruffini. 1994. *Gravitation and Spacetime.* 2d ed. New York: W. W. Norton.

Pauli, W., Jr. 1958. *The Theory of Relativity.* Oxford: Pergamon; original German edition published in 1921.

Planck, M. 1910. "Die Einheit des physikalischen Weltbildes." Reprinted in M. Planck, *Wege zur Physikalischen Erkenntnis*, 1–32 (Leipzig: S. Hirzel, 1934).

Prugovečki, E. 1995. *Principles of Quantum General Relativity.* Singapore: World Scientific.

Reichenbach, H. [1920] 1965. *Relativitätstheorie und Erkenntnis a Priori*. Berlin: J. Springer. Trans. M. Reichenbach as *The Theory of Relativity and a Priori Knowledge*. Berkeley: University of California Press, 1965.

———. 1921. "Erwiderung auf H. Dinglers Kritik an der Relativitätstheorie." *Physicalische Zeitschrift* 22:379–84,

———. 1922. "Der Gegenwärtige Stand der Relativitätsdiskussion." *Logos* 10:316–78; translated with omissions by M. Reichenbach as "The Present State of the Discussion on Relativity: A Critical Investigation," in *Hans Reichenbach: Selected Writings, 1909–1953*, vol. 2, ed. M. Reichenbach and R. Cohen, 3–47. Dordrecht: D. Reidel, 1978.

———. [1928] 1958. *Philosophie der Raum-Zeit Lehre*. Berlin: De Gruyter; translated with omissions by M. Reichenbach and J. Freund as *The Philosophy of Space and Time* (New York: Dover).

Rovelli, C. 1997. "Halfway through the Woods: Contemporary Research on Space and Time." In *The Cosmos of Science: Essays of Exploration*, ed. J. Earman and J. Norton, 180–223. Pittsburgh: University of Pittsburgh Press/Universitätsverlag Konstanz.

Ryckman, T. A. 1996. "Einstein *Agonistes:* Weyl and Reichenbach on Geometry and GTR." In *Origins of Logical Empiricism*, vol. 16 of *Minnesota Studies in Philosophy of Science*, ed. R. Giere and A. Richardson, 165–209. Minneapolis: University of Minnesota.

———. Forthcoming. "The Reign of Relativity."

Schlick, M. 1918. *Allgemeine Erkenntnislehre*. Berlin: J. Springer, 1918; trans. of 2d ed. (1925) by A. E. Blumberg as *General Theory of Knowledge* (Berlin: Springer, 1974; LaSalle, Ill.: Open Court, 1985).

———. 1921. "Kritische oder empirische Deutung der neuen Physik?" *Kant-Studien* 26:96–111; trans. P. Heath as "Critical or Empiricist Interpretation of Modern Physics?" *Philosophical Papers*, vol. 1, ed. H. L. Mulder, B. Van de Velde-Schlick, M. Schlick, 311–28. Dordrecht: D. Reidel, 1979.

Stachel, J. 1986. What a Physicist Can Learn from the Discovery of General Relativity." In *Proceedings of the Fourth Marcel Grossmann Meeting on General Relativity*, ed. R. Ruffini. Amsterdam: Elsevier Science Publishers, 1857–1862.

———. 1989. "Einstein's Search for General Covariance, 1912–1915." In *Einstein and the History of General Relativity*, ed. D. Howard and J. Stachel, 63–100. Boston: Birkhäuser. Based on a paper circulating privately since 1980.

———. 1992. "Einstein and Quantum Mechanics." In *Conceptual Problems of Quantum Gravity*, ed. A. Ashtekar and J. Stachel, 13–42. Boston: Birkhäuser.

———. 1994. "The Meaning of General Covariance: The Hole Story." In *Philosophical Problems of the Internal and External Worlds: Essays on the Philosophy of Adopf Grünbaum*, ed. J. Earman et al., 13–42. Pittsburgh: University of Pittsburgh Press.

Torretti, R. 1984. *Relativity and Geometry*. Oxford: Pergamon.

Trautman, A. 1966. "The General Theory of Relativity." *Soviet Physics Uspekhi* 89: 319–36.

Weinberg, S. 1972. *Gravitation and Cosmology*. New York: John Wiley.

Vienna Indeterminism II

From Exner's Synthesis to Frank and von Mises

MICHAEL STÖLTZNER

Institut für Wissenschafts- und Technikforschung,
University of Bielefeld, and Institute Vienna Circle

> With the concepts of cause and effect one cannot run a tramway.
> —*Ludwig Boltzmann*

In an interview with Thomas S. Kuhn, the physicist-philosopher Philipp Frank recalls his early years at the Institute of Physics of the University of Vienna, where he had studied under Boltzmann and where he became a "Privatdozent" in 1909. "Also, strange as it was, in Vienna the physicists were all followers of Mach *and* Boltzmann. It wasn't the case that people would hold any antipathy against Boltzmann's theory because of Mach. And I don't even think that Mach had any antipathy. At least it did not play as important a role as is often thought. I was always interested in the problem, but it never occurred to me that because of the theories of Mach one shouldn't pursue the theories of Boltzmann" (Blackmore 1995a, 128). In a letter to Arthur Eddington written in 1940, Erwin Schrödinger gives a similar testimony. Schrödinger, who was three years younger than Frank, had begun his studies by the time of Boltzmann's death, and he stayed at the Institute of Physics as an assistant of the experimental physicist Franz Serafin Exner until Exner became emeritus in 1920.

> Filled with a great admiration of the candid and incorruptible struggle for truth in both of them, we did not consider them irreconcilable. Boltzmann's ideal consisted in forming absolutely clear, almost naively clear and detailed "pictures"— mainly in order to be quite sure of avoiding contradictory assumptions. Mach's ideal was the cautious synthesis of observational facts that can, if desired, be

traced back till the plain, crude sensual perception. . . . However, we decided for ourselves that these were just different methods of attack, and that one was quite permitted to follow one or the other provided one did not lose sight of the important principles . . . of the other one. (Moore 1989, 41)

Quite contrary to this synthesis whose elaboration shall be a major objective of the present essay, most German physicists shared Arnold Sommerfeld's view that during the famous polemics on the 1895 *Naturforscherversammlung*, "Mach's natural philosophy stood at the back" of Helm and Ostwald's energeticism and against Boltzmann's atomistic (Broda 1955, 12). Max Planck, who would become the explicit and implicit counterpart of the tradition of Vienna Indeterminism studied in the present chapter, by then entertained a neutral position emphasizing that both the first and the second law of thermodynamics were independent principles that were not reducible to molecular motions. Although his discovery of the law of radiation in 1900 chiefly contributed in turning the tide in favor of Boltzmann, Planck needed some time to fully reconcile himself with the probabilistic nature of the second law.[1] One aspect, however, he tried to avoid at least until it became almost inevitable with the advent of quantum theory, to wit, the idea that the most basic laws of nature are indeterministic.

In his 1908 Leiden lecture, which launched the heavy polemics with Mach, Planck praises as Boltzmann's lifework "the emancipation of the concept of entropy from the human art of experimentation" (Planck 1908, 14)—that is, from the impossibility of a perpetuum mobile of the second kind. The price of this seminal step was to render the second law a merely probabilistic regularity that admitted exceptions—at least in principle.

> *Boltzmann* has drawn therefrom the conclusion that such strange events contradicting the second law of thermodynamics could well occur in nature, and he accordingly left some room for them in his physical world view. To my mind, this is, however, a matter in which one does not have to comply with him. For, a nature in which such events happen . . . would no longer be our nature. . . . *Boltzmann* himself has formulated that condition for gas theory [which excludes these phenomena], it is generally speaking the 'hypothesis of elementary disorder'. . . . By introducing this condition the necessity of all natural events is restored. (15)

Even after quantum mechanics had become generally accepted, Planck emphasized the importance of neatly separating between necessity (dynamical laws) and probability (statistical laws). Physical science, so Planck declared, cannot content itself with statistical explanations that are not in turn explained in terms of more fundamental dynamical laws, because any science needs a solid foundation. Planck thus listed Boltzmann as an ally in his polemics against Mach because the former achieved the de-anthropomorphization of the entropy concept and made an important step toward a stable and unified physical worldview. In his lecture, Planck depicted Mach's epistemology almost as a brand of sensationalism that is—albeit logically coherent—entirely fruitless for physical science. Instead, progress in that discipline is grounded in our belief in the reality of physical objects, such as atoms. In his reply to Mach's rejoinder (1910), Planck (1910) criticized that the principle of economy represented a practical rule that had been elevated to a metaphysical principle. Moreover, he charged Mach of ignorance in thermodynamical matters, in particular, by conflating both laws of thermodynamics—an error that was quite common among energeticists. As John Heilbronn rightly observes, "the exchange with Mach gave Planck the status of a philosopher" (2000, 59). Taking into account his increasing influence within almost all scientific organizations and learned societies, Planck can be safely considered as the philosophical opinion leader among German physicists at least until the end of the 1920s.

Where do these serious divergences concerning Boltzmann's philosophical legacy originate? How deep is the rift between Mach and Boltzmann, and why could it be conceived so differently in Vienna and in Berlin? Typically, comparisons between Mach and Boltzmann start from the notorious fight about the existence of atoms, emphasize Mach's antirealism, and subsequently elaborate on Boltzmann's rather intricate form of realism. John Blackmore (1995b), who seems to be somewhat perplexed about Frank's record,[2] has recently diagnosed a substantial shift toward metaphysical realism in Boltzmann's late works from 1900 until his death in 1906. Yet in his lecture notes on natural philosophy (Fasol-Boltzmann 1990) and in other documents of the time, Boltzmann explicitly endorsed several aspects of Mach's epistemology. Precisely one of them, or so I will argue, constitutes the reason why such different readings of Boltzmann's major scientific achievement could arise.

Mach's replacing the concept of causality by the notion of functional dependencies between sensory elements made it viable to contemplate indeterminism at the most basic level of reality. While Boltzmann endorsed Mach's improvement of the Humean approach, Planck and most other German physicists treated causality still within a Kantian framework as an a priori precondition of scientific knowledge. On their account, only such objects that fell under this category—that is, which obeyed deterministic laws—could ultimately be considered as an element of empirical reality. This reality criterion did not entail that concepts of probabilistic nature, such as entropy, were downgraded to mere auxiliary concepts, but scientists' quest for explanation could not halt at this point. This Kantian approach yielded a rigid and permanent connection between causality and realism because there was no way to obtain the basic concepts of reality independently of the category of causality. Mach's wider notion of causality severed this bond between causality and realism on the *general* level and, accordingly, permitted a whole group of physicists, who had been educated at the Vienna Institute of Physics, to seek reality criteria according to the needs of the theory they intended to formulate. Thus, on the *special* level—that is, for a single theory—the separated issues of causal laws and ontology had to be mutually adjusted.

The important consequence of this separation was that it became possible to define a satisfactory ontology for a genuinely indeterministic theory long before the advent of quantum mechanics. In a recent paper (Stöltzner 1999), I have called this philosophical tradition Vienna Indeterminism and I have given an account of its first phase, which comprises Mach, Boltzmann, and Exner.[3] The temporal starting point of Vienna Indeterminism lies between 1896, when Boltzmann's *Lectures on Gas Theory* (1964) appeared, and 1903, when he started his philosophy lectures at the University of Vienna. Due to his central position within the Vienna Institute of Physics and because of his captivating personality, Exner could convey to the younger generation that rather Machian reading of Boltzmann's late philosophy that is expressed in the quotations from Frank and Schrödinger. This second phase of Vienna Indeterminism contains two branches: (1) In the 1920s Schrödinger follows rather closely Exner's way of thinking, but he subsequently develops a unique philosophy of quantum mechanics (compare Bitbol 1996); (2) more than Schrödinger, Philipp Frank and his close friend, the applied mathematician Richard von Mises, were oriented to French conventionalism and advocated the logical analysis of scientific language. Their views are the main topic of the present chapter, and I introduce Schrödinger's position only insofar as it is discussed by Frank and von Mises. But, of course, the linchpin of my thesis is the continuity and integrity of Vienna Indeterminism. Thus, I first of all give a brief outline of the whole tradition and its context before I discuss Exner's synthesis (section 2) and the positions of Frank and von Mises until the formal foundation of the Vienna Circle in 1929 (section 3).

This temporal restriction excludes both Frank and von Mises's seminal books (Frank 1932 and von Mises 1939), but not primarily for lack of space. I rather intend to locate the tradition of Vienna Indeterminism and the discussion with its critics within one particular journal that—in a very general sense—was the forum for the pro-scientific part of the German-speaking intellectual community. In 1913 the weekly magazine *Die Naturwissenschaften* was founded as a German analog to *Nature* mainly on the initiative of the physicist and retired manager Arnold Berliner. From 1924 it also became the official organ of the venerable Gesellschaft Deutscher Naturforscher und Ärzte and of the Kaiser-Wilhelm Gesellschaft. Like these societies, *Die Naturwissenschaften* emphasized the integrity of all natural sciences and rejected both the antiscientific cultural tendencies prevailing among many German intellectuals and antimodernist trends within science, such as Lenard's *Deutsche Physik*. Apart from survey articles on the progress of various disciplines that were often written by the most renowned German-speaking scientists, *Die Naturwissenschaften* also published papers on philosophy of science. Logical empiricists broadly used this journal as a medium until 1935, when Berliner was forced to resign on racial grounds. For the scientists within the Vienna Circle, it was even the most important philosophical forum before the foundation of *Erkenntnis* in 1930 (Stöltzner 2000, chap. 4). Thus, if one succeeds to establish the integrity of the tradition of Vienna Indeterminism within this forum, one can conclude that it appeared so for a rather broad audience. The same holds for the continuity in the philosophical views of the members of this tradition from the end of the energetics controversy until 1929, when quantum mechanics necessitated a radical shift in attitude toward causality. Another reason for this sociological contextualization is the notorious Forman thesis (1971) claiming a causal influence of the sociocultural milieu of the early Weimar republic on physicists' suddenly converting to indeterminism. After three decades of controversy, an argument as the one outlined here, which contradicts Forman's results by establishing a far-reaching continuity of the philosophical discourse, cannot relapse into internalist considerations only.

A Synopsis of Vienna Indeterminism, Its Critics and Limits

Vienna Indeterminism was made possible by Mach's redefinition of causality in terms of functional dependencies between sensory elements. Mach's ontology was based on facts constituted by relatively stable complexes of such functional relations. Going beyond Hume, Mach expressed them in terms of concrete physical equations—for instance, "Fourier's equations which comprise all conceivable facts of heat conduction" (Mach 1919, 461/415). He calls these laws *direct* descriptions and opposes them to *indirect* descriptions, such as atomistic theories, which are only of hypothetical validity. But in order to guarantee the integrity of those functionally constituted facts, Mach had to posit a principle of unique determination of the actual fact in comparison to all variations of its functional dependencies. Mach also introduced another core tenet of Vienna Indeterminism by emphasizing that for the empiricist it is impossible to finally decide between determinism and indeterminism on the metaphysical level. Nevertheless, he still favored determinism as a regulative principle because only by way of this hypothesis could probabilities make sense. While Mach thus agreed with his opponent Planck that all probabilities required a determinist foundation, Boltzmann was surprisingly vague with respect to the concept of probability. He simultaneously clung to the old concept of equiprobability—which is either based on causal relations or on their absence due to our ignorance—and emphasized against Planck—though mostly in private communications—that the highly improbable entropy-decreasing events really occur. Boltzmann's main objective was, however, to give a proper ontology to atoms by means of a twofold reality criterion. On the one hand, he conceived of atomism as property reduction to theoretically defined universal entities and their interactions. On the other hand, atomism was already implied by humans' finite reasoning powers that made it impossible to actually assess the continuum. At this point, Boltzmann surprisingly endorsed Mach's definition of mathematics as *"economically ordered experience of counting"* (Mach 1919, 68/70).[4] Moreover, he skillfully integrated Mach's empiricism into his struggles against energeticism.[5]

Viewing Boltzmann's conceptual difficulties with probability and atomistic ontology, it is rather surprising that he never adopted or even cited Gustav Theodor Fechner's frequency interpretation of probability published in 1897. Shortly after Boltzmann's death this interpretative move was accomplished by Exner in his 1908 inaugural speech as rector of the University of Vienna, and it became henceforth pivotal for Vienna Indeterminism. As Exner built physical ontology upon collectives, he had to defend a rather firm empiricism in Mach's footsteps because only in this way could he jettison as meaningless all speculations as to whether there exist some unobservable deterministic laws at the most basic level of physical reality. In his polemics against Planck, Exner emphasized that all apparently deterministic laws could well be the macroscopic limit of indeterministic basic laws valid for the single particles or events. Exner's synthesis between Mach and Boltzmann paved the way to accept genuine indeterminism in physics without any reference to quantum mechanics. Exner's reliance upon the second law of thermodynamics did not halt at the boundaries of physics proper. By the end of his life he had completed a comprehensive physicalist and indeterministic theory of culture (Exner 1923), which remained unpublished but gives vivid testimony of the cultural discussions in the large circle around Exner (compare Stöltzner 2002b).

The reception of Exner's philosophical ideas was typically limited to those who had closer contacts to the Vienna physics community. Among them was Exner's longtime assistant, Schrödinger, who constantly stressed Exner's priority in contemplating genuinely indeterministic laws of nature, in particular in his 1922 Zurich inaugural lecture, "What Is a Law of Nature?" Schrödinger also followed Mach's neutral monism by developing a pronounced unease against the Copenhagen interpretation's dualism between observations and an objective particle reality unknowable in principle. Quite in line with Vienna Indeterminism, Schrödinger was searching for a realist but not metaphysically realist ontology for his wave equation, which came close to Boltzmann's universal atomistic pictures; yet neither the wave function nor—in later years—unified field theory turned out to be satisfactory. In 1927 Schrödinger took Planck's chair at the University of Berlin and in 1929 he became a member of the Prussian Academy of Sciences. In his inaugural speech to the academy (1929) he continued Exner's debate with Planck, but whereas in 1922 he had considered the alternative between determinism and indeterminism as an empirical question—as had Exner— he now took a conventionalist tack and considered it a matter of practicability.

French conventionalism was, on the other hand, rather the starting point for the Vienna Circle members Frank and von Mises. I will discuss their positions together not only because of their common intellectual background, but particularly because of the manifold of affirmative cross-references in their works. This intimate connection justifies enrolling—at least for the scope of the present essay —von Mises in the Vienna Circle. Back in 1907 Frank had considered the general law of causality a mere convention, a position he largely revoked in his 1932 book, *The Law of Causality and Its Limits*, where he investigates the conditions under which the general law of causality attains an empirical content. As a matter of fact, the earlier position was hardly reconcilable with von Mises and Frank's constant adherence to Mach. In the introduction to his book, Frank emphasized that his change of mind was caused by quantum theory and by von Mises's "conception of statistical laws and their relation to dynamic laws" (Frank 1932, 12). To von Mises's mind, the two types of law did not actually compete with one another because they simply concerned different observational facts. The law of causality obtains empirical content only once it has been specified by means of certain axioms, such as the differential equations of Newtonian mechanics. Just as the Newtonian dynamical laws govern the motions of point particles, statistical laws deal with mass phenomena that are represented by statistical collectives. Von Mises (1922) explicitly criticizes Boltzmann's formulation of the second law as a blend of microdeterminism and macroprobabilism and advocates a purely probabilistic approach instead. Von Mises and Frank gained this freedom in choosing collectives as a proper ontology by applying the idea that all concepts in physical theories are coordinated to specific observations or measurements. Thus, Frank could argue that the only modification in quantum mechanics was the statistical character of this coordination. When he demanded that, nevertheless, coordination and statistical law had to be unique, it can be viewed as an heir of Mach's principle of unique determination. Finally Exner, Frank, and von Mises never opted for a final decision between determinism and indeterminism.

Vienna Indeterminism ends in 1939 when von Mises's textbook on positivism was published. Moreover, at the same time, a far-reaching convergence between logical empiricists and Ernst Cassirer, the heir of Marburg neo-Kantianism, became manifest. In many respects Cassirer's 1937 book, *Determinismus und Indeterminismus in der modernen Physik,* can be considered as a chronicle of the epoch studied here and as a justification of bestowing on Mach the honor of having given birth to Vienna Indeterminism. Cassirer emphasizes that the gist of the matter lies in the distinction between causality and the object figuring in the laws. He traces this thesis back to his 1910 book, *Substanzbegriff und Funktionsbegriff,* which had focused on the dissolution of substantialism in modern philosophy, a tendency that also provides the background of Mach's reinterpretation of causality. Although, in the end, Cassirer, like the late Planck, opts for maintaining a strongly relativized a priori notion of causality, Frank's review of the book is very laudatory, and he spots there the core tenet of Vienna Indeterminism. "A further principal feature of Cassirer's account is that the form of the law of causality and the concepts of what one calls an *object* mutually presuppose each other. Also this is a basic thesis defended by logical empiricism which has been taken over from positivism. Today's positivism just gives this thesis a more formal turn" (Frank 1938, 73).

This convergence of Cassirer's neo-Kantian position to logical empiricists' epistemology is important because Vienna Indeterminism cannot be considered as a well-entrenched Lakatosian research program let alone as a fixed set of philosophical assertions. It lacks a precisely defined philosophical core that is tenaciously defended throughout the years, because apart from their separation from causality the respective reality criteria are starkly different. After all, the tradition extends over half a century, during which two major scientific revolutions took place, relativity and quantum theory. There are, however, further cohesive traits, such as a firm empiricist attitude and a high esteem for statistical physics in general.[6] But in order to fully prove the existence of such an adaptable tradition, a historical contextualization is wanted that permits one to circumscribe Vienna Indeterminism by its limits, in particular its opponents and allies, and by investigating its institutional and personal basis.

To start with the second question, basically two institutions were carrying the tradition's continuity from the first to the second phase. Founded already in 1850 for Christian Doppler, the Institute of Physics of the University of Vienna was constantly suffering from scarcity of funds. Moreover, neither Josef Stefan nor Josef Loschmidt ever traveled abroad to attend scientific conferences. Both factors—so Boltzmann judges in his obituaries—prevented some possible experimental breakthroughs and lessened their international recognition. But, the spirit was remarkable: "Never did they attempt to express their intellectual superiority in academic conventions. Albeit a student at first and then a long-time assistant, I have never

Stöltzner / From Exner to Frank and von Mises

heard from them any word other than the friend addresses the friend" (Boltzmann 1905, 102).[7] When Stefan and Loschmidt died in 1893 and 1895, this intellectual atmosphere continued under their successors Boltzmann and Exner. Exner, in particular, exhibited an extraordinary understanding for younger people. By the time of Boltzmann's death, he "was surrounded by a bevy of pupils who respected him like a father" (Benndorf 1927, 403). In this way, Exner became "during one generation the center of Austria's physical life" (Sommerfeld 1927, 27). Thus, it seems reasonable to suppose that this climate strongly fostered the spread of Exner's synthesis of Mach and Boltzmann among the students of the Institute of Physics. Let me illustrate this by two appraisals of Exner's personality and achievements written by members of the second phase of Vienna Indeterminism.

In his inaugural address to the Prussian Academy of Sciences, Schrödinger asserts: "Franz Exner (to whom I am personally indebted for his exceptionally great support) was the first who contemplated the possibility of an acausal conception of nature" (1929, 732). Also Frank acknowledges Exner's priority in this respect: "Franz Exner has already drawn attention to the possibility that elementary processes do not follow the pattern of celestial mechanics with their Laplacian causality but that perhaps for an individual event, for example the collision of two molecules, no causal law can be established at all, and that nevertheless, with the formation of averages, laws can be derived by which some causal determination is expressed" (1932, 70–71). In a footnote Frank mentions Schrödinger's 1922 lecture, where "the significance of Exner's thoughts for our time is very correctly characterized" (284). Two chapters further he extends this line of thought to von Mises:

> The statistical conception . . . suggested that the statement of Newton's equations of motion for each individual molecule [in a gas] . . . is not a statement about the real world at all. However this would mean, as Exner already has declared possible, that the proposition that mechanical causality exists for each individual particle of arbitrarily small size is not a statement about reality but can have concrete meaning only as a tautology . . . Perhaps Richard von Mises was the first who has pointed out, in his lecture "On the present crises of mechanics" that in the field of mechanics in the narrower sense, there are observable processes in liquid and solid bodies that also cannot easily be presented with the help of causal laws. (72)

The more detailed discussion to come will show that Exner was indeed the connecting link between both phases of Vienna Indeterminism.

The second institution of relevance is of a rather informal kind. In *Modern Science and Its Philosophy,* Frank related that from 1907 until his departure for Prague in 1912 he met with Neurath, Hahn, and others every Thursday night in a

Vienna coffeehouse. Friedrich Stadler (2001, 143) also lists von Mises among this group, which Rudolf Haller (1986b) has baptized as the "First Vienna Circle." Without entering the debate of whether its importance for the history of logical empiricism suggests this or another name, I view in this circle a well-documented cohesive factor for Frank and von Mises's philosophical formation that was close to the Institute of Physics and that was embedded into other Viennese intellectual forums. One can safely assume that from 1903 to 1906 not only Frank attended Boltzmann's lectures on natural philosophy. Apparently, then, their content did not quite appeal to the coffeehouse circle; Frank curtly remarks that "the effect of the course was slight, because of a lack of coherent approach" (Frank 1961, 244). According to his account, they were mainly interested in French conventionalism and Mach's historical-critical analyses of the physical science, which, so one should add, better suited relativity theory, the field which after the end of the energetics controversy had become pivotal in philosophy of science. Moreover, special relativity was Frank's area of research that would earn him the Prague chair in 1912.[8] Before that, Frank had done mathematically oriented work in the calculus of variations, a topic to which he had most probably been introduced by Hahn as early as 1905, when he was writing his Ph.D. dissertation (Frank 1906). Interestingly, Exner was one of two experimentalists who wrote the opinion about Frank's thesis instead of the deceased supervisor, Boltzmann (compare Stöltzner 2002a). As to the wider context of the First Vienna Circle, Thomas Uebel (2000, 2003) rightly emphasizes the importance of Adolf Höfler's *Philosophical Society at the University of Vienna*, where Frank, Hahn, and Neurath started their philosophical careers and where they received a very specific reading of Kant.

Haller places the First Vienna Circle within "Austrian philosophy" (Haller, 1986a), an intellectual tradition prevailing in the Habsburg monarchy since the days of Bernard Bolzano. One of its major characteristics was the rejection of Kant's transcendental philosophy, a point that also appears in the 1929 manifesto of the Vienna Circle and figures quite prominently in Neurath's later historical writings. Whatever stand one takes with respect to the Neurath-Haller theses in general, one point is essential to properly locate Vienna Indeterminism within this general context. Together with modern logic, general relativity became crucial to the philosophical identity of the Vienna Circle and of logical empiricism in general. By the early 1920s, the Germans Moritz Schlick and Hans Reichenbach—general relativity's most prominent philosophical defenders—who had grown up in a milieu shaped by neo-Kantianism, also arrived at the rejection of any aprioristic conceptions of space and time however relativized. Yet the causality debate developed in a rather different fashion, and the Kantian category of causality enjoyed a surprising longevity despite the general ban against the synthetic a priori prevailing within the Vienna Circle. In particular, the main defender of deterministic Kantian causality in physics, Max Planck, was at the same time mainly responsible for getting both relativity theories accepted by the German physics community. As a matter of fact, Planck's interpretation of relativity was plainly anti-Machian, since he believed that outdated absolute concepts are relativized just in order to find deeper absolute concepts. "Yet when space and time have been denied the character of being absolute, the absolute has not been blotted out, it has just been moved more backward, to wit, into the metric of the four-dimensional manifold" (Planck 1925, 154). Planck's convergent realism manifest in this passage was fundamentally at odds with the highly flexible reality criterion used by the Vienna Indeterminists. Roughly speaking, the front line on causality went right through the logical empiricist camp separating the Viennese from the Germans. The situation changed, however, after quantum mechanics had made it close to inevitable to consider irreducibly statistical microlaws. When both sides came closer to one another in the 1930s, the convergence of position also embraced neo-Kantians such as Cassirer. At the end of this introductory section, I shall sketch why Schlick and Reichenbach, albeit two of the most prominent logical empiricists, did not belong to Vienna Indeterminism. This permits me to draw the philosophical border line of Vienna Indeterminism without invoking the Austrian philosophy thesis in a determinative fashion.

Schlick's first paper on causality (1920) was almost exclusively directed at relativity theory, which did not force him to accept genuine indeterminism. On this account, the statistical character of the second law was not situated in the laws themselves, but in the initial conditions—quite analogous to the relation between the dynamics (the time evolution) and the initial value hypersurface in general relativity.[9] Still, in 1925 Schlick argued quite in line with his teacher Planck that "the validity of causality is thus a presupposition, not an object, of the natural sciences" (1925, 429/31). Moreover, "it is clear . . . that only in utmost extremity will the scientist or philosopher resolve to postulate purely statistical micro-laws" (461/61). When in 1926 this emergency case had happened, Schlick renounced his earlier attempts to provide an explicit characterization of causal laws and turned to scientific practice. "Verification as such, the fulfillment of prediction, confirmation in experience, is therefore the criterion of causality *per se*" (1931, 151/188). Unlike the Vienna Indeterminists, Schlick admitted a logical notion of probability when describing human judgments, but he clearly set this apart from the objective notion of probability occurring in physics.

What separates Reichenbach from Vienna Indeterminism is precisely that he did not assume such a distinction and in this way claimed to possess a probabilistic solution to the problem of induction. Here I cannot enter into this vast topic that led to polemics with his Berlin colleague von Mises, which ultimately estranged him from the movement in the 1930s and which may be one of the motivations why he—unlike Frank and von Mises—turned to quantum logic. But I do not pass over the fact that in the first footnote of his *Philosophic Foundations of Quantum Mechanics,* Reichenbach lists Exner as "perhaps the first" to have criticized the assumption of strict causality (Reichenbach 1965, 1). The author knew well what he was talking about because he had reviewed the first edition of Exner's *Lectures on the Physical Foundations of Science,* published in 1919. The reviewer endorsed "*Exner's* unequivocally advocating the objective meaning of the probabilistic laws in which he rightly conceives a very general regularity of nature" (Reichenbach 1921, 415). This was, of course, also Reichenbach's own position (1920a, 1920b). It is high time now to look into Exner's works themselves and to study their premises and intellectual roots.

Exner's Synthesis and Its Roots

On September 8, 1906, the morning edition of the *Neue Freie Presse* published two obituaries of Boltzmann on its front page. While Mach praised the deceased as an unparalleled experimenter, Exner focused on the atomistic worldview, "in which he found the best mainstay in the struggle against the lately popular, but unclear ideas of energeticism. . . . Against all these theories which signify, in effect, a step backward, Boltzmann fought a stubborn, but righteous and meritorious struggle in which his sharp mathematical weapons always led him to victory." In Boltzmann's philosophical armory one finds Mach's antisubstantialism, too:

> As regards Ostwald's energetics, I think it rests merely on a misunderstanding of Mach's ideas. Mach pointed out that we are only given the law-like course of our sense impressions and ideas, whereas all physical magnitudes, atoms, molecules, forces, energies and so on are mere concepts for the economical representation and illustration of these law-like relations of our sense impressions and ideas. The last are thus the only thing that exists in the first instance, physical concepts being merely mental additions of our own. Ostwald understood only one half of this proposition, namely that atoms did not exist; at once he asked: what then does exist? To this his answer was that it was energy that existed. In my view this answer is quite opposed to Mach's outlook. (Boltzmann 1905, 368/175)

These mentally added concepts, on Boltzmann's account, enjoyed much more freedom than within Mach's adaptive epistemology. Theories could well reach beyond the *known* phenomena, they were not just their economization.[10] By separating clearly the facts and the theories, Boltzmann could better avail himself of the

Machian conception of functional dependencies as an ontological basis for physical theory. He even defended atomism on this line by considering atoms as the result of a reduction of properties to universal basic entities. Boltzmann's ideas about causality largely agreed with Mach's, and he treated the classical metaphysical question as a pseudo-problem. We are "free to denote [the law of causality] either as the precondition of all experience or as itself an experience we have in conjunction with every other" (163/75). When Mach diagnosed our "desire for causality," Boltzmann discerned a general tendency of all our mental habits to "overshoot the mark" by still seeking explanation or definition of the inexplicable elementary concepts. "Indeed people racked their brains over the question whether cause and effect represent a necessary link or merely an accidental sequence, whereas one can sensibly ask only whether a specific phenomenon is always linked with a definite group of others, being their necessary consequence, or whether this group may at times be absent" (354/166).

This language-critical motive would become a core tenet of the Vienna Circle. In a fragment for the philosophy lectures bearing the heading "Cause and Effect," Boltzmann linked causality to probability: "Before any experience takes place, each [an accidental sequence or a causal link between phenomena] is equally probable. But my repeated experiences render it infinitely improbable that all observed *regularity* would be accidental, and infinitely probable that actual actually takes place" (Fasol-Boltzmann 1990, 282). Still, at this time, Boltzmann considered probability as degree of certainty, and he seems to have approved the logical interpretation of Johannes von Kries, whose seminal book (Kries 1886) he quoted in the same year but never afterward. Applying logical probability, however, requires "that the mechanical conditions of the system are known" (Boltzmann 1905, 37/22). As Martin Klein (1973) has convincingly shown, Boltzmann made several major changes in his use of the concept of probability.

But despite these modifications, developing statistical mechanics as a "special science" that studies "the properties of a complex of very many mechanical systems starting from the most varied initial conditions" (Boltzmann 1905, 360/171) was constantly aggravated by obtaining a proper concept of equiprobability. "However, this being the fundamental concept, it cannot in turn be derived and must be regarded as given" (361/171). On January, 31, 1906, Boltzmann's notes for the philosophy class read as such: "Knowledge by the law of causality not in the same way from experience. Source of experience. We stand under its influence.[11] One seeks probability from a priori probability. [This] only [makes] sense, if equally possible cases. Necessarily subjective from our classifications or after known causal law" (1990, 145). To my mind, Boltzmann here argues that in the same way as we necessarily order experiences by (functional) causality, we pose equiprobabilities in order to base probabilistic laws. Both are achieved either by classifications—for example, the symmetry of a die—or according to already empirically known laws, such as "We can infer from experience that in lotto every move is equally probable" (1905, 163/75).

Yet Boltzmann never made the final step in basing probability on experience, although in a letter to Felix Klein in 1899 he expressed his misgivings about Emanuel Czuber's abstract definition of the object of probability calculus (compare Höflechner 1994, 2:318). He evidently was not acquainted with Fechner's relative frequency interpretation of probability posthumously published in the *Kollektivmaßlehre* of 1897 that would fit so neatly to his own definition of statistical mechanics as a special science quoted above. Shortly after Boltzmann's death, "about 1908 Fechner's theory of collectives apparently was standard knowledge for everyone working on probability theory and statistics in the German-speaking area" (Heidelberger 1993, 376).

Why did Boltzmann never read Fechner? An argument with Mach, who conceived Fechner's global tendency to stability as the origin of the second law, teaches that this neglect was a generic one on Boltzmann's side.[12] Thus, instead of referring to Boltzmann's weak sight in his later years, I think Heidelberger is quite right to argue that Fechner's thoughts about probability were too much embedded into his general and often hermetic outlook to be quickly accessible for someone who was—in stark contrast to Mach—unfamiliar with their philosophical context.[13]

Exner, to the contrary, was familiar with this context. In his curriculum vitae for the Austrian Academy of Sciences he wrote, "It was perhaps an unconscious tradition [of Franz Serafin's deceased father Franz Exner (1802–53), who was a professor of philosophy at Prague] that I felt the wish to occupy myself with purely philosophical problems, such as, e.g., in particular with Herbart's system, especially with his psychology and metaphysics" (1917, 3). It is safe to conclude that Exner also studied Fechner's manifold criticisms of Herbart.[14] Moreover, his elder brother Sigmund was a world-renowned physiologist and held a chair at the University of Vienna as well.

Franz Exner did not miss the best opportunity to set forth his philosophical outlook. On October 15, 1908, he delivered his inaugural address as rector of the University of Vienna "On Laws in Science and Humanistics."[15] Like Mach and Boltzmann, Exner denies any irreconcilable difference between science and the humanities that would justify different methods; instead, the two only study different kinds of objects. Still, while the sciences possess mathematically formalizable and universal laws, the humanities obtain weak regularities at best. Exner tackles this classical problem from an unusual perspective: Why do natural laws exist at all? He asserts that all processes in nature fall under the laws of physics. Yet he combines Boltzmann's physicalism with Mach's antisubstantialism; "these laws do not exist in nature, only man formulates them and avails himself of them as linguistic and calculatory means" (Exner 1909, 7).

Looking around us, we do not at first discern any lawlike regularities, but rather the fact that all natural processes are directed. Here Exner pays tribute to Boltzmann, who "was the first to give a definite and clear interpretation of this direction . . . showing that the world ceaselessly develops from less probable into more probable, and hence more stable, states" (9). In the molecular dynamics of a gas we "observe regularities produced exclusively by chance" (13). But highly probable states—namely, stable laws—are only possible for an extremely large number of individual events. All such laws are thus average laws that only hold with high probability, such that there can never be mathematically exact laws.

Throwing two dice sufficiently, we often see the *more probable* numbers of spots to occur *more frequently*. This manifests the law of large numbers, which is "unprovable but taken from the thousandfold experience of men and constitutes the basis of probability calculus. As the one and only law it indeed governs all happenings in nature" (19). But this meta-law only holds if external conditions do not change until many individual events have occurred sufficiently for the exact (but not exceptionless) average laws to stabilize.

Adopting as early as 1908 the relative frequency interpretation of probability represents Exner's most important contribution to Vienna Indeterminism. In the fourth chapter of his 1919 *Lectures,* titled "On Natural Laws," Exner gives an introduction to probability calculus and emphasizes that only if we consider chance not as based on "imperfect knowledge," but as an objective feature of nature, can we reconcile chance and causality by considering the law of causality as expressing that "on average the course of the phenomena is lawful" (1922, 675). Exner adopts Mach's redefinition of causality in terms of functional dependencies and writes: "Ernst Mach to whom one surely must attribute an influential voice, says: 'There is no cause nor effect in nature; nature has but an individual existence'"(675, quoting Mach 1988, 496/580). His staunch empiricism even led Mach to state the limits of determinism almost like Exner: "No fact of experience repeats itself with absolute accuracy. . . . Therefore even the extreme theoretical determinist must in practice remain an indeterminist, especially if he does not wish to speculate away the most important discoveries" (Mach 1991, 282/208).

But Exner's insight that all laws we experience hold only on average gives the empiricist law of causality a new twist insofar as it "expresses nothing else but the fact that natural processes, to the extent we can observe them macroscopically, that is on average, are lawful" (Exner 1922, 674). Nevertheless, Exner objects to Mach's narrow conception of theory and advocates Boltzmann's program of explanation instead. "The kind of natural studies which had as its final aim only a *description* of nature in terms of systems of equations is unsatisfactory. And even if this was in place for a while, today research is directed toward a molecular-mechanical understanding of natural processes" (721). Facts, according to Exner, are the laws that have "objective reality" (724), while theories change drastically in the course of time.

Exner's stance in the realism issue differs significantly from Boltzmann's language-oriented "empirical realism," which commences from specifying the meaning of the existence of unicorns, etc. Boltzmann ultimately considered "the realist mode of expression more purposive than the idealist one" (Boltzmann 1905, 186/75). Exner, on the other hand, does not invoke linguistic considerations and holds that without the assumption "that our sensations are correlated to certain objective processes in the external world . . . all human research would have to appear superfluous" (Exner 1922, 287). But the fact that we cannot entirely turn down metaphysical questions does not justify us in insisting that the question of how clocks "really" tick is meaningful despite relativity theory. Whenever, as in the latter case, we enter a new field of experience, our common habits of thought are not automatically valid, and they can easily become an impediment to science.

Exner focuses on theory reduction and strengthens Boltzmann's conception of atomism as property ascription to more fundamental entities and their interactions.[16] As no measurement is absolutely precise, the firm empiricist can only impose the following condition upon the elementary mechanisms that underlie an average law. "If physical phenomena result from many identical, mutually independent single events, then the causes assumed by the determinist act just as if there were no causes at all, but mere chance ruling" (681).

As Exner is seeking an ontological foundation for average laws, he needs a reality criterion that is independent of the particular nature of the single events and consistent with the law of large numbers. His only option here is to accept the collectives of Fechner's frequency interpretation. Since these are only realized in the limit of infinitely many events, even all apparently deterministic laws admit exceptions—as long as their probability renders them inaccessible to experiment —and they cease to hold below a certain number of single events. On the other hand, the second law of thermodynamics, the probabilistic law par excellence becomes meaningless in microscopic domains.

More generally, Exner's fundamental indeterminism intends to justify all types of regularities found in the world. Depending on the number of events studied by the respective science, the degree of probability varies between zero (in most humanities) and one (in physics). All descriptive sciences lie in between, in particular because the external conditions change too rapidly for exact laws to stabilize.

Already in his inaugural address, Exner discusses some regularities outside physics that are expressions of the second law. While, for instance, primitive societies are very homogeneous, "we find an abundant variation in physical, intellectual, and social respects on the part of civilized peoples that increases with the age of the culture" because uniformity is a very improbable state (Exner 1909, 26).[17]

On August 3, 1914, Planck delivered a rectorial address "On Dynamical and Statistical Regularities" in which he calls for a strict distinction between both types of regularities because of the sharp contrast between reversible—that is, dynamical—and irreversible processes.

> This dualism which has inevitably been carried into all physical regularities by introducing statistical considerations, to some may appear unsatisfactory, and one has attempted to remove it ... by denying absolute certainty and impossibility at all and admitting only higher or lower degrees of probability. Accordingly, there would no longer be any dynamical laws in nature, but only statistical ones; the concept of absolute necessity would at all be abrogated in physics. But such a view should very soon turn out to be a fatal and shortsighted mistake. (1914, 63)

In these lines Planck reports and rejects precisely Exner's views, although he does not cite him by name; except for the open polemics with Mach, Planck hardly ever did mention names on similar occasions.

Exner responded to Planck point by point in his ninety-fourth lecture. While Exner studies how probabilistic macroscopic laws *emerge*, Planck assumes a priori the existence of an absolute causality as a necessary precondition for understanding nature. "But nature does not ask whether man understands her or not, nor are we to construe a nature adequate to our understanding, but only to reconcile ourselves as much as possible with the given one" (Exner 1922, 709). Exner also criticizes Planck's unjustified trust in our habits of thought, which makes it likely "to fall into a sort of physical mythology" (709).

His empiricist approach prompts Exner to reject Planck's distinction between reversible and irreversible processes because in nature we only encounter "irreversible processes which can come, however, arbitrarily close to reversibility" (710). Between the extremes there are many intermediate cases. "Whether a process is reversible or irreversible in fact only depends upon whether the recurrence of a certain state is practically observable" (711).

Exner and Planck also disagree about probability theory. "It is claimed that in its applications probability calculus cannot dispense with the assumption of absolutely dynamical laws for the elementary processes"—here Exner almost quotes Planck (1914, 64)—"[but] the assumption suffices that the elementary processes be equally characterized by average laws" (Exner 1922, 712). While Planck calls for a dynamical explanation of statistical laws, Exner, on the contrary, asserts the following: "Nothing prevents us from regarding the so-called dynamical laws as the ideal limit cases to which the real statistical laws converge for the highest degrees of probability" (713). And Exner even contemplates that gravitational force might be replaced by a statistical process in which the falling body moves along by fits and starts or even on a zigzag path. "Boltzmann has in conversation entirely agreed to this opinion and has considered it not only possible, but even very probable" (670). Indeed, in later years Boltzmann studied whether the entropy curve was a nondifferentiable function and pondered whether there could be "deviations from the principle of energy, perhaps only of the second law, also from the area law or from the center of mass law" (Fasol-Boltzmann 1990, 106).[18] The idea of such a discontinuous dynamic was based on his extension of atomism to time that lapses like the pictures in a cinematograph.[19] In a letter to Brentano, he even estimated the number of atoms in a second as $10^{10^{10^{10}}}$—a number which grossly exceeds its counterpart for matter, Loschmidt's or Avogadro's number $6 \cdot 10^{23}$.[20] "The number of points of time can be made so great that the probability becomes great that a very improbable condition can occur in the whole world" (282). Thus, in the (presumably finite) universe, there could be regions in which the entropy decreases and time flows backward. Moreover, the "force law must differ at different times" (282). Exner generalizes this point and ponders that it would be "presumptuous to claim that any law, for instance, gravity, as it appears to us today, had also been valid in all earlier epochs of the World, or will be valid in all subsequent ones" (Exner 1922, 667). This had been an old cosmological speculation of Fechner.

Frank and von Mises: On the Way to Statistical Causality

One of the first philosophical papers of a core member of the later Vienna Circle is Philipp Frank's "On the Law of Causality and Experience." "The law of causality, the foundation of every theoretical science, can neither be confirmed nor disproved by experience; not, however, because it is an a priori true necessity of thought, but because it is a purely conventional definition" (Frank 1907, 444/63). In order to prove his main thesis, Frank adapted an argument that the biologist-philosopher Hans Driesch had devised to establish the a priori character of the law of energy conservation. This already suggests that Frank to a certain extent discusses the issue of causality within a Kantian frame of thinking in which the a priori category, however, has been replaced by a convention. And, indeed, Frank contends that "the latest philosophy of nature revives in a striking way the basic idea of critical idealism, that experience only serves to fill a framework which man brings along with him as a part of his nature. The difference is that the old philosophers considered this framework a necessary outgrowth of human organization, whereas we see in it a free creation of human arbitrariness" (447/66). Mach and Boltzmann, who are not mentioned in the paper, had been considerably more empiricist in that respect. After a short correspondence in October 1893, both had agreed that the law of energy conservation "has no other evidence than an empirical law" (Höflechner 1994, 2:204).

In his paper, Frank discusses causality in the following form that also involves an induction argument that is rather irrelevant for his line of reasoning. "If, in the course of time, a state of the universe A is once followed by the state B, then whenever A occurs B will follow it" (Frank 1907, 444/63). The crucial point is the arbitrariness in the definition of "state."

> If the law of causality is not valid according to one definition of the state, we redefine the state simply in such a way that the law is valid. If that is the case, however, the law, which appeared to be stating a fact, is transformed into a mere definition of the word "state." (446/65)
>
> If I wish, I can provide all bodies with state variables that are all qualitatively different, in order to fulfill the law of causality. I can regard heat, electricity, magnetism, as properties of bodies, essentially different from one another, just as is done in modern energetics, and as Driesch does. On the other hand, if I wish, I can get along with less qualitatively different properties. For example, I can introduce only the motion of masses; but then, in order to obtain the necessary diversity, I must take refuge in uncontrollable hidden motions. This leads to the purely mechanical world view, which Democritus dimly conceived as an ideal, and which occurs mostly in the form of atomism. (448/66)

In the interpretation of quantum mechanics, arguments similar to Frank's have become popular. They state that causality could be rescued by adding so-called hidden variables that permit a realist ontology for the quantum particles at the price of adding in-principle unmeasurable quantities. Frank, on the contrary, avoids any ontological commitment—either to Boltzmann's property ascription to fundamental entities or to Mach's (qualitatively different) facts. Although he remarks that some worldview will be more simple than another, he opts neither for Boltzmann's ontological simplicity nor for Mach's principle of economy.

Frank's 1919 paper on "The Statistical Approach in Physics," published in *Die Naturwissenschaften*, further elaborates on the notion of state;[21] however, now within a probabilistic setting. He starts from a definition of the principle of causality that is less contaminated with the problem of induction. "The present state of a closed system of bodies uniquely determines its future state, that is, whenever the systems reaches the state A, a particular state B follows . . . If one understands by state the sum of all *physically measurable* properties of the system, the law of causality has no validity. In the sense of molecular theory one must rather add to the description of the state also the positions and velocities of all molecules, by means of which the law of causality is saved, but its actual application becomes impossible" (1919, 727). Frank's main justification for the actual invalidity of the law of causality, revoking his earlier radical conventionalist views, comes from Brownian motion. Recall that this phenomenon had also been Boltzmann's and Exner's case in point for a genuinely indeterministic physical world. In the theory of gases the number of molecules is typically so large that highly improbable events, such as a spontaneous departure from equilibrium, practically never occur. While for gases the invalidity of causality is only theoretically inferred, in Brownian motion we observe these spontaneous density fluctuations, so that Frank concludes: "in the realm of the empirical-physical, the experimentally measurable quantities . . . there exists no causality" (728). It remains, however, possible to establish an average law, Smoluchowski's law of diffusion for the Brownian particle.

Frank bases the law of causality exclusively on the prediction of a future state and he interestingly does not require that it refer to genuine laws. This becomes clear when he extends his considerations—as Exner had done—to the humanities, which study properties considerably less detailed than those of statistical mechanics. "The law of causality does not require at all the existence of historical laws. It might well be that the properties by which the historian describes groups of nations do not suffice in principle to fulfill the law of causality, and that there exists no historical, but only an individual psychological causality" (728). Frank's outline of the "statistical conception of nature" shares the core tenet of Vienna Indeterminism, the separation of causality and ontology (701). One should not be led astray by terminology. As does Planck, he classifies only dynamical laws admitting unique predictions as causal, but he does not demand that statistical laws should be reduced to dynamical ones, because, after all, Brownian motion empirically teaches that this is impossible. Yet, this evidence had been precisely Exner's empiricist rationale in the polemics with Planck. Exner had, more generally, considered the very fact that all natural processes are directed as the starting point of our understanding of nature and the second law of thermodynamics as the basic law of nature. Right at the beginning of his paper, Frank characterizes "the tendency of assimilating all distinctions" as the most characteristic trait of natural processes and as a "brazen law" that originates in a game of chance (701). For this result he credits Boltzmann and Marian von Smoluchowski, a former student of Exner's. Despite these similarities, Frank does not subscribe to so radical a probabilism as Exner. And although he obviously applies the frequency interpretation, he does not yet derive clear-cut ontological consequences from that.

Let me turn to the role of Mach within Frank's early thinking. In 1910, the same year in which he had visited Mach to explain to him Minkowski's formulation of special relativity, he reviewed for the *Monatshefte* Planck's Leiden lecture.[22] There he holds that Planck's attack essentially has arisen from various misunderstandings. In particular, one could simultaneously maintain that our worldview is an arbitrary creation and that this, nonetheless, reflects natural processes independent from us. The real conflict, according to Frank, lies in the fact that—unlike Mach—Planck assumes that our present physical worldview possesses some lasting traits. The reviewer rightly holds that even Mach would consider energy conservation as real, once the quantities of all the single energies have been specified. As a general law, however, it is merely a convention whereas Planck considers it as the most important guiding principle. In a review of a new edition of Planck's classical historical and systematic study on *The Principle of Energy Conservation* (Planck 1913), Frank is even more direct. To an argument of Planck in favor of the empirical character of the general law, he retorts: "There is still a breach through which the skillfully expelled 'conventionalism' can intrude into this [general] form of the energy law and this lies in the concept of 'the same state.'" (1916, 18). The reviewer also rejects introducing metaphysical realism as a guiding postulate; "it is even less admissible to repeat now what had happened with God, freedom, and immortality in favor of atoms and electrons" (1910, 47).

Both reviews form the basis of Frank's commemorative article on "The Importance for Our Times of Ernst Mach's Philosophy of Science," which lies, according to the author, in Mach's having adapted the great project of Enlightenment to the present. As a main tenet of Enlightenment philosophy Frank considers "the protest against the misuse of merely auxiliary concepts" as an absolute foundation of physics and philosophy, because this bears the danger of conceiving any change in the foundations of physical theory as a bankruptcy of the scientific world conception as such (1917, 70/80). Of course, each epoch creates its own auxiliary concepts that may in turn transcend their own domain of definition. "The work of Mach is therefore not essentially destructive, . . . but on the contrary it is an attempt to create an unassailable position for physics" (68/75) despite the constant change of theories. Not entire parts of theories, as Planck held, will become lasting truths, but only the functional dependencies between the phenomena will remain—the direct descriptions in Mach's terminology. "The known connections among phenomena form a network; the theory seeks to pass a continuous surface through the knots and threads of the net. Naturally, the smaller the meshes, the more closely is the surface fixed by the net. Hence, as our experience progresses the surface is permitted less and less play, without ever being unequivocally determined by the net" (66/72). A later quotation from Poincaré might suggest that Frank is heading toward a sort of structural realism based on Mach's functional dependencies. Yet he does not posit any analog of Mach's principle of uniqueness in order to guarantee the integrity of the facts constituted by this network. Instead, he affirms that all our theories are empirically underdetermined and contain an irreducible conventional element.

In his article, Frank also takes a stand on atomism. Emphasizing that Mach above all strove after concepts that were applicable in all sciences, he concludes: "I will not deny that Mach allowed himself to be misled by this argument into attacking the use of atomistics in physics more sharply than can be justified. After all, the usefulness of the atomic theories in this limited realm is certainly indisputable. His followers, as is generally the case, often saw in this weakness of the master his greatest strength . . . I believe that one can completely free the nucleus of Mach's teachings from this historically and individually conditioned aversion to atomistics" (68/77).

Let me now turn to von Mises. In his 1922 article "On the Present Crises of Mechanics," he still appeared almost as an orthodox Machian. Von Mises starts out by distinguishing two strands of mechanics that differ in content, not just in method. "Bound mechanics" contains all mechanical problems in the classical physical sense that can be subsumed under a single variational principle. "Free mechanics" denotes all theories that are consistent with the wider framework of Newton's axioms and that are specified by arbitrary force functions. Also general relativity, von Mises continues, seems to be expressible as a part of free mechanics, but the force functions admitted are of a highly restricted type. "It seems to us that the mechanics of relativity is much more absolute or 'absolutistic' than the usual one, 'more bound' in our words . . . Perhaps here one finds part of the reasons which have induced *Ernst Mach* in his posthumous 'Optics' to reject relativity theory so firmly from the standpoint of experience" (1922, 26).

Whether Mach really wrote the infamous preface or not, von Mises touches upon the absolute character of the metric which for Planck marks a main virtue of relativity theory.[23] Von Mises also stubbornly sticks to Mach's classical terminology concerning atoms, while Frank already in 1917 had attempted to detach Mach's epistemology from that. "I want to clearly emphasize that I do not think of hypothetical molecules, electrons, a-particles and the like, but that I have in mind only phenomena of motion and equilibrium at sensorily perceptible masses" (28).

Indeed, von Mises studies the main question of his article—to wit, whether the framework of "free mechanics" suffices to explain all observable phenomena of motion and equilibrium—at purely classical examples. Both turbulence phenomena in liquid media and Brownian motion teach that no satisfactory result can be obtained unless one resorts to statistical methods that, in the first case, yield a phenomenological theory whose degenerate system of equations "provides the welcome opportunity to adapt the theory to observations" (27). While in this case von Mises argues almost like an engineer, in the case of Brownian motion he takes the position of a methodological purist and charges the theories of Einstein and Smoluchowski for

> the intolerable contradiction that the course of events at one time was considered as uniquely determined by physical or mechanical laws, while one subsequently believed to be able to reach results about this course from a completely different angle. This contradiction particularly comes to light in Boltzmann's version of the kinetic theory of gases (which, however, deals with the hypothetical molecules and not with observable masses, so that it can serve only as an analogy here) where one calculates first the velocity changes according to the laws of elastic scattering and then thwarts these calculations by purely statistical considerations. (29)

In kinetic gas theory this connection between the deterministic microlevel and the probabilistic macrolevel is established by "the notorious ergodic hypothesis" (29). It is instead more coherent to pursue a thoroughly probabilistic approach. As in Exner's case, the frequentist account of probability permits von Mises even to furnish the probabilistic laws with a suitable ontology—to wit, mass phenomena that become an independent object of physical theorizing in the same vein as Newtonian point particles. "Probability calculus is part of *theoretical physics* in the same way as classical mechanics or optics, it is an entirely self-contained theory of certain phenomena, the so-called mass phenomena, irrespective of whether they are of mechanical, electric or other nature" (28). This calculus maps initial probabilities, which play the combined role of the force functions and initial values of "free mechanics," into other probabilities without ever yielding deterministic results about single processes. The burden of finding and verifying these probability distributions remains with the empirical sciences. Thus, statistical physics "never directly competes with a result of mechanics or of the rest of deterministic physics" (28).

By attributing to probability calculus its own domain of facts, mass phenomena, von Mises can maintain the strict separation between deterministic physics and statistical physics. This avoids Exner's radical outlook that, presumably, all deterministic laws are really indeterministic. Von Mises's compartmentalization of physical ontology seems to be the price paid for his staunch rejection of atomism and his continuous criticism of Boltzmann. Despite his friend Frank's references to Exner, von Mises even in later years reads Boltzmann much more in the Berlin fashion than as a Vienna Indeterminist. This makes him overlook many Machian traits in Boltzmann's and Exner's radical probabilism. Nevertheless, von Mises employs all pertinent arguments, firm empiricism and the rejection of any a priori category of causality foremost. As did the empiricist Exner, he considers Brownian motion as a decisive case in point for the unavoidability of indeterminism in this factual domain. "It is entirely irrelevant whether we stick to the assumption that the orbits *would be* determined if we knew the exact initial conditions and all influences; since we have no prospect of ever achieving this knowledge, this is an assumption of which it can never be decided whether it is true or not, hence an unscientific one" (29). Von Mises quite generally believes that such unanswerable questions could be excluded in the course of scientific progress. This already hints at how centrally the issue of language will figure in his later philosophy (1939).

In his inaugural address Exner had considered the law of large numbers as the empirical meta-law basic to all science. This law indeed represents a characteristic trait of the frequentist theory of probability, and in von Mises's first rigorous formulation of the theory it becomes the first axiom about collectives. During the same year he published the mature version of his theory, von Mises took the occasion to apply his theory to refute the philosopher Karl Marbe's claim that probabilistically distributed events harbor an inherent tendency of equilibration. Von Mises commences his 1919 paper by distinguishing concept formation in philosophy, which starts out from everyday language, and in the sciences, which rest upon exact but arbitrary definitions within a partially or fully axiomatized theory. Thus "probability" in everyday parlance, our subjective degree of certainty, is sharply distinct from its mathematical homonym.

Von Mises's definition of probability as the limit of the relative frequency of a property within an infinite series presupposes that this series forms a collective. There are two conditions for a collective: (1) The relative frequencies of the occurrence of the property converge to a limit. (2) "*If out of the whole series of elements one forms a subseries without using the differences between the properties in the subseries to be selected, then within this subseries the relative frequencies for the occurrence of the properties possess the same limits as for the whole series*" (von Mises 1919, 171). This second condition is called the "irregularity of coordination" or the "impossibility of a gambling system."

According to von Mises, the first condition is based on our manifold experience that in lotteries, birth rates, etc., the relative frequencies become more and more stable as the observed series gets longer. In those days empirical investigations into such simple phenomena were still very common; Frank (1919, 704), for instance, reports in detail a statistical investigation of the number of pedestrians within a small strip of the sidewalk. Since Poisson, the empirical fact of the convergence of relative frequencies is often called the law of large numbers. But, as von Mises demonstrates in a later paper that is largely identical with a part in *Probability, Statistics, and Truth*, this terminology is ambiguous because Poisson also used it for a particular mathematical theorem that generalizes a result of

Jacob Bernoulli.[24] This states that the probability p for an experiment repeated n times to lie within $[pn-\varepsilon n, pn+\varepsilon n]$ (ε is a small positive number) converges to 1 as $n \to \infty$. Von Mises now shows that if one stays within the realm of classical a priori probabilities, this theorem is of purely algebraic nature and does not permit any conclusion about actual experiments. Adopting the frequency interpretation, however, it yields a valuable statement about "the *order* of the experimental results" or "about the course of the phenomena" that transcends the empirical law of large numbers, which had only concerned the existence of the limit (1927, 501, 502). In order to derive Poisson's theorem, one has to assume the irregularity condition. While in his 1919 paper he had argued that this axiom is hardly accessible to direct empirical observations—but derives its empirical support mainly from the manifold experimental corroborations of the multiplication rule of probability, which can be derived from it—he now provides a direct analogy from physics: "As modern physics has deduced from the failed attempts to construe for centuries a perpetuum mobile the valuable energy law or the principle of the excluded perpetuum mobile, so we have to avail ourselves of the experiences of the system players in the casinos" (von Mises 1927, 501).[25] The analogy presupposes a Machian reading of the principle of energy conservation as empirical and contradicts Frank's 1907 conventionalist account. Moreover, it runs counter to Planck's idea of the de-anthropomorphization of scientific concepts according to which both perpetua mobilia had been replaced by the respective principles of energy conservation and entropy increase.

While in the "Crises of Mechanics" von Mises (1922) considers probability theory on a par with mechanics and attributes to it its own domain of facts, in 1919 he was still holding that in the application to theoretical physics "the connection between probability theory and reality is not so immediate [as in games of chance or population statistics] because theories of physical nature lie between them" (1919, 173). Instead, he compares probability calculus to geometry because probabilities are calculated from given probabilities; but "the determination of the initial collectives of the calculation does not belong to the tasks of probability calculus in the narrow sense" (175). Similarly, the procedures of determining the base length and the angles of the triangles do not belong to geodesy itself. Pure geometry, he continues, corresponds to the games of chance.

Von Mises's analogy was not so far-fetched because, since Hilbert's sixth problem (1900, 272), geometry was considered as the pattern of axiomatizing an empirical theory, and Hilbert, already there, had explicitly suggested the axiomatization of probability theory in order to attain a rigorous formulation of the theory of gases. As for both Mach and Hilbert, geometry was undoubtedly an empirical science, it was only a short step for von Mises to subsequently consider mass phenomena as the ontology suitable for not only societal but also physical probabilities.[26] His 1919 paper still envisages probability theory predominantly from the mathematical side and leaves the specification of the probability distributions and statistical collectives to the empirical sciences. But in 1922 he would find that this question was decisive for the scientific import of probability calculus and for an ontology suitable to statistical physics—even more after quantum mechanics had won favor by the end of the decade.

When on September 16, 1929, Frank and von Mises opened the Prague biennial meeting of the German Physical Society, most physicists of the younger generation were already deeply convinced that quantum mechanics required a final farewell to well-entrenched methodological convictions. As president of the regional society and as chair of the local organizing committee,[27] Frank arranged in conjunction with the conference a meeting on "Epistemology of the Exact Sciences" co-organized by the Vienna Circle and the Berlin Society for Scientific Philosophy. At this meeting the Vienna Circle went public with its famous manifesto. A major topic of this meeting and of the first volume of *Erkenntnis* was probability theory; papers by Reichenbach, Friedrich Waismann, and Herbert Feigl were followed by an extensive discussion that clearly showed the rift between Reichenbach and von Mises.[28] Since these contributions contain various intermediate positions as to whether our degrees of certainty represent genuine probabilities, I briefly mention only one major criticism of Reichenbach's that is of relevance for the demarcation of Vienna Indeterminism because it touches upon von Mises's comparison between probability calculus and geometry. Reichenbach argues that

> in the coordination of a physical body to a mathematical theory the notion of approximation occurs which contains the concept of probability . . . : within certain limits these physical objects correspond *with high probability* to the mathematical axioms. Thus, the problem of coordination itself contains the concept of probability. It is true, in geometry one is allowed to separate the coordination problem from the mathematical theory because the coordination problem does not contain any *geometrical* concept; in probability theory however the concept constituted by this theory enters itself into the coordination problem: this is the logical particularity of the problem of probability. (*Erkenntnis* 1:275)

Von Mises, to the contrary, considers the collective as an ideal concept, and the question "whether an empirically given series represents a collective . . . does not constitute a problem within probability calculus" (272). He insists "that approximation and statistics are not to be confused with one another" (280). Strictly in line with Hilbert's program of the axiomatization of the sciences he even extends his analogy and asserts that by "modifying the axiom of disorder . . . one can obtain another probability calculus in the same sense as there is an Euclidean and a non-Euclidean geometry" (280).

The main source of disagreement seems to me von Mises's firm empiricism, according to which there cannot be any difference between the observed and the existing that would require a probabilistic theory of approximative correspondence. On the ontological side, Mach, Boltzmann, and Exner's insistence upon the individual existence of the world and, accordingly, the rejection of possible-world arguments blocks—or at least makes very unattractive to frequentists such as Frank and von Mises—Reichenbach's probabilistic reasoning concerning coordination. Against this backdrop it is not surprising that Mach's principle of unique determination resurges in Frank's broad insistence on the uniqueness of coordination. In both cases uniqueness does not contradict conventionalism because coordination is not one-to-one: also, other theories could uniquely map experiences into experiences.

According to Frank's recollections (1961, 49), his opening address to the Prague congress was "intended to give the scientists a kind of preview of our ideas and to prove that the new line of philosophy [logical empiricism] is the necessary result of the new trends in physics." If after these changes physicists still refuse to address anew those philosophical questions, they will almost certainly relapse into "school philosophy," which does not pose any limit on questions about the "real" position and momentum of quantum particles. But, "beside the relativity theory and quantum mechanics there cannot exist a philosophy that contains a fossilization of the earlier physical theories" (Frank 1929, 991/119)—to wit, the classical worldview of Newtonian mechanics that yielded both the exalted optimism expressed in Laplace's demon and the unjustified pessimism of Emil du Bois-Reymond's *Ignorabimus*. Frank in particular criticizes the naive correspondence theory of truth in which he conceives the common tenet of the various heavily conflicting directions of "school philosophy." This conception yields properties that cannot be measured or experienced but nevertheless exist. "Since, on the other hand, the doctrines of the "school philosophy" in the field of mechanical phenomena require strict determinism, one is forced to assume for the motion of the electron some mystical vital causes" (973/102). Against all this, Frank asserts:

> The task of physics is only to find symbols among which there exist rigorously valid relations, and which can be assigned uniquely to our experiences. This correspondence between experiences and symbols may be more or less detailed. If the symbols conform to the experiences in a very detailed manner we speak of causal laws; if the correspondence is of a broader sort we call the laws statistical. I do not believe that a more exact analysis will establish a definite distinction here. We know today that with the help of positions and velocities we cannot set

up any causal laws for single electrons. This does not exclude the possibility, however, that we shall perhaps some day find a set of quantities with the help of which it will be possible to describe the behavior of these particles in greater detail than by means of the wave function, the probabilities. (992/123)

Let me elaborate on two aspects of Frank's summary. First, what conclusion can we draw from the fact that the values of Planck's constant h observed in black-body radiation and in atomic spectra agree? To Planck's mind, this marks a clear sign that we have successfully moved up one step in the ladder, from the relative to the absolute, because, after we had given up simultaneously precise positions and momenta, we gained a new absolute constant. On Frank's account, this agreement of various determinations of Planck's constant does not warrant any metaphysical inference, but it permits us to uniquely define h as a symbol in physical theory. But since every measurement is the comparison of an object theory with the theory of the measurement apparatus, the experience of uniqueness in a measurement has consequences. "Every verification of a physical theory consists in the test of whether the symbols assigned by the theory to the experiences are unique" (987/111).

Second, Frank's account of experience and theory could be pictured like a commuting diagram in geometry between symbols at t_0 and t_1 and the respective experiences. This suggests that statistical features enter in two places that are, however, strongly correlated: in coordination (or assignment) and in the law. In the above quotation, Frank takes Exner's stand against Planck and asserts that the distinction between deterministic and statistical laws is at best a gradual one. This, however, does not entail Reichenbach's point of view, because Frank rejects the idea that experiences were not unique, but only an ensemble over which we put a probability distribution. Instead, Frank uses the distinction between the macro- and the microlevel to argue that collectives (or, more precisely, objects derived from them) correspond to single experiences and, accordingly, represent a possible ontology for physical laws that map probabilities into probabilities. "For these probabilities (the squares of the absolute values of the wave functions) Schrödinger in his wave mechanics, sets up rigorous causal laws. To the probabilities that occur in these laws and define the state of the system one can therefore assign definite experiences" (1929, 992/122).[29] In his 1932 book on causality, Frank expressed this importance of the coordination rule in a different wording that meanwhile had become current in quantum mechanics: "Each formulation of the law of causality even contains 'interpretation' as an essential part" (1932, 235). So, as Exner held, strict macrolaws for collectives emerge from chance. But Frank emphasizes that quantum mechanics might not be the final word because the idea of "absolute chance," or of an irrational element in science, presents itself only to the advocate of "school philosophy" and does not make any sense within the scientific world conception.

Following Frank's talk, von Mises (1930) explained to the Prague congress that this change toward a statistical ontology is rooted in a modified attitude to causality. It is rather easy, von Mises begins, to rephrase any statistical law in such a manner that it conforms to Kant's very general definitions of causality. Thus, the principle of causality is not a necessity of thought, "but *changeable*, and it will *subordinate itself to the demands of physics*" (1930, 146). For this reason, causality does not provide an adequate basis to assess the more relevant distinction between determinism and indeterminism, or between the description of nature by means of differential equations and by means of probabilities. Von Mises returns to his earlier considerations about classical mechanics and states that Laplace's demon could act properly only as long as the force laws are not too complex.

> Newtonian mechanics only provides a useful means of causal explanation of nature as long as *relatively simple force laws entail more complex motions* ... Explanation just means reducing to something more simple. (146)
>
> The deterministic approaches of classical physics can be maintained *formally*, or better: *ideally*, in the entire realm of directly observable phenomena, but in many cases ... they become *idle*, they lose *the character of a causal explanation*, they do not contribute to our knowledge, to describing or predicting the course of phenomena. ... For those who comprehend these concepts [occurring in physical theories] only as means introduced in the approaches based on differential equations in order to jointly enable an orientation in the phenomenal world, the limits of applicability and the limits of determinism itself coincide. (147)

Once again we find Mach's empiricism at the root of the indeterministic approach. More precisely than in his earlier papers, von Mises studies the difference between the macro- and the microlevel. Hydrodynamics, Brownian motion, and Boltzmann's various attempts to provide a mechanical foundation of the kinetic theory all show that "the transition between the physics of the single elementary body, atom, proton, electron, etc. to the macroscopic phenomena simply is *obtained only by statistics*" (148). If one consequently adopts a purely statistical approach the notorious ergodic hypothesis becomes a solvable mathematical problem. Although the time evolutions themselves do not form a collective, and, accordingly, the original concept of probability cannot be carried over to them, the law of large numbers (in the general sense) can be applied to the time evolutions.

The statistical approach—as any scientific investigation, to the empiricist's mind—tries "to find out observable processes which are limited in space and time and which reoccur to a reasonable approximation" (151). Thus those who equate the idea of causality to naive determinism must assume that the precision of measurements can be increased beyond any limit. But this contradicts the atomistic hypothesis that, accordingly, limits the determinist in a second respect. But von Mises maintains his earlier empiricist position that determinism and indeterminism do not contradict each other. Recalling the failure of the 1924 Bohr, Kramers, and Slater theory, which had contemplated a merely statistical validity of energy conservation in the atomic realm, he concludes: "The systematic theory, as I have pursued it for more than a decade, has never known of any failure of deterministic physics other than that it becomes *idle* in certain cases" (152). Absolute chance, once again, does not make sense to a Vienna Indeterminist. Von Mises also emphasizes the continuity between quantum mechanics and prequantum indeterminism in another respect. Born's interpretation of Schrödinger's wave function and Heisenberg's theory of measurement just teach that "also in microphysics the concrete measurement process does *not represent an elementary process, but a statistical event*" (153). So, ultimately even von Mises had made peace with atomism, and he only had to slightly modify his deeply Machian reading of Vienna Indeterminism.

Notes

I am indebted to Michael Heidelberger, Eckehart Köhler, Merrilee Salmon, and Thomas Uebel for their comments on draft versions of this paper. Thanks go to Paolo Parrini and Wes Salmon for having set up such a beautiful meeting at Florence—a city Wes greatly enjoyed. From Wes's writings and in few but extensive discussions, I was lucky to learn a lot from him, perhaps not the least of which was how to be simultaneously open minded and firm in philosophical matters. Although the present chapter does not explicitly address works of his, Wes Salmon's ideas are behind the scene, and in a certain sense they cross the front lines investigated here. While with respect to statistical causality I would put him into the Viennese tradition, drawing realistic consequences from coinciding values of a natural constant in different domains aligns him with Planck, and his late ideas basing causality in the transmission of invariant quantities might be seen as an attempt in bridge-building.

Throughout the text, quotations are based on German originals and translations are usually mine. Where translations already existed or were published during the authors' lifetime, I have tried to follow them as long as it was not necessary to restore a terminological continuity between different authors in the German originals. References contain both the page number of the German original and (after a slash) the page number of the indicated English translations.

1. Cf. Kuhn (1987). Planck wrote on philosophical matters only after 1908. His own interpretation of the 1900 discovery—to wit, that he had obtained a new constant of nature— did not significantly change thereafter. Hence, the most contested point of Kuhn's book, whether Planck actually thought about a quantum theory from the very beginning, is irrelevant for the scope of the present chapter.

2. Blackmore asserts "that some caution is in order" because Frank was almost eighty years old at the time of the interview (1995a, 133 n. 18).

3. For most remarks concerning Mach and Boltzmann made in the sequel, see this paper for further details.

4. See Fasol-Boltzmann (1990, 159) and the letter to Felix Klein given in Höflechner (1994, 2:270).

5. See the quotation at the beginning of the following section.

6. One may add here that on Exner's suggestion the Austrian Academy of Science founded one of the most important institutes for the early research in radioactivity (see Karlik and Schmid 1982). During the 1920s Prague was also a very active place in atomic and nuclear physics (see Stöltzner 1995).

7. Cf. also Exner's picture of Loschmidt in Exner (1921).

8. Cf. the bibliography in the English translation of Frank (1932, 290–96).

9. Note that before Gödel's rotating universe of 1949 no causality-violating solution of the theory was known, such that one could hope to always find such a hypersurface that admits a causal dynamics.

10. Already in 1873 Boltzmann (1905, 11) praised Maxwell's electrodynamics for providing a touchstone for further experimentation, while its competitors only reach as far as the phenomena are known.

11. Here I cannot make sense from the shorthand text other than changing "entstehen" (originate, emerge) into "stehen". See Blackmore (1905a, 169) for a translation of the entire note.

12. See Mach (1919, 381/351) and Boltzmann (1905, 154/53 n. 9), where he frankly admits his ignorance. Mach's interpretation that the maximization of probability in the second law arises from the tendency toward stability represents also a main target when Planck (1910) rightly charges Mach of conflating both laws of thermodynamics.

13. Michael Heidelberger, private communication. Indeed both contexts of Fechner's indeterminism listed in Heidelberger (1993, 338–53), "Freedom and Physiology" and "Epigenesis and Philosophy of History," were quite extraneous for Boltzmann. His preparatory notes for the philosophy lectures do not mention Fechner either; cf. Fasol-Boltzmann (1990, 13).

14. On the multifarious relation between Fechner and Herbart, which was by no means a simple opposition, see Heidelberger (1993).

15. All Vienna Indeterminists avoided the terms "Geisteswissenschaften" and later "humanities," the meanings of which do not coincide but equally suggest a fundamental methodological division of the scientific disciplines. The title of Exner's speech "Humanistik" apparently tries a way out by a very uncommon term referring to the tradition of the classics. I have recently (Stöltzner 2002b) translated this—admittedly as queer as the German original—as "humanistics"; for the origin of the word "humanities," see Hiebert (2000, 10), who places the inaugural address within a broad tendency "in which probability and chance, as generated from within the social and humanistic disciplines, came to inspire and motivate investigators in the physical science to take a deeper look (deeper than classical mechanics allows) at processes that occur in nature" (7).

16. Paul A. Hanle holds that on the basis of Exner's indeterminism "we cannot in principle apply any mechanistic program of physics to molecular processes" (1979, 256), while this was the case with Boltzmann's. But Hanle simply misunderstands Boltzmann's concept of atomism as if it relapsed into mechanical or deterministic explanation. As Exner discusses his fundamental indeterminism in connection with the example of Brownian motion, Hanle criticizes his "failure to distinguish between indeterminacy in principle and the practical inability to analyze the determinate causes in an aggregation of micro-physical events" (227). Here Hanle does not appraise the ontological consequences of the frequentist interpretation of probability (see below) and Exner's downright empiricism.

17. By the time of his death, Exner was working over a compendious study on the history of culture titled *From Chaos to the Present* that would elaborate his physicalist and indeterminist approach to the humanities. See Stöltzner (2002b) for a detailed discussion of this theory.

18. See his correspondence with Felix Klein (Höflechner 1994, 2:277–80) and Boltzmann (1898).

19. Cf. Fasol-Boltzmann (1990, 105).

20. See Höflechner (1994, 2:384) and (Blackmore 1995a, 125), where one finds translations of many letters and a large part of the lectures.

21. As stated in the introduction, the following papers have all appeared in this magazine.

22. The *Monatshefte für Mathematik und Physik* was the house organ of the Institute of Mathematics of the University of Vienna; so it is quite natural that Frank and Hahn published several "Literaturberichte" (reviews) there, but apparently Uebel (2000) was the first to notice this.

23. There exists an extensive literature as to whether the preface of the *Optics* was forged, as Gereon Wolters claims, and, more generally, Mach's late position on relativity theory; see Wolters (1987) and Holton (1989).

24. To wit, the first half of the fourth lecture in the German original (von Mises 1936, 129–43).

25. This analogy also appears in von Mises (1930, 148).

26. Interestingly, von Mises (1939) was very critical about Hilbert's axiomatic method. In a recent paper (2002a) I have argued that Hahn and Frank identified Hilbert's axiomatization program as professing the faith of a Leibnizian preestablished harmony between mathematics and the empirical sciences, a bridge that contradicted their rigid separation between analytical and empirical statements. To my mind, a similar case can be made with respect to von Mises. Recent work on Hilbert (Corry 1997; Majer 2002) shows that such an account misrepresents Hilbert's intentions, which were of a methodological kind and not tied to ontological reductionism. This is not to say that he did not overstate reductionism and universality at some places.

27. As outlined in Stöltzner (1995), the German University at Prague and Frank personally were well embedded into the physical life of the German-speaking world. Additionally, it was quite common to accept philosophically oriented talks on meetings of the society, but there were never two such talks placed so prominently in an opening session. Frank's dramaturgy ended the morning session on the "Kleine Bühne" with a talk of Arnold Sommerfeld—not quite a positivist—on "Some principal remarks concerning wave mechanics." This underlines to what extent quantum mechanics already dominated the scene.

28. Together with von Mises's congress speech, all papers were published in *Erkenntnis* 1:159–285.

29. In the German original, Frank writes "frequencies" instead of "probabilities".

References

Benndorf, H. 1927. "Zur Erinnerung an Franz Exner." *Physikalische Zeitschrift* 28: 397–409.
Bitbol, M. 1996. *Schrödinger's Philosophy of Quantum Mechanics*. Dordrecht: Kluwer.
Blackmore, J., ed. 1995a. *Ludwig Boltzmann: His Later Life and Philosophy, 1900–1906. Book One: A Documentary History*. Dordrecht: Kluwer.

—. 1995b. *Ludwig Boltzmann: His Later Life and Philosophy, 1900–1906. Book Two: The Philosopher.* Dordrecht: Kluwer.

Boltzmann, L. 1898. "Über die sogenannte H-Kurve." *Mathematische Annalen* 50:325–32.

—. 1905. *Populäre Schriften.* Leipzig: J. A. Barth. Partially translated in *Theoretical Physics and Philosophical Problems,* ed. Brian McGuinness. Dordrecht: Reidel, 1974.

Broda, E. 1955. *Ludwig Boltzmann: Mensch, Physiker, Philosoph.* Vienna: Franz Deuticke.

Cassirer, E. [1910] 1994. *Substanzbegriff und Funktionsbegriff: Untersuchungen über die Grundfragen der Erkenntniskritik.* Darmstadt: Wissenschaftliche Buchgesellschaft (originally published in Berlin).

—. [1937] 1957. *Determinismus und Indeterminismus in der modernen Physik.* In *Zur modernen Physik,* 129–376. Darmstadt: Wissenschaftliche Buchgesellschaft (originally published in Gothenburg).

Corry, L. 1997. "David Hilbert and the Axiomatization of Physics (1894–1905)." *Archive for History of Exact Sciences* 51: 83–198.

Exner, F. S. 1909. *Über Gesetze in Naturwissenschaft und Humanistik.* Vienna and Leipzig: Alfred Hölder.

—. 1917. Handwritten curriculum vitae in the archive of the Austrian Academy of Sciences.

—. 1921. "Zur Erinnerung an Josef Loschmidt." *Die Naturwissenschaften* 9:177–80.

—. 1922. *Vorlesungen über die physikalischen Grundlagen der Naturwissenschaften.* 2d ed. Leipzig and Vienna: Franz Deuticke.

—. 1923. "Vom Chaos zur Gegenwart." Unpublished typescript.

Fasol-Boltzmann, I. M., ed. 1990. *Ludwig Boltzmann: Principien der Naturfilosofi. Lectures on Natural Philosophy.* Heidelberg and New York: Springer.

Forman, P. 1971. "Weimar Culture, Causality, and Quantum Theory, 1918–1927: Adaption by German Physicists and Mathematicians to a Hostile Intellectual Environment." *Historical Studies in the Physical Sciences* 3:1–114.

Frank, P. 1906. "Über die Kriterien für die Stabilität der Bewegung eines materiellen Punktes in der Ebene und ihren Zusammenhang mit dem Prinzip der kleinsten Wirkung." Ph.D. diss. (handwritten), University of Vienna.

—. 1907. "Kausalgesetz und Erfahrung." *Ostwald's Annalen der Naturphilosophie* 6: 443–50. English translation in Frank (1961, 62–68).

—. 1910. Review of Planck's "Die Einheit des physikalischen Weltbildes." *Monatshefte für Mathematik und Physik* 21:46–47.

—. 1916. Review of the third edition of Planck's *Das Prinzip der Erhaltung der Energie. Monatshefte für Mathematik und Physik* 27: 18.

—. 1917. "Die Bedeutung der physikalischen Erkenntnistheorie Machs für das Geistesleben der Gegenwart." *Die Naturwissenschaften* 5: 65–72. English translation in Frank (1961, 69–85).

—. 1919. "Die statistische Betrachtungsweise in der Physik." *Die Naturwissenschaften* 7:701–5, 723–29.

—. 1929. "Was bedeuten die gegenwärtigen physikalischen Theorien für die allgemeine Erkenntnislehre." *Die Naturwissenschaften* 17: 971–77, 987–94. Also in *Erkenntnis* 1:126–57. English translation "Physical Theories of the Twentieth Century and School Philosophy," in Frank (1961, 96–125).

—. [1932] 1998. *The Law of Causality and Its Limits.* Dordrecht: Kluwer (translation of *Das Kausalgesetz und seine Grenzen* [Vienna: Springer]).

————. 1938. "Bemerkungen zu E. Cassirer: Determinismus und Indeterminismus in der modernen Physik." *Theoria* 4: 70–80.

————. 1961. *Modern Science and Its Philosophy.* New York: Collier Books.

Haller, R. 1986a. "Gibt es eine Österreichische Philosophie?" In *Fragen zu Wittgenstein und Aufsätze zur Österreichischen Philosophie*, 31–43. Amsterdam: Rodopi.

————. 1986b. "Der erste Wiener Kreis." In *Fragen zu Wittgenstein und Aufsätze zur Österreichischen Philosophie*, 89–107. Amsterdam: Rodopi.

Hanle, P. A. 1979. "Indeterminacy before Heisenberg: The Case of Franz Exner and Erwin Schrödinger." *Historical Studies in the Physical Sciences* 10: 225–69.

Heidelberger, M. 1993. *Die innere Seite der Natur. Gustav Theodor Fechners wissenschaftlich-philosophische Weltauffassung.* Frankfurt am Main: Vittorio Klostermann.

Heidelberger, M., and F. Stadler, eds. 2002. *History of Philosophy of Science. New Trends and Perspectives.* Dordrecht: Kluwer.

Heilbronn, J. L. 2000. *The Dilemmas of an Upright Man. Max Planck and the Fortunes of German Science.* Cambridge, Mass.: Harvard University Press.

Hiebert, E. N. 2000. "Common Frontiers of the Exact Sciences and the Humanities." *Physics in Perspective* 2: 6–29.

Hilbert, D. 1900. "Mathematische Probleme." *Nachrichten von der Königl. Gesellschaft der Wissenschaften zu Göttingen (Mathematisch-physikalische Klasse)*: 253–97. English translation in *Bulletin of the American Mathematical Society* 8 (1902): 437–79 (reprinted in the new series of the *Bulletin* 37 (2000): 407–36).

Höflechner, W., ed. 1994. *Ludwig Boltzmann: Leben und Briefe.* Graz: Akademische Druck- und Verlagsanstalt.

Holton, G. 1989. "More on Mach and Einstein." *Methodology and Science* 22:67–81

Karlik, B., and E. Schmid. 1982. *Franz Serafin Exner und sein Kreis. Ein Beitrag zur Geschichte der Physik in Österreich.* Vienna: Verlag der Österreichischen Akademie der Wissenschaften.

Klein, M. J. 1973. "The Development of Boltzmann's Statistical Ideas." *Acta Physica Austriaca* Suppl. X (*The Boltzmann Equation: Theory and Applications*, ed. E. G. D. Cohen und W. Thirring): 53–106.

Kries, J. von. 1886. *Prinzipien der Wahrscheinlichkeitsrechnung.* Freiburg i.B.: Mohr.

Kuhn, T. S. 1987. *Black-Body Theory and the Quantum Discontinuity.* With a new afterword. Chicago: Chicago University Press.

Mach, E. 1910. "Die Leitgedanken meiner naturwissenschaftlichen Erkenntnislehre und ihre Aufnahme durch die Zeitgenossen." *Scientia* 7:225–40. English translation in Blackmore (1992, 133–40).

————. 1919. *Die Principien der Wärmelehre: Historisch-kritisch entwickelt.* Leipzig: J. A. Barth. English translation *Principles of the Theory of Heat: Historically and Critically Elucidated.* Dordrecht: Reidel, 1986.

————. [1883] 1988. *Die Mechanik in ihrer Entwicklung: Historisch-kritisch dargestellt.* Ed. R. Wahsner and H. H. von Borzeszkowski. Berlin: Akademie-Verlag. Authorized English translation: *The Science of Mechanics: Account of Its Development* (La Salle, Ill.: Open Court, 1989).

————. 1991. *Erkenntnis und Irrtum: Skizzen zur Psychologie der Forschung.* Darmstadt: Wissenschaftliche Buchgesellschaft. English translation: *Knowledge and Error* Dordrecht: Reidel, 1976.

Majer, U. 2002. "Hilbert's Program to Axiomatize Physics (in Analogy to Geometry) and Its

Impact on Schlick, Carnap, and Other Members of the Vienna Circle." In Heidelberger and Stadler (2002, 213–24).

Moore, W. 1989. *Schrödinger—Life and Thought*. Cambridge: Cambridge University Press.

Planck, M. 1908. "Die Einheit des physikalischen Weltbildes." In *Wege zur physikalischen Erkenntnis*, 1–24. Leipzig: S. Hirzel, 1944.

———. 1910. "Zur Machschen Theorie der physikalischen Erkenntnis: Eine Erwiderung." *Physikalische Zeitschrift* 11:1180–90.

———. 1913. *Das Prinzip der Erhaltung der Energie*. 3d ed. Leipzig and Berlin: Teubner.

———. 1914. "Dynamische und statistische Gesetzmäßigkeit." In *Wege zur physikalischen Erkenntnis*, 54–67. Leipzig: S. Hirzel, 1944.

———. 1925. "Vom Relativen zum Absoluten." In *Wege zur physikalischen Erkenntnis*, 142–55. Leipzig: S. Hirzel, 1944. (Originally in *Die Naturwissenschaften* 13:52–59).

Reichenbach, H. 1920a. "Die physikalischen Voraussetzungen der Wahrscheinlichkeitsrechnung." *Die Naturwissenschaften* 8:46–55 and "Nachtrag," 8:349.

———. 1920b. "Philosophische Kritik der Wahrscheinlichkeitsrechnung." *Die Naturwissenschaften* 8:146–53.

———. 1921. Review of "Exner, Franz, Vorlesungen über die physikalischen Grundlagen der Naturwissenschaften." *Die Naturwissenschaften* 9:414–15.

———. 1965. *Philosophic Foundations of Quantum Mechanics*. Berkeley: University of California Press.

Schlick M. 1925. *Naturphilosophie*. In *Lehrbuch der Philosophie: Die Philosophie in ihren Einzelgebieten*, ed. M. Dessoir, 397–492. Berlin: Ullstein. English translation in *Philosophical Papers*, ed. H. Mulder and B. F. B. van de Velde-Schlick, 2:1–90. Dordrecht: Reidel.

———. 1931. "Die Kausalität in der gegenwärtigen Physik." *Die Naturwissenschaften* 19:145–62. English translation in *Philosophical Papers* 2:176–209.

Schrödinger, E. [1922] 1929. "Was ist ein Naturgesetz?" *Die Naturwissenschaften* 17:9–11. English translation by J. Murphy and W. H. Johnston in *Science and the Human Temperament*, 133–47. New York: W. W. Norton & Co.

———. 1929. "Aus der Antrittsrede des neu in die Akademie eintretenden Herrn Schrödinger." *Die Naturwissenschaften* 17:732. English translation of the unabbreviated text in the introduction to *Science and the Human Temperament*, xiii–xviii.

Sommerfeld, A. 1927. "Franz Exner." In *Jahrbuch der Bayerischen Akademie der Wissenschaften 1926*, 27. Munich: Oldenbourg.

Stadler, F. 2001. *The Vienna Circle: Studies in the Origins, Development, and Influence of Logical Empiricism*. Vienna and New York: Springer.

Stöltzner, M. 1995. "Philipp Frank and the German Physical Society." In *The Foundational Debate (Vienna Circle Institute Yearbook 3)*, ed. W. DePauli-Schimanovich, E. Köhler, and F. Stadler, 293–302. Dordrecht: Kluwer.

———. 1999. "Vienna Indeterminism: Mach, Boltzmann, Exner." *Synthese* 119: 85–111.

———. 2000. "Kausalität in den *Naturwissenschaften*: Zu einem Milieuproblem in Formans These." In *Wissensgesellschaft: Transformationen im Verhältnis von Wissenschaft und Alltag*, ed. H. Franz, W. Kogge, T. Möller, and T. Wilholt, 85–128 (accessible via http://archiv.ub.uni-bielefeld.de/wissensgesellschaft/publikationen/Michael%20Stoeltzner%20Wissensgesellschaft.pdf).

———. 2002a. "How Metaphysical Is 'Deepening the Foundations'? Hahn and Frank on Hilbert's Axiomatic Method." In Heidelberger and Stadler (2002, 245–62).

Philosophy of Physics

———. 2002b. "Franz Serafin Exner's Indeterminist Theory of Culture." *Physics in Perspective* 4:267–319.

Uebel, T. E. 2000. *Vernunftkritik und Wissenschaft: Otto Neurath und der erste Wiener Kreis im Diskurs der Moderne.* Vienna and New York: Springer.

von Mises, R. 1919. "Marbes 'Gleichförmigkeit der Welt' und die Wahrscheinlichkeitsrechnung." *Die Naturwissenschaften* 7:168–75, 186–92, 205–9.

———. 1922. "Über die gegenwärtige Krise der Mechanik." *Die Naturwissenschaften* 10:25–29.

———. 1927. "Über das Gesetz der großen Zahlen und die Häufigkeitstheorie der Wahrscheinlichkeit." *Die Naturwissenschaften* 15:497–502.

———. 1930. "Über kausale und statistische Gesetzmäßigkeit in der Physik." *Die Naturwissenschaften* 18:145–53. Also in *Erkenntnis* 1:189–210.

———. 1936. *Wahrscheinlichkeit, Statistik und Wahrheit.* Vienna: Springer. English translation *Probability, Statistics, and Truth.* London, 1939.

———. [1939] 1990. *Kleines Lehrbuch des Positivismus. Einführung in die empiristische Wissenschaftsauffassung.* Den Haag. Reprint ed. by F. Stadler. Frankfurt am Main: Suhrkamp, 1990.

Wolters, G. 1987. *Mach I, Mach II, Einstein und die Relativitätstheorie—Eine Fälschung und ihre Folgen.* Berlin: de Gruyter.

V

The Mind-Body Problem

The Mind-Body Problem in the Origin of Logical Empiricism

Herbert Feigl and Psychophysical Parallelism

MICHAEL HEIDELBERGER

Universität Tübingen

It is widely held that the current debate on the mind-body problem in analytic philosophy began during the 1950s at two distinct sources: one in America, deriving from Herbert Feigl's writings, and the other in Australia, related to writings by U. T. Place and J. J. C. Smart (Feigl [1958] 1967). Jaegwon Kim recently wrote that "it was the papers by Smart and Feigl that introduced the mind-body problem as a mainstream metaphysical Problematik of analytical philosophy, and launched the debate that has continued to this day" (Kim 1998, 1). Nonetheless, it is not at all obvious why these particular articles sparked a debate, nor why Feigl's work in particular came to play such a prominent part in it, nor how and to what extent Feigl's approach rests on the logical empiricism he endorsed.

Following the quotation cited, Kim offers an explanation backed by a widespread (mis)conception of logical empiricism. He claims that work concerning the mind-body relation done prior to Feigl and Smart dealt either with the logic of mental terms—as Wittgenstein's and Ryle's work had—and therefore missed the point, or lacked the sophistication of our modern approaches. One exception, C. D. Broad's laudable work, could not alter this, for it "unfortunately . . . failed to connect with the mind-body debate in the second half of this century, especially in its important early stages" (Kim 1998, 1). Kim seems to extend his verdict on

Ryle and Wittgenstein to include all authors writing on the mind-body problem throughout the decades preceding Feigl and Smart.

If we ask what distinguishes the young mind-body dispute of the late 1950s from older debates on the topic, we are told that Feigl and his friends and precursors of the Vienna Circle introduced new methods of logical analysis for solving or dissolving the mind-body problem. Feigl himself would probably have given that very answer. Others might say that the debate grew out of general frustration with Cartesian dualism and that it acquired its own specific character in dealing with the problems created by refuting that position (Jackson 1998, 395; see also Bieri 1997, 5–11). The story goes that reflection on the mind-body relation was horribly wrapped in Cartesian obscurity and confusion until Feigl and the Australian materialists entered the scene. Their "brain state theory," writes Kim, "helped set basic parameters and constraints for the debates that were to come—a set of broadly physicalist assumptions and aspirations that still guide and constrain our thinking today" (Kim 1998, 2).

Interpreting the difference between the older and younger mind-body debate in the United States in this way may contain a grain of truth, but from the perspective of German-speaking scholars, it is entirely wrong. Seen against the backdrop of nineteenth-century German and Austrian philosophy, Feigl's approach was neither novel nor audacious; he merely revived a tradition that had once been a mainstream topic turned unfashionable; to be exact, he modified and spelled out one specific traditional position.

It is time to readjust our appraisal of Feigl. I intend to show that Feigl's treatment of the mind-body problem upheld an active anti-Cartesian tradition; it follows a pattern in philosophy that was widespread in German-speaking countries throughout the nineteenth century and well into the twentieth, even after World War I.[1] According to Thomas Kuhn's categories, not only Feigl, but almost all the scholars who discussed the mind-body problem within the Vienna Circle and similar movements, were doing "normal science," guided by one single paradigm, so there was nothing revolutionary about Feigl's endeavors. Clearly, Feigl's solution is characterized by the particular twist he gave to the dominant paradigm—an originally neo-Kantian attitude passed on to him by his mentor Moritz Schlick.

In order to understand Feigl's project, we need to first take a look at how the mind-body relation was discussed from mid-nineteenth century onward. So sections 1 and 2 of this chapter will deal with psychophysical parallelism, its popularity during the second half of the nineteenth century and how it was subsequently treated up to the late 1920s. My aim is to capture the setting in which young Herbert Feigl must have encountered the issue when he took up his university studies in Vienna in 1922. The third section deals with the special twist that Moritz Schlick and Rudolf Carnap gave the issue in their writings from that period. Finally, in the fourth section I analyze Feigl's fundamental essay "The 'Mental' and the 'Physical'" (1958) and discuss how it compares to positions he had advocated prior writing it.

Psychophysical Parallelism Dates Back to the 1850s

In order to properly understand twentieth-century mind-body debate, we must turn our attention to the 1850s.[2] At that time German-speaking scientists were engaged in a quarrel over materialism that came to be known as the "materialism dispute." In reaction and opposition to German idealism's metaphysical and speculative post-Kantian philosophy, authors like Carl Vogt, Ludwig Büchner, and Jacob Moleschott propagated a very radical, albeit philosophically indigent, materialism identifying mental processes with physical processes. Vogt, for example, stated that any astute scientist must come to the conclusion "that all those capacities that we consider to be activities of the soul are merely functions of brain substance; or, put in simple terms, that thoughts issue from the brain just as gall is produced by the liver or urine by the kidneys" (Vogt 1847, 206). Büchner refers to Virchow, who wrote: "An expert on nature acknowledges only (material) objects and their properties; whatever goes beyond that is transcendental, and transcendence means intellectual confusion" (Virchow, as quoted by Büchner 1855, 274).

As Büchner's reference suggests, at that time the materialistically motivated movement also aimed to weaken religious dominance; it contributed notably to political liberalism prevalent in 1848 and afterward. The outcome was that hardly a natural scientist or otherwise educated person dared risk seriously adhering to Christian or Cartesian dualism in solving the mind-body enigma. In one famous case the physiologist Rudolph Wagner gave a lecture at a congress for German natural scientists and physicians that started the whole materialism dispute. He insisted that for ethical reasons science must maintain belief in a personal God and in immortality, not only when scientific proof is lacking but even when science seems to disprove it. Needless to say, materialists scoffed.

As uncouth and simple as both the materialists' and their opponents' opinions were, combined with turbulent progress in physiology and gradual alienation from idealistic philosophy of nature *(Naturphilosophie)*, the dispute over materialism aroused more scientific interest in the question of how mind is possible in a wholly physical world. Any solution offered for the mind-body puzzle that violated scientific conceptions or was reminiscent of substance dualism was strictly rejected. The latter hypothesis never had many devotees among scholars in Germany anyway.

The introduction of Darwinism pressed the relevance of finding the mind's place in physical nature and increased support for materialism. Slowly this move-ment became "monism"—led initially by Ernst Haeckel, Darwin's advocate in Germany, and then by the founder of physical chemistry, Wilhelm Ostwald. In fact, it is most likely that in their youth all our (German-speaking) heroes of logical pos-itivism devoured the monists' books, as Carnap himself admitted (Carnap 1963, 11). It is also known that Moritz Schlick played a prominent role in a monistic organi-zation.

In terms of providing a serious philosophical position, the second edition of Friedrich Albert Lange's history of materialism, published in 1873–75 (Lange [1873–75] 1974), offered the most effective and sophisticated criticism of early pop-ular materialism. While Lange defended Büchner against the claim that materialism terminates in a loss of morals, and also admitted that as a *method* materialism was not only feasible but also necessary for scientific work, he went to great lengths to analyze the difficulties, weaknesses, and contradictions inherent in materialism, if taken as a serious philosophical position. This critique in turn decidedly raised momentum for neo-Kantianism and contributed considerably to a revival of phi-losophy in general in Germany after 1860.

Lange ventured beyond offering a mere critique of materialism by affirming a doctrine that was to determine the course of the mind-body debate well into the twentieth century, namely the theory of "psychophysical parallelism." Along with many other scientists and philosophers of the period, Lange viewed psychophysi-cal parallelism as compatible with science and science's materialistic inclination, without necessitating recourse to crude materialism of the type disseminated by Büchner and others. Simultaneously, psychophysical parallelism promised to pro-vide a sophisticated program of empirical scientific research into the mind-body relation.

Psychophysical parallelism had been established and developed by the physi-cist, philosopher, and psychologist Gustav Theodor Fechner. First mention of his theory dates in the 1820s, but the contents became well known through his mature work, *Elements of Psychophysics*, in 1860.[3] This work marks a turning point in the history of experimental and quantitative psychology, and, I claim, also marks a crucial moment in the history of the mind-body debate and the history, or—if one prefers—the prehistory of scientific philosophy in general. Fechner himself did not use the term "psychophysical parallelism" to designate his standpoint. My guess is that this designation has been taken from Alexander Bain's book *Mind and Body* (1874), published in an authorized German translation fourteen years after Fech-ner's main work; but it may equally be to the merit of the unremitting psychologist Wilhelm Wundt (see Mischel 1970, 10).

A widespread misconception pervading pertinent English literature confuses this type of parallelism with forms of Cartesian doctrine of two noninteracting substances, such as doctrines of occasionalism or preestablished harmony.[4] Psychophysical parallelism means the exact opposite: It declines the Cartesian division of the world into extended substance (matter) and nonextended substance (mind).[5] While this conception is congruous with Leibniz's notion of noncausal "conformity of the soul and the organic body" (Leibniz 1714, §78), at the same time it entirely rejects the theological and metaphysical explanation that Leibniz offered for it. Psychophysical parallelism has an entirely different explanation. It propounds a kind of aspect dualism that must be strictly distinguished from what should preferably be called "Cartesian parallelism."

In fact, it is best to distinguish three different kinds of psychophysical parallelism (not only regarding Fechner, but in general), each built upon the other.[6] The *primary* form of psychophysical parallelism is an *empirical postulate*—a methodical rule for researching the mind-body relation, claiming that there is a consistent correlation between mental and physical phenomena. In the living human body, mental events or processes are regularly and lawfully accompanied by physical events and processes in the brain; or, as Fechner put it, they are "functionally dependent" on them. A particular physical state corresponds to every mental state; for every mental event there is a correlated brain state.

It is important to emphasize that functional dependence between the mental and physical says nothing about the causal nature of the relationship; causal influence is neither claimed nor denied. This type of psychophysical parallelism refrains from all causal interpretation of the mind-body relation. Fechner said that it is neutral regarding every imaginable "metaphysical closure" compatible with it. This sort of parallelism constitutes the factual foundation for any and every ambitious explanation of the relation holding between the body and the mind, whether or not such explanations ultimately turn out to be causal and interactive.

As a maxim for research, psychophysical parallelism is not only neutral in terms of any causal interpretation that may later seem necessary, it is also neutral regarding the exact nature of the correlation holding among mental and bodily phenomena—namely, whether it is one to one or one to many—and also neutral in terms of precisely which mechanism physically manifests the mental. Understood this way, psychophysical parallelism presupposes nothing about the exact nature of the mental and the physical and how these relate. It is to be taken as a metaphysics-free description of phenomena on which any advanced and scientifically acknowledged mind-body theory must be founded. In his endeavor to clearly state—without any recourse to metaphysics—just how the mental depends on the physical, Fechner came quite close to what we today call "supervenience" (see Heidelberger forthcoming, chap. 2).

Many scholars, who were skeptical in other respects, found this type of parallelism thoroughly agreeable. William James, for instance, confined himself—as he said—to "empirical parallelism," although he rejected all stronger forms of parallelism (see below). "By keeping to it," he wrote in *Principles of Psychology*, "our psychology will remain positivistic and non-metaphysical; and although this is certainly only a provisional halting-place, and things must some day be more thoroughly thought out, we shall abide there in this book" (James 1891, 182).

The *second*, stronger, form of psychophysical parallelism is a *metaphysical theory* about the relationship between the body and the mind. It adds to the primary form of parallelism a certain interpretation, or enhances it, by providing a metaphysical explanation for the alleged correlation. Fechner called his own interpretation the "identity view" of the body and soul. It provides philosophical underpinnings for functional dependence, including the following theses: (1) A living human being is not to be considered a conglomeration of two substances—a human being is one single entity; (2) the properties of this entity are considered mental when they are perceived inwardly, meaning from the perspective of the entity itself; and (3) the entity is considered something physical, when it is viewed from the outside, meaning from a perspective that is not the perspective of the entity itself. The mental and the physical are therefore two different aspects of one and the same entity. This position is also sometimes called double aspect theory, or—more correctly—the "doctrine of two perspectives."

The theory suggests that each human being has double access to, or has two perspectives of, himself: When I am aware of myself in a way in which no one else can be aware of me, I am aware of mental processes. When I am aware of myself in a way in which other persons can also perceive me (for example, when I see myself in a mirror), then I see the same processes in a physical, objective form; I appear to myself as a physical, material being.

This second form of psychophysical parallelism abandons the neutrality implied by the primary form and takes a stand on the true nature of the mind-body relation. It is defined as noncausal and therefore noninteractionist. But this noncausal interpretation is not merely postulated *per fiat,* as is the case for Cartesian parallelism; instead, it results from the definition of the psychical and the physical in terms of the perspective in which something is given. Viewing the physical as something that causes the mental, or vice versa, results from scrambling differing perspectives. Wherever causality may be found in the world, that will not be within the mind-body relation. We can demonstrate that distinguishing perspectives is nothing mysterious by considering a bent coin. It would be ridiculous to say that a dent on the head's side causes a bulge on the tail's side. While both sides of the coin are intimately connected, their joint occurrence has nothing to do with causality; they are merely two sides of one underlying substrate—two aspects that appear parallel to each other when the coin is damaged.

Obviously, the metaphysical identity view is not the only logically possible improvement on empirical parallelism. Reductive materialism and Cartesian interactionism can also be seen as being augmentative. Fechner finds all these theories that build upon empirical parallelism metaphysical, not because they lack empirical significance or because they are speculative, but because, ultimately, no finite experience can prove them. In Fechner's opinion, any meaningful interpretation of empirical parallelism must be conceivable as something that anticipates future experiences. The status of an improvement on empirical parallelism achieved by amending parallelism with metaphysical interpretation is—evaluated epistemologically—in principle no different than the status of a normal law of nature: Based on inductive generalization, both refer hypothetically to future experiences.

Some of the benefits of psychophysical parallelism presented as an empirical postulate can also be found in psychophysical parallelism presented as an identity view. First and foremost the identity view provides a nonarbitrary way of defining those claims of materialism that are reasonable, as well as imposing limits upon it. It allows for nonreductive materialism and dismisses crude reductive materialism, without reverting to antimaterialism. Materialism can thus be upheld as a research avenue while being dropped as a universal metaphysical doctrine. Another important benefit is that this stance confers upon psychology the autonomy it requires for explaining the mental and its phenomenal reality without colliding with the causality of physical reality. And, finally, the notion offers the additional benefit that it does not infringe on the autonomy of philosophy. Philosophy is not condemned to skepticism, but it can work on a reasonable explanation for the mind-body relation, one that goes beyond neutral scientific description.

It is noteworthy that Ernst Mach, one of the earliest and most enthusiastic devotees of Fechner's psychophysical parallelism, ultimately abandoned Fechner's own amendment to the empirical postulate and instead tried to do without any explanation whatsoever—not only in terms of the psychophysical relationship but also for all relations among phenomena in the whole of science (see Heidelberger forthcoming, chap. 4). Mach wanted to restrict natural science exclusively to those neutral functional dependencies among phenomena, which Fechner had only meant to be a provisional stage of psychophysics. In doing so, Mach desired to banish causal claims not only from psychophysics but also from physics and psychology. This indicates that Mach's prime motive for rejecting causal explanation and scientific realism originated in his preoccupation with mind-body theory, rather than in his work on physics or from some basal animosity to atoms. It also shows that Fechner actually (if perhaps unintentionally) headed an antimetaphysical movement skeptical of causation that Mach picked up and furthered, and that ultimately led to logical empiricism and beyond.[7]

Basically, the identity view form of psychophysical parallelism was supported by four arguments: First, none of our experiences compels us to acknowledge the reality of a thinking substance independent of a material bearer of mental properties. Second, the realm of physical phenomena and processes is causally closed; this means that each event is caused by another physical event and in physics there are no "gaps" in which the mental could "intervene" with the physical. The same holds for phenomena in the psychical realm: they, in turn, can only be explained in mental terms. Third, the law of the conservation of energy shows that physical energy can only be transformed into or derived from other physical energy. Therefore, the physical can neither affect the mental nor vice versa. And the fourth argument for the identity view—and Fechner considered this one the most important —is that it is simple and frugal. All other amendments to the basic empirical fact of the psychophysical relation are metaphysically stronger than the identity view because, for the purpose of explanation, they involve more causality than the identity version does.

In its *third* form, psychophysical parallelism is a cosmological thesis stretching beyond the range of human life. It claims that even inorganic processes have a psychical side to them. Fechner was convinced that we can, by reasoning from analogy, plausibly assume in a scientifically respectable way that there exists a psychical dimension other than the realm of inner human experience. He believed that his identity view applies not only to humans and perhaps also to animals, but also to plants, the earth, planets, and the whole universe. His argument rested on the premise that the mental must not necessarily correlate to a nervous system; it could also be realized in other material systems. This notion became popular in our times under the banner of functionalism. Fechner elaborated the idea several times beginning around 1848, but it met with resistance and ridicule—even as late as 1925, brought forth by Moritz Schlick (see Heidelberger forthcoming, chap. 3, §3.2).

The way that Fechner heightened psychophysical parallelism in this third type of parallelism (to become full-blown panpsychism) led many of his contemporaries to also dismiss his identity view—I feel, unjustifiably—as entirely speculative and inappropriate. But even extending the view into cosmology was not simple nonsense; it actually represents the origin of what later came to be called "inductive metaphysics," as opposed to dogmatic metaphysics. In order to prevent being mistaken for panpsychists and to explicitly limit psychophysical parallelism to living human beings, many authors preferred the term "psychophysiological" to "psychophysical" parallelism (see, for example, Erdmann 1907).

At first glance one would think that a cosmological type of parallelism could be interpreted as pure Spinozism. But Spinoza saw the difference between mental and material attributes as something ontological and objective, something that refers to real intrinsic properties; whereas Fechner and many of his followers viewed the distinction as epistemological, based on the perspective from which the substance is investigated. This difference between Fechner and Spinoza demonstrates that Spinozism is more strongly tied to Cartesian dualism than is Fechner's parallelism.

Another difference is how Fechner treats teleology. While Spinoza rejected all teleological assumptions, Fechner chose to do the exact opposite and used psychophysical parallelism to argue *for* a teleological view of nature. According to this interpretation, the purposiveness of the mental inner side as seen from the outer perspective is completely compatible with mechanistic, nonteleological natural necessity, including the Darwinian version of it. Leibniz would have agreed to a similar type of reconciliation, one stating that causal laws to which bodies are subjected are compatible with the laws of final causes that hold for activity of the soul (see Heidelberger forthcoming, chap. 7; Leibniz 1714, §79).

Another difference from Spinoza is Fechner's treatment of the concept of substance. Very early on, Fechner noticed that letting psychophysical parallelism depend unquestioningly on the concept of substance is very problematic, both for the identity view version and for the cosmological version. That sort of a substance would be a strange metaphysical entity, neither purely mental nor purely material and thus even worse than the notion of noumenon, a concept he opposed energetically. In order to dispense of this undesirable entity, he suggested a phenomenalistic conception of substance: a substance is nothing but a bundle of lawfully connected appearances. And since physical appearances are nomologically connected to other physical appearances as well as often connected to psychical appearances, we end up with it being entirely admissible to speak of material substances that also possess mental properties. Readers may recognize that this is precisely the source of Mach's view that substance is nothing but a "complex of (sensory) elements."

But most scholars failed to notice Fechner's early phenomenalistic modification of the identity view, so that for a long time it was regarded as faulty and obscure metaphysics. This was particularly the case when, in the last years of his life,

Fechner came to consider all appearances, whether mental or material, to be appearances in the mind of God, thus landing in "objective idealism" of a sort similar to that of Charles Sanders Peirce.

Psychophysical Parallelism from Fechner to Feigl

From a philosophical point of view the most pressing problem for psychophysical parallelism was the question of the precise role attributable to causality. It was one thing to dismiss causal interaction between the body and the soul, but to determine which role legitimately remains for the causality of nature and the causality of the mind—without forgoing psychophysical parallelism—was quite another matter. It seems that Fechner favored various options at different times: When he was young he tended to think that there are two different sorts of causality and that these are neither exclusive of one another nor intolerant of one another: physical causality in the realm of physical phenomena and psychical causality in the realm of inner experience. (Thus, on this issue Fechner also tended toward Leibniz's interpretation, which says that "bodies are active as if souls did not exist . . . and souls likewise, as if there were no bodies, and yet both move as if one had influenced the other" [Leibniz 1714, §81]). But throughout the phase represented by his major works, Fechner limited causal efficacy to that realm of reality that underlies all appearances of both types of aspect. In his old age, as mentioned previously, he adhered to objective idealism, which says that the correct place for causality is within the sphere of the mental. The distinctions separating these three views are subtle and tend to vanish if we take Fechner's phenomenalist dissolution of the concept of substance seriously. "Neutral monism" as it was later propagated by William James and taken up by Ernst Mach and Bertrand Russell became the logical outcome of these ruminations (James [1904] 1976; Mach [1886] 1922, 14, 35, 50; Russell 1921, chap. 1).

As the controversy continued, several other variations were suggested and many new distinctions were introduced, raising the complexity of the issue.[8] The discourse centered mainly around the role played by causality in the second type of psychophysical parallelism. Some scholars limited causal efficacy—and thus also reality—to the realm of the physical, ending up with "materialistic" parallelism implying epiphenomenalism for the mental. For many, this seemed to be the price to be paid for psychophysiological parallelism absent of panpsychism. The result implies a discontinuity in causality for the realm of the mental.

Others assumed the opposite—namely, that the realm of the mental is primary —and this led to causal inefficacy on the material side. Besides materialistic and idealistic parallelism, a third type was suggested, occasionally called "realistic monism" or "monistic parallelism." It held both the psychical and the physical sides for equally causally inefficacious epiphenomena of an underlying and causally efficacious actual reality. So, in the end, parallelism itself encompassed all those philosophical positions that it had originally intended to conquer! We have already noticed Ernst Mach's reaction to this confusing situation. He cut the Gordian knot by entirely forgoing causality and permitting solely functional dependence.

But the most significant form of psychophysical parallelism of interest here is not Ernst Mach's, but rather what is called "critical realism." The main advocate of this interpretation during Fechner's and Mach's time was the Austrian philosopher Alois Riehl. He wrote on the mind-body problem in 1872 and then extensively again in 1887 (Riehl [1872] 1925, 128; 1887, 176–216; 1894, 167–205).[9] Riehl defended the second type of the monistic form of psychophysical parallelism, which assumed that the reality underlying physical and psychical aspects of our perception is identical with Kant's noumenon. Since he shared this and other concepts with Kant, he is usually considered a neo-Kantian. But contrary to the other —for the most part Marburger scholars—he interpreted noumena as objective and causally effective reality independent of human consciousness, and he defended, in contrast to Kant, the notion that noumena are to a certain degree recognizable. Riehl labeled this mind-body conception "identity theory" and "realistic monism," thereby idiosyncratically constricting the traditional meanings of those terms (This contradicts Place's claim that in 1933 the American psychologist E. G. Boring may well have been the first to use the term "identity theory." See Place 1990, 22.)

Let us now briefly discuss Wilhelm Wundt's opinion. Wundt, the principal representative of "new psychology" in Germany advocated an interesting form of partial parallelism. On the one hand he was an outspoken opponent of the "theory of reciprocal effect," and therefore there are many passages in his writing where he unrestrictedly endorses psychophysical parallelism at least as a research maxim. But on the other hand he wants parallelism confined to those physical and mental events for which we have actual proof that they are parallel (Wundt 1894, 42). In his opinion parallelism applies "*only to those elementary* psychical processes (sensations), *to which alone* certain limited movements run parallel." Parallelism is "merely the parallel running of elementary physical and mental events, never parallel movements amongst complex performances on both sides" (Wundt [1863] 1990, 1:509, 513; see also 487). Wundt does not claim that there is thought without brain activity. It would seem, rather, that he struggles with a distinction familiar to present-day mind-body study—namely, the distinction between type identity and token identity. Elementary mental processes, Wundt maintained, are type identical with corresponding physiological processes (each occurrence of a specific sensation always corresponds to a specific physical event), while one and the same complex or higher mental event can, at different times, also be accompanied by differing physiological processes. While it is possible on the elementary level to know the psychical meaning of physical events, this is no longer possible on the sophisticated level (Wundt 1894, 42). It is true that the perceptual contents of our mental life are linked to physiological events, but the "mental configuration" of these contents, "being what links them according to logical and ethical standards," can no longer be bound to physiological events (Wundt 1880, 67). The outcome of Wundt's partial parallelism is a very complex theory of volition.

In spite of the complications and modifications, psychophysical parallelism was endorsed by the majority of both psychologists and physiologists well into the twentieth century. To them it seemed to be a scientific and philosophically respectable doctrine that honored the autonomy of psychology, permitting it to peacefully coexist alongside physiology and science in general. It also gave philosophers enough room to exercise sagacity in criticizing deviating positions and to discover new ways to fill in the outline provided. Toward the end of the nineteenth century idealistic notions gained more significance. Charles Sanders Peirce said: "The new invention of Monism enables a man to be perfectly materialist in substance, and as idealistic as he likes in words" (Peirce 1960, §15–126). Of course, an author not at home in the ongoing philosophical discussion had less interest in and awareness of the distinctions separating the various forms of psychophysical parallelism. But the result was a widespread diluted type of psychophysical parallelism that obscured many of the important distinctions that Fechner and subsequent thinkers had introduced into the debate.

Fechner's adherents included many prominent names. For instance, a letter from 1922 addressed to a Swiss journal and dealing with the theory of relativity shows that Albert Einstein adhered to Fechner's ideas: "To guard against the collision of the various sorts of 'realities' with which physics and psychology deal, Spinoza and Fechner invented the doctrine of psychophysical parallelism, which, to be frank, satisfies me entirely" (Bovet 1922, 902).[10] Although Niels Bohr apparently never mentioned it in print, he adopted Fechner's psychophysical parallelism as taught by his philosophy mentor, his father's close friend Harald Höffding.[11] In his successful book *Psychology*, Höffding had discussed the identity theory at length and praised Fechner as the first "to construct the theory of the relation between the mind and body based on the consequences ensuing from the axiom of the conservation of energy" (Höffding 1893, 92).[12] It is no coincidence that as late as 1932 John von Neumann phrased the distinction made in quantum mechanics between the observer and the system under observation in terms taken from the "principle of psychophysical parallelism." He made reference to Bohr, who was the first to have pointed out that "the, in formal respects unavoidable, duplicity in describing nature in quantum mechanics" is related to psychophysical parallelism as a fundamental principle of the scientific worldview (Neumann [1932] 1968, 223–24, 262 n. 207).[13]

Of course, psychophysical parallelism also had opponents. Many natural scientists were not particularly interested in philosophical controversies and did not wish to be involved in anything resembling philosophy of nature in the post-Kantian tradition, with which Fechner initially had been closely associated (see Heidelberger 1994a). Simply using the expression "identity theory" or "identity view" suggested proximity to F. W. J. Schelling's "identity philosophy," or at least it seemed so for scholars like Hermann von Helmholtz. This criticism could be easily refuted by noting that the first form of psychophysical parallelism was limited to being a mere empirical postulate. Helmholtz was not even willing to concede that and opposed even the more or less harmless form of psychophysical parallelism. He argued the incompatibility of free will and determinism. In his opinion, the realm of the mental, with all its voluntary and spontaneous activity, should not be mixed with nomological and necessary processes of nature, as psychophysical parallelism mixes them, and that even in natural science, for the time being, one must tolerate interactionism (see Heidelberger 1994b, 493; 1997, 43–47). His student Heinrich Hertz advocated a similar opinion in the introduction to *Principles of Mechanics* in 1894.

It was entirely natural for other critics to reject parallelism's pan-psychical implications, arising in its generalized third form. But it is surprising to discover that a willingness to adopt such an idiosyncratic and highly speculative consequence of Fechner's doctrine was much greater then than it would be today. For example, in a private letter, the physicist H. A. Lorentz admitted in 1915 that he believed in Fechner's psychophysical parallelism and came to the conclusion that "the mental and the material are inviolably tied to one another, they are two sides of the same thing. The material world is a way in which the *Weltgeist* appears, since the smallest particle of matter has a soul, or whatever one chooses to call it. This is all closely tied to Fechner's views . . . and I think that we have to assume something similar."[14]

By way of reviving a bit of the atmosphere in which parallelism enjoyed such widespread recognition over such a long period of time and in order to also demonstrate the significance that the law of the conservation of energy had for this controversy—even for philosophers—I would like to quote a passage taken from a letter of 1875 from the philosopher Hans Vaihinger to Friedrich Albert Lange, mentioned above, written forty years prior to Lorentz's letter. In this letter Vaihinger deals with some of the motives that made the identity theory—at least as an empirical postulate—so attractive for the scientifically enlightened public. (When he writes about "moderate occasionalism" he means something like psychophysical parallelism of the first type. This letter was intended to deny rumors —that Lange had heard—claiming that Vaihinger had converted to occasionalism.)

> I made the following distinction: a scientist has two options: either "moderate *occasionalism*," or *Spinozism* as rectified by Kant. For a scientist may only *either* say: *occasionally* certain brain activity occurs *simultaneously* with certain psychological events; but he may not permit them an inner connection at all; he makes no hypothesis about how they are connected, he states only the *fact* that the totally inclusive cycle of mechanical causality in the brain is *accompanied* in some mysterious way by psychical phenomena. If he were not satisfied by this provisionary and insufficient notion, a notion that serves only those who are anxious and overly careful, then the scientist would have to proceed towards the wider *Spinozian* hypothesis, which says that whatever appears to us to be an *external material event, is*—for us—*inwardly* a *sensation*; and I added that this latter opinion, which after Kant has been advocated by *Fechner, Zöllner, Wundt, Bain*, and others, and which is also your view, seems to me to be the only possible consequence of the *Law of the Conservation of Energy.* So you see, my dear professor, that "occasionalism" is hardly perilous. It is merely a provisionary stopover for those unwilling to address the other conclusion; and for those persons advocating an intermediate position it is at least better than either opposite position, viz., *materialism* or *spiritualism*, both of which violate the law of the conservation of energy by allowing physical things to "become" psychical and psychical things to "effect" physical things and be involved in the "mechanic series of causes" (see Lange 1968, 358).[15]

During the late 1870s the arguments and methods supporting Fechner's psychophysics were more frequently attacked by neo-Kantians (see Heidelberger forthcoming, chap. 6). But that hardly damaged the peaceful and fruitful rule of psychophysical parallelism within German-speaking culture. What abruptly ended that rule was a new chapter that the philosopher Christoph Sigwart added to the second edition of his *Logic* in 1893 in an attempt to refute psychophysical parallelism and demonstrate its intolerable conclusions (see Sigwart 1911, 2:542–600). (We must remember that opposing psychophysical parallelism and subsequently adopting a form of psychophysical interactionism did not necessarily mean that one embraced Cartesian substance dualism.) Sigwart tried to show that neither the concept of causality nor the principle of the conservation of energy encompass parallelism and that only the doctrine of reciprocal effect between the mental and the physical is philosophically permissible and valid.

As if Sigwart had opened the locks, a flood of refutations against parallelism broke through. The author of a dissertation in Vienna in 1928 noted dryly that the ensuing dispute over parallelism was surpassed only by the Trojan War (Kronstor-fer 1928, 173; see also 95). The most influential critiques of parallelism after Sigwart were written by Wilhelm Dilthey, Carl Stumpf, and Heinrich Rickert, but there were also many other authors who spoke out against parallelism, who were of lesser importance for academic philosophy or less interested in the relationship between philosophy and natural science. Most critics were bothered by parallelism's proximity to materialism, which robbed the human soul of causal efficacy and subjected the mind to determinism.

Dilthey was one of the founders of an antinaturalistic movement that came to be called *Lebensphilosophie* and sought an autonomous fundament for the sciences of the spirit—that is, the humanities. In the 1880s he had already come to the conclusion that a "correlation" between the mind and body that was understood as being noncausal was "the worst of all metaphysical hypotheses" and that the various attempts of his coevals to establish empirical psychology were nothing more than "poor metaphysics" (Dilthey 1982, 281; see also 279). In 1894 Dilthey read two essays to the Prussian Academy of the Sciences, to which he had belonged since 1887, contrasting two types of psychology: One was descriptive and analytic, a type to which he himself subscribed. This type of psychology strove to describe and analyze real psychological experience. The other was explanatory and constructive psychology, a method used in contemporary scientific psychology, going beyond actual experience and postulating an abstract psychological reality in an entirely hypothetical and deductive manner.

In this context Dilthey branded psychophysical parallelism an essential but unfounded and hypothetical construct for new psychology. He reproached its advocates for their "refined materialism" that reduces the "most powerful mental facts" to "mere accompaniments of our bodily life." Its deterministic consequences, he thought, had already begun to disintegrate "political economics, criminal law, and constitutional law" (Dilthey [1894] 1974, 142). The experimental psychologist Hermann Ebbinghaus, who adhered to psychophysical parallelism, replied skillfully to these heavy attacks and defended new psychology against Dilthey's accusations. Debates over the Dilthey-Ebbinghaus controversy lasted well into the Weimar Republic period and left traces that are more or less noticeable to this very day.[16]

Although William James, as we saw earlier, provisionally advocated "empirical parallelism," he criticized identity theory as early as 1879 in a way similar to Dilthey. Simply by calling it *automaton-theory* he made it clear that in his opinion parallelism degrades man to a mere automaton, such that "whatever mind went with it would be there only as an 'epiphenomenon', an inert spectator . . . whose opposition or whose furtherance would be alike powerless over the occurrences themselves." But mind, according to James, must have some effect on the body, otherwise it would not have been able to outlive the "struggle for survival." James came to the conclusion that "to urge the automaton theory upon us, as it is now urged, on purely *a priori* and *quasi*-metaphysical grounds, is an *unwarrantable impertinence in the present state of psychology*" (James 1891, 129, 138).[17] Thirty-two years after publication, James's critique was echoed by Edmund Husserl, who protested that parallelism treats the psyche as a "merely dependent modification of the physical, at best as a secondary parallel accompaniment" and that it interprets all beings as having "a psychophysical nature unequivocally determined by fixed laws" (Husserl [1911] 1987, 9).

Dilthey arranged Carl Stumpf's appointment to a chair for psychology in the department of philosophy in Berlin in order to prevent parallelists like Ebbinghaus, Wundt, or Benno Erdmann from attaining this position, to which they claimed rights. (What he could not prevent, however, was the call for Friedrich Paulsen for another chair in Berlin. Paulsen's interpretation of psychophysical parallelism, however, did not tend toward materialism, but in the exact opposite direction, toward panpsychism.) Although Stumpf was a leading psychologist at the time, who emphasized experimental and scientific methods, he defended interactionism vehemently in a well-received opening speech at the third congress for psychology in 1896 in Munich (Stumpf [1896] 1997, 154–82; 1910, 65–93). It is possible that he was then still influenced by his teachers Franz Brentano and Rudolph Hermann Lotze, who belonged to a small group of older scientists cum philosophers of the nineteenth century refusing to follow the fashion of psychophysical parallelism and advocating an interactionist position instead.[18] Stumpf found parallelism obscure and ambiguous, a theory that, if examined carefully, actually represents a concealed form of dualism, since it assumes two different realities. He also claimed that since according to Darwin's theory of evolution all reality must be causally efficacious—then causal efficacy must also be attributable to the mental.

The third prominent antiparallelist was Heinrich Rickert, a leading advocate of decidedly antinaturalistic neo-Kantianism in southwest Germany. In a contribution to a *Festschrift* for Sigwart in 1900, Rickert claimed with astute elegance that any concession made to parallelism that weakens the relation of psychophysical causality inevitably leads to intolerable panpsychism (Rickert 1900). He tried to show that the mind-body problem is a pseudo problem, originating in unqualified attempts to reunite the sciences of physics and psychology after such great effort had been made to distinguish and divorce them—the former as the science of quantity and the latter as the science of quality. Rickert believed in a special type of causality holding for the realm of qualities that is different from "mechanical causality" as found in the realm of quantity and therefore not subject to the law of the conservation of energy. He emphasized that both determinism and

The Mind-Body Problem

parallelism are useless categories for historians and that the discipline of history, dealing as it does with real human activity, must assume psychophysical causality and interactionism. If we start with this concept, instead of some kind of parallelism, human action appears to be an exception to determinism. In history, individual actions of civilized humans have nothing in common with mechanical causality of the kind found in natural science.

Since this kind of antinaturalistic neo-Kantianism surged mainly in Germany, throughout the 1890s resistance to psychophysical parallelism was greater in Germany than in Austria. There were frank interactionists in Austria too, however, and among their most eminent advocates were Franz Brentano, Wilhelm Jerusalem and—less obviously—Alois Höfler, although their motives for resisting parallelism differed from those of Germany's neo-Kantians.[19] The Viennese academics and Hapsburgian culture in general seem to have been more favorable for identity theory, a fact attested by the work of Ernst Mach, Friedrich Jodl, Ewald Hering (who later taught in Prague), as well as Josef Breuer and—at some distance—Sigmund Freud.[20] There were even parallelists among the followers of Brentano, the dualist. (A late member of this group was Gustav Bergmann, who subsequently was to become a member of the Vienna Circle. After Feigl helped him get a job in Iowa in 1939 he tried to enhance psychophysical parallelism by combining it with Brentano's concept of intentionality and with methodological behaviorism (compare Natsoulas 1984). Like Herbert Feigl, he was involved in bringing psychophysical parallelism to the United States, although to a lesser degree.)

As early as 1896 Friedrich Jodl had phrased identity theory in terms of a two-language theory, a development for which both Feigl and Schlick later claimed the credit (and which, incidentally, came up again with Donald Davidson's "anomalous monism"). Jodl thought that physiological and psychological descriptions for a state or process in a living organism are identical and refer to the same event, although they take on different forms (Jodl 1896, 74). We can probably trace this early two-language theory back to Hippolyte Taine, who in 1870 had already compared the relation of descriptions for the mental and the physical with the relation of two languages that mutually augment and elucidate one another (Taine 1870, pt. 1, bk. 4, chap. 2, §§4, 5). Höffding also advocated a two-language theory when writing that brain processes and processes of consciousness refer to one another "as if one and the same fact were expressed in two different languages" (Höffding 1893, 85).

It is only a small step from Riehl's and Jodl's identity-theoretic interpretations of parallelism to Moritz Schlick and his Viennese colleague Robert Reininger. In 1916 the latter dedicated an entire book to *The Psychophysical Problem* (Reininger 1916) and taught a course on Gustav Theodor Fechner during the summer term at the University of Vienna, almost two years after the young Herbert Feigl (1902–88) had come to Vienna—it was perhaps the one single course ever to be given dealing solely with Fechner (see Kronstorfer 1928, p. iv of the bibliography).

Schlick and Carnap Enter the Scene

Considering all that has been said, it is not surprising that philosophers well educated in natural science, as Moritz Schlick and Rudolf Carnap were, stood squarely within the tradition of psychophysical parallelism when it came to dealing with the mind-body problem. In *General Theory of Knowledge*, published in 1925, Schlick referred to himself explicitly as an advocate of that doctrine (Schlick [1925] 1979, 336). He stressed, however, that his own position is more radical than that of common parallelism and surpasses it in two respects: First, his position includes the "reduction of psychology to brain physiology" in the sense that there is an "identity" of reality such that "two different systems of concepts"—psychological concepts and concepts of physics—refer to it, and, second, his parallelism is not of a metaphysical but of a purely epistemological nature (351, 335, 336). In a letter to Ernst Cassirer in 1927, Schlick wrote the following: "The psychophysical parallelism in which I firmly believe is not a parallelism of two 'sides' or indeed 'ways of appearing' of what is real, rather, it is a harmless parallelism of two differently generated concepts. Many oral discussions on this point have convinced me (and others) that this way we can really get rid of the psychophysical problem once and for all.[21]

Schlick's solution for the mind-body problem reflects two components of his philosophy, which originate from diverging traditions and therefore appear at first contradictory. On the one hand we have Schlick's critical realism, which (besides naive realism) rejects positivism and every other form of "immanence philosophy" while simultaneously accepting a reality that transcends the given. On the other, we have Schlick's positivistic inheritance that views reality as consisting of qualities, whether or not they are actually given for consciousness. Schlick explicitly used a positivistic strategy adopted from Richard Avenarius claiming that the riddles of the mind-body connection (and other challenges) can be seen as an inappropriate use of "introjection" (more on this later).

Schlick's realism rests on a threefold distinction: First, there is a realm of noumena, consisting of complexes of qualities, which must not necessarily be given to any consciousness. Second, there exists reality characterized by the quantitative concepts of natural science; it results from eliminating (secondary) qualities in the course of scientific progress. And, third, there exist our intuitive perceptual

The Mind-Body Problem

events with which reality (in the second sense) is represented in consciousness—namely, experience. In understanding reality we must learn to distinguish "knowing" *(kennen)* from "recognizing" *(erkennen)*. In this sense, noumena can never be directly known—they are never given to consciousness—but we can at least partially recognize them by their causal effects and thus determine their place in the network of objective relationships by characterizing them with scientific and quantitative concepts. Recognition consists of assigning systems of symbols to circumstances. However, objectively recognized reality, which to a certain extent also encompasses what is not given, is represented through our acquaintance with our perceptual subjective experience, for only in this way can we have access to the realm of noumena. But since these qualities and complexes of qualities themselves are part of reality, they can in turn be described using scientific concepts.

Now, these distinctions imply a very specific meaning for "psychical" and "physical." For Schlick, the concept of the psychical refers to what is at all given (The Given), meaning what is identical to "content of consciousness." Reality is called physical, "inasmuch as it is described by the spatio-temporal quantitative system of concepts provided by natural science" (Schlick [1925] 1979, 324, 329). It is important to note the special role of spatial extension in this distinction. Schlick insists that space appears in two ways that must be kept strictly separate: one is perceptually imaginable space as we know it in sight, from touch, and from our kinesthetic sensations and so on; the other is physical space as conceptually construed by natural science. If we do not make this distinction and use introjection instead, meaning that we locate mental properties inside the brain or that we attribute experiential extension to the physical, we suffer from the fundamental confusion that Schlick considered to be the source of the mind-body problem. Schlick remarked with amazement that—in spite of all their differences—Avenarius and Kant nonetheless were both able to avoid that kind of unfounded introjection.

So we must now pose the crucial question regarding how Schlick intends to avoid introjection in the relationship of the psychical to the physical. This can best be done using Schlick's example of person *A* looking at a red flower and that person's brain processes ("with *A* having an open skull exposing the brain") being observed by another person, person *B* (Schlick [1925] 1979, 348–50). *B* is interested in those cerebral processes that are necessary and sufficient for *A* to see a flower. *A* is not acquainted with the noumena, the flower, at all, but she can comprehend it using scientific terms; she can employ botanical and physical terms of classification, she can describe its molecules, and so forth. Thus *A* can recognize the flower in a scientific way. But *A* also undergoes a perceptual event; she experiences the flower in a way that can be described as "red," "which in the same sense is something very real in itself, just as the transcendental object 'flower' is real." Person *B* does this: His experience shows him that the same reality that *A* describes as "red" can—using a physical term—be described as a brain process of such and such a kind. But *B* cannot only know about *A*'s brain and her mental world, he can also have a perceptual experience of *A*'s brain.

Schlick felt that this example shows well just how the mind-body problem arises by confusing characterized reality with the terms used in that very characterization, or confusing it with its perceptual representative. The first mistake is to think of the actual brain process in *A* as a physical concept of the brain process. The result is an unwarranted reduplication of reality: Instead of assuming just one reality, which is either described as physical or mental, a distinction is made between the reality of *A*'s brain and that of her consciousness. It is this sort of confusion that encourages the question of how both realities are related.

Another mistake is to confuse the concept that a physicist might have of *A*'s brain for the real intuitive experience that *B* has of that brain. All three realities, says Schlick, the flower itself and the contents of *A*'s and *B*'s consciousness, are equally valid and must each be understood for itself. And for all three of these realities it is clear from the start that they are causally tied to one another: The first causes the second and this causes the third. For what *A* and *B* know, we have a "parallelism of ways of description: both psychological and physical concepts can be applied to them" (Schlick [1925] 1979, 350).

If we compare Schlick's discussion with other, earlier attempts made at psychophysical parallelism, we inevitably fall back on Alois Riehl. He seems to be the true representative of "Spinozism rectified by Kant," mentioned by Vaihinger in the letter quoted above. As I already noted, Riehl considered noumena to be at least partially recognizable *(erkennbar)*, even if they are not direct objects of experience. In a way similar to Schlick's, he was also convinced that scientific progress consists in increasingly freeing scientific objects from secondary qualities and reducing those objects to primary qualities. He also thought that spatial extension was a property capable of being experienced, therefore deserving a status similar to that of color and taste (Riehl 1887, 38; 1894, 40). In dealing with the psychophysical problem, he stressed, just as Schlick did, the "definite identity of that process which underlies at the same time physical and psychical phenomena." He rejected "the hypothesis so popular today that physical and psychical correspond" because it involves "some hidden dualism" (Riehl 1887, 196; 1894, 185). And, finally, Riehl and Schlick entirely agree on disposing of metaphysics. Neither wants to turn the identity of the physical and mental into a "theory of the universe" but to confine it to those "points at which the objective and the subjective world actually touch"—as Riehl put it (Riehl 1887, 196; 1894, 185).

In commemorating Schlick, Feigl claimed that Schlick's solution to the mind-body problem differed entirely from all traditional metaphysical solutions. "Neither materialism," he wrote, "nor spiritualism is being maintained here, neither monism nor dualism, neither parallelism, the double-aspect theory, nor interactionism in the usual sense." But Feigl did admit that Schlick came closest to identity theory "as found, say, in the 'philosophical monism' of Alois Riehl's." Nonetheless, Feigl hurried to make clear, "even this must first be divested of its metaphysical character." He concluded that "Schlick's solution is best described, no doubt, as a two-language theory" (Feigl 1937–38, 413).[22] Still, I find Feigl's attempt to detach Schlick from the tradition of psychophysical parallelism exaggerated. The obvious conformity of Schlick's and Riehl's views demonstrates Schlick's involvement in that traditional debate. If we compare Schlick to his predecessors, we see that he did not express a more radical, effective antimetaphysical or materialistic attitude, nor did he make a progressive "semantic ascent" (Quine), meaning a linguistic analysis of the problem in the manner of twentieth-century analytic philosophy.

Let us now consider how Carnap dealt with the problem prior to turning toward physicalism. There is not much to say, since his treatment of the problem was relatively brief. But, naturally, that does not mean that he found it insignificant. In *The Logical Structure of the World* he even called the psychophysical relation the central problem of metaphysics (Carnap 1928, §22). He said that the "essence problem" of the psychophysical relation lies in the difficulty of understanding and explaining the surprising parallelism of such heterogeneous phenomena as that of the mental and the body (Carnap 1928, §166). In his opinion, only three different metaphysical solutions need be considered seriously: the hypothesis of mutual effect, the identity thesis, and the thesis of parallelism without identity. However, none of the three hypotheses is better than any of the others, for strong arguments refute all three. Carnap's most important argument was the standard objection to identity theory, stating that identity is an empty term, as long as it is not entirely clear what it means to "underlie an inner and outer side" (Carnap 1928, §22).

Carnap's radical solution to the "essence problem" of the psychophysical relationship is well known and follows the pattern provided by his general critique of metaphysics. The fact that the given can be ordered in two parallel series should be accepted without reserve. If the issue of "interpreting" or "explaining" parallelism persists, this can only be seen as an unqualified inclination toward metaphysics. Within the means provided by the system of constitution such issues can no longer be stated seriously or meaningfully. "The question of how to interpret the finding [that the Given can be ordered in parallel series] lies beyond the scope of science. This can be seen in the very fact that it cannot be expressed in constitutive terms. ... The question of interpreting that parallelness belongs to metaphysics" (Carnap 1928, §169). Science investigates functional dependencies, not "essence relationships." Carnap mentions Ernst Mach as the chief advocate of this interpretation.

Compared to the previous development discussed thus far in these investigations, neither Carnap's nor Schlick's interpretation of the mind-body problem appears to be particularly revolutionary. We can certainly say that Schlick's and Carnap's solutions convey and focus a tension that was already prevalent in Fechner's treatment of the problem and that ruled the whole ensuing discourse. It is a tension and a dilemma, if you will, between the antimetaphysical and empiristic tendency of psychophysical parallelism of the first type and the realism of the second type of parallelism's identity theory. The problem with which the Fechner tradition struggled was the following: If we want solely to deal with the facts, then parallelism can only be understood as research heuristics. But this would mean that we dismiss any explanation for a very strange regularity that obviously seems to suggest some underlying causal mechanism that would make it understandable. But accepting the simplest imaginable explanation for this regularity—namely, the identity theory—means to transcend the direct realm of facts and invite panpsychism or similar metaphysics.

Confronted with this dilemma, Schlick—properly following the tradition set forth by Riehl and other "critical realists"—opted for realism and tried to modify the concept of the physical object as much as necessary in order to make undesirable metaphysical consequences vanish. The best elucidation of the parallelism between mind and body is seen as the two-language theory. The unobservable realm underlying the different conceptual constructs is—as it were—tamed by realism. Carnap, though, was more willing to follow up Fechner's original solution, which was later radicalized by Ernst Mach, William James, and Bertrand Russell, thus dealing with the other side of the dilemma. Like many others in logical empiricism, he strove to demonstrate that natural science can describe the world without losing something in the process and without recourse to explanations that transcend the given. If we shun every reference to realms and objects that are inaccessible to experience and restrict natural science to the description of what is observable, we can retain meaningful science without unwelcome metaphysics.

Psychophysical Parallelism in the United States: Herbert Feigl

For a long time Herbert Feigl was a devotee of Schlick's critical realism and its related realistic solution for the mind-body enigma. Since he set up both the subject and the name index for the second edition of the *General Theory of Knowledge* and helped Schlick with corrections, he must have been well acquainted with his teacher's views (see Schlick [1925] 1979, 11). Thus it is not surprising that—as he himself reports—he "opposed Carnap's phenomenalism from the start in Vienna" and that he was involved "in a standing dispute with Carnap ... over the 'realism', 'subjective idealism' or the 'phenomenalism' issues" (Feigl 1964, 231.) His Viennese friends must have made life quite difficult for him. He later recalled being scorned for advocating realism: "You metaphysician! they told me in Vienna. Imagine! This was the worst thing that could happen to a philosopher at that time" (Feigl 1964, 243). To his "great chagrin" he watched his teacher and friend finally give in and under Wittgenstein's influence become "a positivist in terms of the problem of reality." That was too much for young Herbert, for—as he later recalled—he himself was "temporarily overwhelmed" by Carnap, a thinker "tremendously resourceful in discussion . . . who has thought through everything a hundred times more fully than is evident from his publications" (Feigl 1964, 242).

In the period between the late 1920s and 1958, when Feigl's essay was published, the Vienna Circle turned to embrace physicalism. We are uncertain whether Feigl went along with this change in every respect. In retrospect he described the first phase of physicalism as an error: In his opinion it soon became obvious that mental states could be neither identified with overt behavior nor reduced to neurophysiological states. But in Carnap's retreat from the principle of verification and his concern after 1956 with bilateral reduction laws as a method for introducing mental terms Feigl saw a revision of the original view of the *Logical Structure of the World* leading back to critical realism and a two-language theory for mind and body. He considered two factors responsible for reinstating "clarified critical realism": One was Tarski's "pure semantics," which Carnap further developed, and the other, the "pure pragmatics" demonstrated in Wilfried Sellars's work. These developments encouraged Feigl to return to his own previous interpretations and those of his mentor, Schlick.

In his first publication on the mind-body problem, in 1934, after the general turn to physicalism, Feigl held the relationship between the physical and the mental for a *logical* identity between two descriptions of the given, a description in psychological and a description in physical vernacular (Feigl 1934, 436).[23] It was a more radical version of Schlick's notion interpreting the identity spoken of in identity theory as a relation holding between realities.[24] This focus can be understood as a concession made to logical behaviorism, which at that time was quite popular within the Vienna Circle. But by 1958 at the latest, Feigl returned to Schlick's views and those he himself had held prior to 1934. (Incidentally, in the foreword to the essay he remarks that he was initially introduced to "philosophical monism" by reading Alois Riehl and that he found "essentially the same position again in Moritz Schlick" [Feigl [1958] 1967, v; see also 79 n]). Growing criticism regarding behaviorism at that time might have encouraged Feigl ([1958] 1967, 62, 109). The most significant change was that now Feigl no longer saw the identity of the mental and the physical as a necessary, but as an empirical identity.

In several passages of the essay Feigl asks what distinguishes his own identity theory from parallelism and concedes that the distinction is not of an empirical nature. "The step from parallelism to the identity view is essentially a question of philosophical interpretation." Thus, deciding for one of the two positions is similar to making a choice between phenomenalism and realism or between the regularity theory and the modality theory of causality—things that cannot be decided empirically. The principle of frugality, or "inductive simplicity," demands that we forgo the doubling of realities, such as parallelism assumes, in favor of identity theory (Feigl [1958] 1967, 94).[25] The advantage of this theory is that it "removes the duality of two sets of correlated events and replaces it with . . . two ways of knowing the *same* event— one direct, the other indirect." If we admit a synthetic element in the psychophysical relation, then "there is something which purely physical theory does not and cannot account for" (Feigl [1958] 1967, 106, 109). The translation of the mental into terms of the physical still assumed in 1934 has totally disappeared here. In lieu of a two-language theory we now have Feigl's "double-knowledge, double-designation view" (Feigl [1958] 1967, 138), which is nothing other than a revival of the second form of psychophysical parallelism.

The preceding discussion shows once again that even twentieth-century logical empiricism is more firmly rooted in tradition than has often been assumed and more so than the logical empiricists themselves conceded. And we have also seen which wealth of options, distinctions, and arguments lay and still lie waiting in the tradition of identity theory, which unfortunately have frequently been lost in contemporary discourse. In the present tiny renaissance of the identity theory it is time to recall forgotten history. In this exposition I limited myself to the tradition leading up to Herbert Feigl and neglected the Australian version of identity theory (for the latter, see Place 1988, 1990). An excellent comparison of both schools of thought, which elaborates the differences separating them, is available, however, and it confirms my view of Feigl's proximity to parallelism (Stubenberg 1997; see also Sturma 1998).

But the story told here also shows (and I intend to suggest it to my readers repeatedly) that the mind-body problem is not simply one problem amongst many, nor one that logical empiricism could have skipped elaborating. The twists and turns in discussing this very problem has formed logical empiricism in essential ways. Do not forget that the prehistory of logical empiricism roots not only in logic and physics but also particularly in psychology and physiology. At times, these two latter disciplines were of greater interest in the nineteenth century than the former two. I mentioned earlier that a tendency of many empiricists in the tradition of logical positivism to replace causality with functional dependency can be explained by reviewing the discussion of the mind-body problem. Their antimetaphysical inclination also originated not (only) in physics, but was due to the efforts made to find a serviceable scientific basis for emerging empirical psychology. Even the preference for "description" over "explanation" resulted from psychophysics' neutrality regarding causality. Something similar can be said for the origins of phenomenalistic critique of the concept of substance in the discussion on parallelism. And the early logical empiricists' antirealism, as expressed at the dawn of the twentieth century, has roots not only in physics, but also in the endeavors of psychophysical parallelism to prevent from the start any sort of conclusion about a panpsychical side of the world. It seems an irony of history that precisely Feigl's theory, a realistic variation of parallelism, "survived." In light of what I have reported it seems even thoroughly possible that Carnap's philosophical neutrality in epistemological matters, as he elaborated it in the system of the *Logical Structure of the World*, and that pervades all of his work, in the end is obligated to Fechner's demand for neutrality in psychophysical parallelism as a maxim of research and to the two-language theory related to it.

Translated by Cynthia Klohr

Notes

I wrote and read an earlier version of this essay in 1998–99 as fellow at the Center for Philosophy of Science at the University of Pittsburgh. I am grateful for the support I experienced there. Further versions were read at universities in Hannover, Mainz, Florence, and Tunis, as well as at the Humboldt University in Berlin and in a work group on "Psychological Thinking and Psychological Praxis" at the Berlin-Brandenburg Academy of the Sciences. I would like to thank all my audiences for helpful discussions and especially Cynthia Klohr for translating this paper. The German version appeared as "Wie das Leib-Seele Problem in den Logischen Empirismus kam," in *Phänomenales Bewusstsein—Rückkehr der Identitätstheorie?* edited by M. Pauen and A. Stephan, 40–72 (Paderborn: Mentis 2002).

1. To my knowledge, the only author writing in English aware of identity theory's long anti-Cartesian prehistory is Milič Čapek. Cf. Čapek (1969), which provides valuable information.

2. More recent portrayals of nineteenth-century materialism are given in Wittkau-Horgby (1998) and Heidelberger (1998).

3. The most important source for Fechner's psychophysical parallelism is the foreword and introduction to *Elements of Psychophysics*. See Fechner (1860, 1:vii–xiii, 1–20).

4. This is probably partly due to Bertrand Russell's incorrect portrayal of psychophysical parallelism in *Analysis of Mind* (1921). Therein he claimed that modern psychophysical parallelism is hardly distinguishable from Cartesian theory (35). For later authors expressing similar views, see Armstrong (1993, 8) and loci mentioned in that book's index; Heil (1998, 27); and (lacking all knowledge of the German tradition) Bieri (1997, 7). See also note 18 below.

5. The situation is, in fact, much more complicated. Occasionalism is logically *compatible* with (but not identical to) the first version of psychophysical parallelism, but not

with the second version, which is, to a greater degree, of philosophical interest. (See the following discussion.)

6. A general depiction is given in Heidelberger (forthcoming, chap. 2).

7. For a detailed comparison of Fechner's and Mach's mind-body theories, see Heidelberger (2000a; 2000b; forthcoming, chap. 4, §4.4).

8. A contemporary's survey of the issues is given in Busse (1913), with an excellent appendix.

9. Cf. also Riehl (1921, 112–46).

10. Thomas Ryckman kindly drew my attention to this source.

11. See Favrholdt introduction to Bohr (1999, xliii, 7). Favrholdt is wrong in claiming that Höffding's psychophysical parallelism is indebted to Leibniz (xliv).

12. See also Höffding (1891; 1903). On page 30 of this latter work, Höffding dismisses the term "parallelism" as actually being inappropriate and ambiguous and prefers "identity theory" instead.

13. Neumann ([1932] 1968, 262 n. 208) also feels obligated to discuss this topic with Leo Szilárd.

14. H. A. Lorentz to theologian H. Y. Groenewegen, April 5, 1915. Inv.-No. 27, Archive H. A. Lorentz, Rijks-Archive North Holland, Haarlem, The Netherlands. Private information of Dr. L. T. G. Theunissen, Institute for the History of Science, Utrecht University. I would like to thank Bert Theunissen for permission to quote from Lorentz's letter, which he discovered while working on a project with his colleague Henk Klomp.

15. Vaihinger emphasized using the term "moderate occasionalism" in Lotze's interpretation of it.

16. Recent research on this controversy can be found in Kusch (1995, 162–69) and Gerhardt, Mehring, Rindert (1999, 162–68).

17. Chapter 5 in James's *Principles,* where these quotations are to be found, is titled "The Automaton-Theory" and appeared in almost the same wording in 1879 in *Mind.* A discussion of James's arguments is given in Čapek 1954. It is a serious mistake and highly misleading to characterize James's concept of automaton-theory as "logically identical to the sort of parallelism familiar from the writings of Leibniz and Malebranche," as Owen Flanagan puts it in his article (1997, 32; cf. also note 5 above).

18. See Pester (1997, chaps. 3.3 and 3.5) on Hermann Lotze's sophisticated methodical occasionalism. The Cambridge philosopher James Ward (1902, 66–69), who had studied with Lotze for some time, gives a remarkable defense of Lotze's view, containing one of the very rare presentations of psychophysical parallelism written in English.

19. See Höfler (1897, 57–63). Höfler mentions a discussion he had with Boltzmann on this topic (58 n).

20. In dealing with the mind-body problem, Jodl appears to have been influenced by Alois Riehl. On the relationships of Breuer and Freud to Fechner, see Heidelberger (forthcoming, chap. 1, sect. 1.11; chap. 7, sect. 7.3); for further information on the situation in Austria, see ditto chap. 6; on Mach's relationship to Fechner, see Heidelberger (forthcoming, chap. 4).

21. Schlick to Ernst Cassirer, Vienna, March 30, 1927, Moritz Schlick, Papers, Inv. No. 94. Used with kind permission from the Vienna Circle Stichting, Amsterdam, and the Philosophical Archive at the University of Constance, which let me see copies of the Schlick literary collection.

22. Feigl also wrote that Schlick, "perhaps with greater clarity than all other monistic

critical realists of the times[,] elaborated a physicalistic identity theory" worth being rediscovered "in modern semantic terms" (Feigl 1963, 261, 254). Once again, Riehl is explicitly
mentioned as the scholar to whose views in these matters Schlick comes the closest. See also
Feigl ([1950] 1953, 614), where Schlick's view is called "double-language theory." In this article Feigl repeatedly characterizes his own theory as "identity or double-language view of
mind and body" (617, 624, 626).

23. Cf. Feigl's own elucidation of his standpoint in 1934 in Feigl ([1958] 1967, 23). In
Feigl ([1950] 1953) Feigl reports Felix Kaufmann's insistence that strict identity must be understood logically. Cf. also Sturma (1998).

24. Cf., for instance, Schlick ([1950] [1925] 1979, 347).

25. Cf. also 95–97, 104. See also Feigl (1953, 616–17), where he refers to the "principle of
parsimony."

References

Armstrong, D. M. 1993. *A Materialist Theory of the Mind*. 2d ed. London: Routledge.
Bain, A. 1874. *Geist und Körper. Die Theorien über ihre gegenseitigen Beziehungen*, Leipzig:
    Brockhaus. English original: *Mind and Body: The Theories of Their Relation*. New York:
    Appleton 1873.
Bieri, P., ed. 1997. *Analytische Philosophie des Geistes*. 2d ed. Königstein: Hain.
Bohr, N., ed. 1999. *Complementarity Beyond Physics (1928–1962)*. Vol. 10, *Niels Bohr Col-
    lected Works*, ed. D. Favrholdt, gen. ed. Finn Aaserud. Amsterdam: Elsevier.
Bovet, E. 1922. "Die Physiker Einstein und Weyl. Antworten auf eine metaphysische Frage."
    Wissen und Leben 15, no. 19:901–6.
Büchner, L. 1855. *Kraft und Stoff*. 5th ed. Frankfurt am Main: Meidinger.
Busse, L. 1913. *Geist und Körper, Seele und Leib*. 2d ed. Leipzig: Meiner.
Čapek, M. 1954. "James's Early Criticism of the Automaton Theory." *Journal of the History
    of Ideas* 15:260–79.
———. 1969. "The Main Difficulties of the Identity Theory." *Scientia* 104:388–404.
Carnap, R. 1928. *Der logische Aufbau der Welt*. Berlin-Schlachtensee: Weltkreis-Verlag.
———. 1963. "Intellectual Autobiography." In *The Philosophy of Rudolf Carnap*, ed. Paul
    Arthur Schilpp, 1–84. La Salle, Ill.: Open Court.
Dilthey, W. [1894] 1974. "Ideen über eine beschreibende und zergliedernde Psychologie." In
    Wilhelm Dilthey, *Gesammelte Schriften*, vol. 5, ed. Georg Misch, 136–240. Stuttgart:
    Teubner.
———. 1982. "Ausarbeitungen und Entwürfe zum zweiten Band der Einleitung in die Geis-
    teswissenschaften. Viertes bis Sechstes Buch (ca. 1880–1890)." In Wilhelm Dilthey,
    Gesammelte Schriften, vol. 19, *Grundlagen der Wissenschaften vom Menschen, der
    Gesellschaft und der Geschichte*, ed. Helmut Johach and Frithjof Rodi, 78–295. Göttingen:
    Vandenhoeck.
Erdmann, B. 1907. *Wissenschaftliche Hypothesen über Seele und Leib*. Köln: Dumont-
    Schauberg.
Fechner, G. T. 1860. *Elemente der Psychophysik*. 2 vols. Leipzig: Breitkopf & Härtel.
Feigl, H. 1934. "The Logical Analysis of the Psychophysical Problem: A Contribution of the
    New Positivism." *Philosophy of Science* 1:420–45.
———. 1937–38. "Moritz Schlick." *Erkenntnis* 7:393–419. Reprinted as "Moritz Schlick, a

Memoir," in *Philosophical Papers*, vol. 1, ed. H. L. Mulder and B. F. B. van de Velde-Schlick (Dordrecht: Reidel, 1979), xv–xviii, and in *Rationality and Science: A Memorial Volume for Moritz Schlick in Celebration of the Centennial of His Birth*, ed. Eugene T. Gadol (Vienna: Springer 1982), 55–82.

———. [1950] 1953. "The Mind-Body Problem in the Development of Logical Empiricism." In *Readings in the Philosophy of Science*, ed. H. Feigl and M. Brodbeck, 612–26. New York: Appleton-Century-Crofts. First published in *Revue Internationale de Philosophie* 14:64–83.

———. [1958] 1967. *The "Mental" and the "Physical": The Essay and a Postscript*. Minneapolis: University of Minnesota Press. The essay appeared first in *Concepts, Theories, and the Mind-Body Problem*, vol. 2, *Minnesota Studies in the Philosophy of Science*, ed. Herbert Feigl, Michael Scriven, and Grover Maxwell (Minneapolis: University of Minnesota Press, 1958), 370–497.

———. 1963. "Physicalism, Unity of Science and the Foundations of Psychology." In *The Philosophy of Rudolf Carnap*, ed. P. Arthur Schilpp, 227–67. La Salle Ill.: Open Court.

———. 1964. "Logical Positivism after Thirty-Five Years." In *Philosophy Today* 8 (4):228–45.

Flanagan, O. 1997. "Consciousness as a Pragmatist Views It." In *The Cambridge Companion to William James*, ed. Ruth Anna Putnam, 25–48. Cambridge: Cambridge University Press.

Gerhardt, V., R. Mehring, and J. Rindert. 1999. *Berliner Geist: Eine Geschichte der Berliner Universitätsphilosophie bis 1946*. Berlin: Akademie Verlag.

Heidelberger, M. 1993. *Die innere Seite der Natur: Gustav Theodor Fechners wissenschaftlich-philosophische Weltauffassung*. Frankfurt am Main: Klostermann. English translation in preparation with University of Pittsburgh Press (see Heidelberger forthcoming).

———. 1994a. "Fechners Verhältnis zur Naturphilosophie Schellings." In *Schelling und die Selbstorganisation: Neue Forschungsperspektiven* (= *Selbstorganisation. Jahrbuch für Komplexität in den Natur-, Sozial- und Geisteswissenschaften* 5), ed. Marie-Luise Heuser-Keßler and Wilhelm G. Jacobs, 201–218. Berlin: Duncker & Humblot.

———. 1994b. "Force, Law, and Experiment: The Evolution of Helmholtz's Philosophy of Science." In *Hermann von Helmholtz and the Foundations of Nineteenth-Century Science*, ed. David Cahan, 461–97. Berkeley: University of California Press.

———. 1997. "Beziehungen zwischen Sinnesphysiologie und Philosophie im 19. Jahrhundert." In *Philosophie und Wissenschaften: Formen und Prozesse ihrer Interaktion*, ed. Hans Jörg Sandkühler, 37–58. Frankfurt am Main: Peter Lang.

———. 1998. "Büchner, Friedrich Karl Christian Ludwig (Louis) (1824–1899)." In *Routledge Encyclopedia of Philosophy*, vol. 2, ed. Edward Craig, 48–51. London: Routledge.

———. 2000a. "Der Psychophysische Parallelismus: Von Fechner und Mach zu Davidson und wieder zurück." In *Elemente moderner Wissenschaftstheorie: Zur Interaktion von Philosophie, Geschichte und Theorie der Wissenschaften*, ed. Friedrich Stadler, 91–104. New York: Springer.

———. 2000b. "Fechner und Mach zum Leib-Seele Problem." In *Materialismus und Spiritualismus: Philosophie und Wissenschaften nach 1848*, ed. Andreas Arndt und Walter Jaeschke, 53–67. Hamburg: Meiner.

———. Forthcoming. "Nature from the Inside" (working title), English translation of Heidelberger 1993 in preparation with University of Pittsburgh Press.

J. 1998. *Philosophy of Mind: A Contemporary Introduction*. London: Routledge.

The Mind-Body Problem

Höffding, H. 1891. "Psychische und physische Activität." *Vierteljahrsschrift für wissenschaftliche Philosophie* 15:233–50.

———. 1893. *Psychologie in Umrissen auf Grundlage der Erfahrung.* 2d German ed. Leipzig: Reisland.

———. 1903. *Philosophische Probleme.* Leipzig: Reisland.

Höfler, A. 1897. *Psychologie.* Prague: Tempsky.

Husserl, E. [1911] 1987. "Philosophie als strenge Wissenschaft." In E. Husserl, *Aufsätze und Vorträge (1911–1921)* (= Husserliana 25), 3–62. Dordrecht: Nijhoff.

Jackson, F. 1998. "Identity Theory of Mind." In *Routledge Encyclopedia of Philosophy,* vol. 6, ed. Edward Craig, 395–399. London: Routledge.

James, W. 1891. *The Principles of Psychology.* Vol. 1. London: Macmillan.

———. [1904] 1976. "Does 'Consciousness' Exist? " In *Essays in Radical Empiricism,* vol. 3, *Works of William James,* 3–19. Cambridge, Mass.: Harvard University Press.

Jodl, F. 1896. *Lehrbuch der Psychologie.* Stuttgart: Cotta.

Kim, J. 1998. *Mind in a Physical World: An Essay in the Mind-Body Problem and Mental Causation.* Cambridge, Mass.: MIT Press and Bradford Books.

Kronstorfer, R. 1928. *Drei typische Formen des Psychophysischen Parallelismus (Spinoza, Fechner, Mach).* Ph.D. diss., University of Vienna.

Kusch, M. 1995. *Psychologism: A Case Study in the Sociology of Knowledge.* London: Routledge.

Lange, F. A. [1873–75] 1974. *Geschichte des Materialismus und Kritik seiner Bedeutung in der Gegenwart.* 2d ed. 2 vols. Frankfurt am Main: Suhrkamp. Translated into English as *The History of Materialism and Criticism of Its Present Importance,* 3 vols., trans. E. C. Thomas (London: English and Foreign Philosophical Library, 1877–79).

———. 1968. *Über Politik und Philosophie. Briefe und Leitartikel 1862 bis 1875.* Ed. Georg Eckert. Duisburg: Walter Braun.

Leibniz, G. W. 1714. *Monadologie.*

Mach, E. [1886] 1922. *Die Analyse der Empfindungen und das Verhältnis des Physischen zum Psychischen.* 9th ed. Jena: Gustav Fischer.

Mischel, T. 1970. "Wundt and the Conceptual Foundations of Psychology." *Philosophy and Phenomenological Research* 31 (1): 1–26.

Natsoulas, T. 1984. "Gustav Bergmann's Psychophysical Parallelism." *Behaviorism* 12:41–69.

Neumann, J. von. [1932] 1968. *Mathematische Grundlagen der Quantenmechanik.* Berlin: Springer.

Peirce, C. S. 1960. "Lowell Lectures of 1903." In *Collected Papers,* ed. C. Hartshorne and P. Weiss, vol. 1, 2d ed. Cambridge, Mass.: Harvard University Press.

Pester, R. 1997. *Hermann Lotze: Wege seines Denkens und Forschens.* Würzburg: Königshausen & Neumann.

Place, U. T. 1988. "Thirty Years On—Is Consciousness Still a Brain Process?" *Australasian Journal of Philosophy* 66:208–19.

———. 1990. "E. G. Boring and the Mind-Brain Identity Theory." *British Psychological Society, History and Philosophy of Science Newsletter* 11:20–31.

Reininger, R. 1916. *Das psycho-physische Problem. Eine erkenntnistheoretische Untersuchung zur Unterscheidung des Physischen und Psychischen überhaupt.* Leipzig: Braumüller.

Rickert, H. 1900. "Psychophysische Causalität und psychophysischer Parallelismus." In *Philosophische Abhandlungen: Christoph Sigwart zu seinem siebzigsten Geburtstage 28. März 1900 gewidmet* (no editor), 59–87. Tübingen: Mohr.

Riehl, A. [1872] 1925. *Über Begriff und Form der Philosophie.* Leipzig: Haacke 1872. Reprinted in *Philosophische Studien aus vier Jahrzehnten*, by Alois Riehl, 91–174, 332–39. Leipzig: Quelle & Meyer.

———. 1887. *Zur Wissenschaftstheorie und Metaphysik.* Vol. 2, pt. 2 of *Der philosophische Kriticismus und seine Bedeutung für die positive Wissenschaft*, 2 vols. Leipzig: Engelmann, 1876–87.

———. 1894. *The Principles of the Critical Philosophy: Introduction to the Theory of Science and Metaphysics.* Trans. Arthur Fairbanks. London: Kegan, Paul, Trench, Trübner & Co. This is the translation of Riehl 1887. Chap. 2 of this volume is titled "On the Relation of Psychical Phenomena to Material Processes," 167–205. (In Riehl 1887, the German original reads "Ueber das Verhältnis der psychischen Erscheinungen zu den materiellen Vorgängen," 176–216.)

———. 1921. *Zur Einführung in die Philosophie der Gegenwart.* 6th ed. Leipzig: Teubner. Fifth Lecture: "Der naturwissenschaftliche und der philosophische Monismus" (Scientific and Philosophical Monism), 112–46.

Russell, B. 1921. *The Analysis of Mind.* London: George Allan & Unwin.

Schlick, M. [1925] 1979. *Allgemeine Erkenntnislehre.* 2d ed. Frankfurt am Main: Suhrkamp. Translated as *General Theory of Knowledge* by A. Blumberg. Vienna: Springer, 1974.

Sigwart, C. 1911. *Logik.* 4th ed. 2 vols. Arranged by Heinrich Maier. Tübingen: Mohr.

Stubenberg, L. 1997. "Austria vs. Australia: Two Versions of Identity Theory." In *Austrian Philosophy Past and Present*, ed. K. Lehrer and J. C. Marek. Dordrecht: Kluwer.

Stumpf, C. [1896] 1997. "Leib und Seele." In *Carl Stumpf—Schriften zur Psychologie*, ed. Helga Sprung, 154–82. Frankfurt am Main: Lang. Originally published in *Philosophische Reden und Vorträge*, by Carl Stumpf, 65–93. Leipzig: Barth, 1910.

Sturma, D. 1998. "Reductionism in Exile? Herbert Feigl's Identity Theory and the Mind-Body Problem." *Grazer Philosophische Studien* 54:71–87.

Taine, H. 1870. *De l'intelligence.* 2 vols. Paris: Hachette. (I consulted the German translation L. Siegfried, *Der Verstand*, 2 vols., 3d ed. [Bonn: Emil Strauss, 1880].)

Vogt, C. 1847. *Physiologische Briefe für Gebildete aller Stände.* Stuttgart: Cotta.

Ward, J. 1902. "Psychology." In *The Encyclopaedia Britannica*, 54–70. Edinburgh: Adam and Charles Black. See esp. the part entitled "Relation of Body and Mind: Psychophysical Parallelism," 66–69.

Wittkau-Horgby, A. 1998. *Materialismus. Entstehung und Wirkung in den Wissenschaften des 19. Jahrhunderts.* Göttingen: Vandenhoeck.

Wundt, W. [1863] 1990. *Vorlesungen über die Menschen-und Thierseele.* 2 vols. Ed. Wolfgang Nitsche. Heidelberg: Springer.

———. 1880. "Gehirn und Seele." *Deutsche Rundschau* 25:47–72.

———. 1894. "Über psychische Causalität und das Princip des psychophysischen Parallelismus." *Philosophische Studien* 10:1–124.

Logical Positivism and
the Mind-Body Problem

JAEGWON KIM
Brown University

When the mind-body problem is mentioned along with logical positivism, most philosophers immediately think of logical behaviorism. They think of the verifiability criterion of meaning, the verificationist reduction of psychological expressions to observational/behavioral terms, and the claim that introspectionist psychology, lacking cognitive meaning as it does, is not a genuine science.

None of these associations is entirely wrong. Stereotypes often are not without a foundation. However, the facts are usually quite a bit more complicated and murkier, and often more interesting, than the stereotypes. In this regard, logical positivism and the mind-body problem is no exception. In this presentation I will look at how three important positivists, Schlick, Carnap, and Hempel, dealt with the mind-body problem, paying special attention to the influence of their general positivist commitments in shaping their views on the status of the mental. In doing this, I will work in reverse historical order, taking up Hempel first and then Carnap and Schlick.

I

As far as I know, Hempel's "Logical Analysis of Psychology" (1935) is his only substantive statement on the mind-body problem. In this paper, Hempel states a strong form of behaviorism/physicalism and mounts a vigorous positivist argument in its support—just the kind of argument you would expect from a committed positivist.

The problem Hempel sets for himself concerns the nature of psychology as a science—in particular, how to answer the question "Is psychology a natural science, or is it one of the sciences of mind and culture *(Geisteswissenschaften)*?" It is interesting to note that in presenting his claims and arguments, Hempel is evidently thinking of himself as speaking for the entire group of philosophers associated with the Vienna Circle. For example, he says at the outset, "The present article attempts to sketch the general lines of a new analysis of psychology, one which makes use of rigorous logical tools, and which has made possible decisive advances toward the solution of the above problem. This analysis was carried out by the 'Vienna Circle'" (Hempel 1935, 15). He then goes on to mention specific names, like Schlick, Carnap, Neurath, Waismann, and Feigl. As all of you know, Hempel's position, or "the position of the Vienna Circle," is that, contrary to the traditional conception of psychology as the science of inner mental life accessible solely through introspection, psychology, insofar as it is a significant cognitive enterprise, is a branch of natural science, and that there are no important differences between the subject matter of psychology and that of physics.

The way Hempel begins his argument for this claim is characteristically positivist: he makes a conceptual/linguistic ascent—that is, he transforms the question about the subject matter of psychology and of physics into a question about psychological and physical languages. Since "the theoretical content of a science is to be found in statements" (Hempel 1935, 16), the problem boils down to determining "whether there is a fundamental difference between [the contents or meanings of] the statements of psychology and those of physics" (Hempel 1935, 16). Before Hempel addresses the contents of psychological statements, he sets forth some general principles concerning meaning, of which the following are the main ones (Hempel 1935, 17):

(i) Each meaningful statement has associated with it a class of "test sentences," sentences whose truth values determine the truth value of that statement.

(ii) These test sentences are "physical test sentences" (Hempel 1935, 17) and describe intersubjectively observable physical phenomena and conditions.

(iii) The class of test sentences associated with a statement exhausts the content/meaning of that statement.

(iv) Two statements have identical content/meaning if and only if the same class of test sentences is associated with each.

This of course is a straightforward form of the verifiability theory of meaning, or "cognitive significance"—and, more specifically, a physicalist version of it. And it is condition (ii) that makes it physicalist. Earlier, under the influence of Ernst Mach, most positivists, including Carnap, worked with a psychological/phenomenological basis; for example, in *Der logische Aufbau der Welt* (1928), Carnap's basic entities were *Elementarerlebnisse*. This meant that the basic units of meaning were taken to be sentences that directly report a phenomenological experience, like "I sense a red round thing in front of me." It seems that members of the Vienna Circle made a more or less collective switch to a physicalist basis by the early 1930s, making sentences about middle-sized physical objects and their observable properties the basic units of cognitive meaning. Hempel's article under discussion appeared in 1935.

Given the physicalist version of the verifiability theory of meaning, an argument can proceed quickly and smoothly to the conclusion that in respect of content/meaning, psychological statements cannot differ from physical statements, since all contents and meanings must be captured by intersubjectively and observationally verifiable "physical test sentences." More specifically, Hempel argues for a *translatability* thesis to the effect that every psychological statement can be translated into a set of physical statements in which no psychological concepts appear. For any psychological statement, the class of its physical test sentences will serve as a physical translation of it.

Notice that under a phenomenological version of condition (ii) above, we would again have the conclusion that psychological and physical statements have meanings of a uniform kind, except this: meanings would now be captured not by physical test sentences but by phenomenological test sentences, sentences about subjective psychological contents of experience. This means that, on this approach, meanings of both physical and psychological statements are at bottom anchored in phenomenological statements about experience, giving us a reduction of all cognitively meaningful discourse, including physical language, to phenomenological language. Thus, although a phenomenological version of (ii) would give us part of what Hempel was looking for, namely the thesis that psychology and physics share the same subject matter, it would not have delivered the crucial physicalist component of Hempel's overall position. In particular, it would not have delivered behaviorism.

In any case, the physicalist position defended by Hempel is what is now called logical (or analytic, conceptual) behaviorism, the claim that the meaning of any psychological expression is fully definable in terms of expressions that refer to observable physical behavior and dispositions to emit such behavior. The reason of course is that the engine that drives Hempel's argument is the verifiability criterion of meaning, and the thesis he attempts to establish is the proposition that psychological statements are fully translatable into physical statements. But I believe that things are a little more complicated. Let us look at an example Hempel gives of how the translation from psychological to physical statements works. Consider, Hempel says, the statement "Paul has a toothache." Hempel then lists the following as a partial list of its test sentences:

1. Paul weeps and makes gestures of such and such kinds.
2. At the question "What is the matter?" Paul utters the words "I have a toothache."
3. Closer examination reveals a decayed tooth with exposed pulp.
4. Paul's blood pressure, digestive processes, and the speed of his reactions show such and such changes.
5. Such and such processes occur in Paul's central nervous system.

I list these not to raise the usual difficulties and objections but to see the kind of reduction base Hempel, and perhaps others in the Vienna Circle, had in mind for psychological language. The most surprising thing about this list is that only one of the items, the first, is unproblematically about observable physical behavior. Item 2 concerns linguistic behavior, a highly complex and, many would argue, richly psychologically laden phenomenon (for item 2 to be relevant, Paul must be a *speaker of English*, for one thing, and we must assume that Paul has *understood* the question and that he *desires* to speak the truth). Items 3 and 4 concern not behavior at all but physiological conditions and changes, and, most interestingly, item 5 concerns a neural process in the brain—presumably, the neural substrate of pain. The last three items are not the sort of thing that can be considered to be part of the concept of pain; it is not at all plausible to think that mere "logical analysis" could reveal a toothache's connections to blood pressures or brain states. Most of us would regard such bodily states as only contingently and empirically connected with mental events; only empirical scientific research could establish such connections.

In light of this, what are we to make of the standard view of Hempel's paper as an unambiguous statement of logical behaviorism? This view seems to have been Hempel's as well. In his short introductory note, written in 1977, to the version of the paper reprinted in Ned Block's *Readings in Philosophy of Psychology,* Hempel writes: "Since the article is so far from representing my present views, I was disinclined to consent to yet another republication, but I yielded to Dr. Block's plea that it offers a concise account of an early version of logical behaviorism" (Hempel [1935], 1980, 14). The implication is that Hempel was in agreement with Block's assessment that logical behaviorism was the position advocated in his 1935 paper. The trouble with this view is that it makes Hempel come out badly mistaken, as we have just seen, about a physical translation of the sample psychological state-

The Mind-Body Problem

ment "Paul has a toothache." Only one of the five physical test sentences is unambiguously about observable behavior. It is a little difficult to believe that Hempel, with his famously astute mind, could have been wrong on four out of the five! Perhaps we could discount the second test sentence about linguistic behavior; here, we might say, Hempel made an honest mistake, a consequence of the fatally simplistic view of language to which most philosophers and psychologists with positivist sympathies subscribed at the time. Still, this does not explain the fact that three of the five conditions cited by Hempel have to do not with behavior or behavior dispositions but with physiological/neural conditions, like having a decayed tooth, a rising blood pressure, and a neural state of the brain—things that no amount of "logical analysis" of "toothache" is likely to uncover.

I would suggest that Hempel's inclusion of these items should prompt us to reexamine the concepts of meaning and translation that are operative in his argument. That is, we should try to understand Hempel's talk of "translation" in a way that is consistent with the fact that he takes "Paul is in neural state N" as a partial translation of "Paul has a toothache." Translation, for Hempel as well as for other positivists (as we shall see), was not necessarily a process of turning an expression into one of its *semantical/conceptual equivalents*. It is clear that Hempel, like Carnap, took empirical lawlike connections as sufficient to underwrite translation. For "Paul's blood pressure is rising" to count as a physical test sentence for "Paul has a toothache," it made no difference to Hempel that a rising blood pressure was empirically and contingently, not conceptually or semantically, connected to the onset of a toothache. But then the thesis that psychological statements are fully translatable, in this broad sense of translation, into physical statements no longer amounts to logical or analytical behaviorism. It is possible to read Hempel as a logical behaviorist, but that requires us to think that Hempel was badly confused when he wrote "The Logical Analysis of Psychology." A more reasonable interpretation, I believe, is to view the position Hempel was defending in 1935 as a more broad-based physicalism, not a version of Rylean logical/analytic behaviorism.

But what does broad physicalism of this sort really amount to? Few, including Cartesian dualists, need to quarrel with the claim that psychological events and processes are lawfully associated with behavioral dispositions and physiological processes. Some of these associations may be logical/conceptual and others empirical/contingent; and they may be rich and variegated enough so that for each psychological kind there is some unique physical kind that correlates with it. Consider a Spinozistic double-aspect theorist or Leibnizian parallelist. Such a person is committed to the following thesis: For each psychological property/kind M, there exists a physical property/kind P such that M and P always co-occur. These psychophysical correlations could be empirical or a priori and they could be contingent or necessary, though most serious dualists who would countenance such correlations are likely to take them as contingent and empirical. In any case, it is clear that if a pervasive system of psychophysical correlations holds, then, on Hempel's broad sense of translation, any "atomic" psychological statement of the form "*x* instantiates *M* at *t*" would be translatable into a physical statement of the form "*x* instantiates *P* at *t*." Ultimately, therefore, it follows on this supposition that all psychological statements are translatable into physical statements. This means that Hempel's translatability thesis—the claim that all psychological statements are translatable into physical statements—is fully consistent with the Spinozistic or Leibnizian dualism; in fact, Hempel's translatability claim is logically entailed by these dualisms. The conclusion seems inescapable that the notion of translation used by Hempel in "The Logical Analysis of Psychology" is too broad to support a philosophically significant notion of translation, and that it cannot serve as a basis for formulating a robust and significant form of physicalism. Physicalism, however it is formulated, must exclude dualisms of the sort just considered. We will recur to some of these issues below.

II

Carnap wrote extensively on the mind-body problem, starting with his *Der logische Aufbau der Welt*(1928). His other works addressing the problem include "Die physikalische Sprache als Universalsprache der Wissenschaft" (*Erkenntnis* 2, 1932), later published in English as *The Unity of Science* (London, 1934), "Psychology in Physical Language" (the original German version in *Erkenntnis* 1932–33), and "The Logical Foundations of the Unity of Science" (in *International Encyclopedia of Unified Science*, 1:1938; reprinted in *Readings in Philosophical Analysis*, ed. H. Feigl and W. Sellars [New York, 1949]). As we noted earlier, in the *Aufbau*, Carnap takes as his basic entities *Elementarerlebnisse*, or elementary experiences, and remembered similarity over these items as his single primitive relation. His project is to reconstruct "the world," a global conceptual/linguistic scheme that will be adequate for the entirety of scientific knowledge. Although Carnap's official position was that the ontological issues were metaphysical pseudo-problems, and he studiously avoided the mind-body problem as an ontological issue, the clear thrust of his project in the *Aufbau* is one of phenomenological reduction—reduction of everything to a phenomenological basis. By "everything" I mean both language and things in the world.

I take it that around this time Carnap and others were under the influence of Ernst Mach's hyper-empiricism and phenomenalism. And the kind of phenomenological reduction Carnap attempted in the *Aufbau* had a long history and tradition that started with the early British Empiricists, in particular Berkeley and Hume, and continued on with John Stuart Mill. But isn't such a reduction a form of subjectivism and idealism? Doesn't it reduce everything, including the material world, to perceptions, impressions, and ideas—that is, to the domain of the mental?

Chapter B of part 4 of the *Aufbau*, the final part of the book, is titled "The Psychophysical Problem." Here Carnap discusses the problem in a pretty standard setting, without referring to the fundamentally psychological character of his constructional system. And he says what you would expect him to say, namely that the so-called mind-body problem has two aspects, the scientific problem of ascertaining neural correlates of psychological processes and the metaphysical problem of explaining these correlations. According to him, the first problem is an empirical problem outside the domain of philosophy, whereas the second problem is a metaphysical problem that cannot even be formulated within his constructional system or indeed within science. But we should keep the following in mind: the mind-body problem as Carnap sees it within his *Aufbau* system is a problem concerning the physical as it is constructed within the system and the mental, which also appears as constructions at an upper level, that of "heteropsychological and cultural objects." So elementary experiences at the base of his system are perhaps conscious phenomenal events and states from the first-person point of view, whereas the mental, for which Carnap's mind-body arises, is a third-person psychological domain constructed in the higher reaches of his constructional system. The mind-body problem so construed, as a problem arising at upper levels of his constructional system, seems to be a valid problem that needs to be addressed. We can imagine the mind-body problem arising in a similar way for the phenomenalist/ idealist systems of Hume and Berkeley. Given this approach, however, the following question is yet to be addressed: Isn't there a mind-body problem also for the *Elementarerlebnisse?* That is, what might be *their* physical bases? Or do they float free from physical processes? If these questions *don't* make sense, why? Carnap would have replied: these questions cannot even be formulated within the constructional system and that is why they make no sense. He might have added that at this bottom level we don't even have a mental-physical distinction. This would have been the logical positivist party line, but it is obviously not very satisfying. After all, in constructing the kind of system that he did, Carnap made a series of philosophical decisions, decisions that are not arbitrary or random, and if these decisions are to determine the meaningfulness and legitimacy of certain philosophical questions, they would have to be motivated and rationally defensible decisions. At least, that is what some of us today would be inclined to say in response.

In the pre-Vienna days, Schlick, too, appears to have had a distinctly idealistic tendency in his views on the mind-body relation. In his "Ideality of Space, Intro-jection and Psycho-Physical Problem" (the German original published in 1916), Schlick sees the mind-body problem as consisting in the competition between mental and physical items for spatial location. Consider the white of this sheet of paper. Such a sensory quality, Schlick says, has a location in space. But where? Schlick writes:

> That [sensory qualities] do not exist at the place of the physical object "paper" we have just been emphatically told by natural science; it could find nothing else there but a physical substrate, matter, electrons or what you will, in specific physical states. The only other place that could still come into question is the brain. But the sensory qualities are not there either, for if somebody could ex-amine my brain while I am looking at the white paper, he would never discover the white of the paper anywhere there, . . .
>
> So sensory qualities cannot be localized in either the one or the other part of physical space: the place they need to claim is found everywhere to be already occupied by physical things, which exclude them from being present at the same time. The world of the physicist is absolutely complete in itself, and contains no room for the world of the psychologist, with its sensory qualities. The two are in contention for the possession of space. The one says: "It is white at this point!"; the other: "It is not white at that same point!" It is these contradictions of local-ization, and no others, which constitute the true psycho-physical problem." (Schlick 1916, 195)

These passages are interesting in that, although that is not the way we would now set up the mind-body problem, it does echo, and perhaps anticipate, the so-called causal exclusion problem much debated today in connection with the mind-body problem and mental causation (see Kim 1998). For Schlick the physical world is a closed space-time world that excludes all things that are nonphysical. The current analog of this principle is the thesis that the physical world is a causally closed world that excludes all nonphysical causal agents and forces. In any case, what is Schlick's solution to the spatial exclusion/competition problem for the mental and the physical?

According to Schlick, there are two ways of overcoming these "contradictions": Kant's critical philosophy and the positivism of Richard Avenarius. Although their starting points are different, both solutions, in Schlick's view, arrive at the same conclusion: the ideality of space. Thus, sensory qualities win the battle with physical objects for locations in space, since space itself is not something real that is out there but only a construction based on appearances and experiences. Here is what Schlick says: "We see that with Avenarius the sensory qualities carry off victory in the battle for the possession of space. . . . It is [sensory qualities] in their motley variety that fill space and congregate into bodies and 'I-complexes'. It obviously makes no sense to seek out a place for 'consciousness' among them, since they themselves are ingredients of consciousness. The decision was bound to fall out thus, since the sensory qualities' claim to space is manifestly prior, absolutely given and inalienable, whereas physical objects, electrons and so on do not represent things of comparable immediacy but are arrived at only by inference and theoretical constructions" (Schlick 1916, 197).

According to Schlick, "Kant takes space into consciousness, whereas Avenarius spreads consciousness over space." However, Schlick himself finds neither the Kantian solution nor the strict positivist solution fully acceptable, for reasons that are not wholly clear to me. The final picture that Schlick describes in this paper and with which he seems generally satisfied goes like this: (1) "the data of consciousness" are not mere Kantian appearances but real objects "without qualification" (Schlick takes this to be the positivist line); (2) the mental items are not the only real things in the world; there are also Kantian "things-in-themselves" that are not experienced but inferred; (3) these things-in-themselves, or transcendent objects, are not the familiar "physical bodies of nature," these latter being only a system of conceptual signs constructed by science to represent the nonspatial transcendent objects; (4) there is a coordination, dependence, and interaction between "the real contents of consciousness" and "the real objects beyond consciousness."

I am not quite sure that I have grasped the full significance of this system, and I am not sure I understand how it generates a solution to the space competition problem for the mental and the physical. Perhaps the solution that Schlick had in mind is something like this: the spatial external world inhabited by physical things is a mere "sign system" constructed by science as a way of representing and attaining knowledge of the transcendent objects. This means that, as in Kant's system, space is only ideal, being a merely epistemological construction, not something that pertains to the real objects, namely the transcendent objects of the world. Since space is ideal, not real, the competition for space is not a genuine competition. Hence, the problem is dissipated.

As I said, I am not wholly confident of my reading of Schlick of this pre-Vienna period. However, what is clear is that his thinking on the mind-body problem had a distinctly idealistic cast; much of his philosophical concern and argumentation was squarely rooted within the idealist tradition. But Schlick of this pre-Vienna period was not a logical positivist. Wittgenstein's *Tractatus* (German 1921; English 1922) was yet to be published, and the Vienna Circle was still almost ten years away (Carnap arrived in Vienna in 1926).

III

Let us now turn to the physicalist phase of the Vienna Circle. As noted earlier, the crucial transition from phenomenalism to physicalism seems to have been made around 1930. In the early thirties, we see numerous publications by Carnap vigorously promoting the physical language as "the universal language of science"—that is, the language for all cognitively significant discourse. Hempel's "The Logical Analysis of Psychology," discussed earlier, belongs to this period (although he was not officially a member of the Vienna Circle). Let us begin with Carnap, focusing on his "Psychology in Physical Language" (Carnap 1932–33).

Carnap's main aim in this paper is, as he says, "to explain and establish the thesis that every sentence of psychology may be formulated in physical language," or in the material mode, "all sentences of psychology describe physical occurrences, namely the physical behavior of humans and other animals." This, Carnap says, is a subthesis of physicalism—that is, physicalism applied to the psychological language—which is the claim that physical language is the universal intersubjective language for all significant discourse. As with Hempel, the main idea used by Carnap to make precise the physicalist claim about psychology is translation: the thesis comes to the claim that every psychological sentence is translatable into a physical sentence, a sentence in the physical language. Thus, Carnap's project so characterized seems exactly identical with Hempel's program in "The Logical Analysis of Psychology."

Carnap's statement of his project brings up two questions: What is a physical language? How does translation work? In "Psychology in Physical Language" Carnap doesn't say anything specific about physical language. However, this issue is explicitly raised and addressed in his "Physical Language as the Universal Language of Science" (1932). He writes: "The physical language is characterized by the fact that statements of the simplest form (e.g., the temperature of such and such a place at a specified time is so much): [in the formal mode] attach to a specific set of co-ordinates (three space and one time co-ordinate) a definite value or range of values of a coefficient of physical state; [in the material mode] express a quantitatively determined property of a definite position at a definite time" (Carnap 1931, 404). Note the occurrence in the formal mode of the expression "physical," which is in need of further explanation if circularity is to be avoided. In the material mode version, the quantitative character of the properties seems to carry the main burden in defining the physical. Whether or not such an approach will ultimately work is debatable, but Carnap's characterization is clear enough in outline, and in its intent, to be amply serviceable.

The Mind-Body Problem

What then is translation? Carnap has this to say: "We say of a sentence P that it is translatable ... into a sentence Q if there are rules, independent of space and time, in accordance with which Q may be deduced from P and P from Q; to use the material mode of speech, P and Q describe the same state of affairs ... The definition of an expression 'a' by means of expressions 'b', 'c', ... represents a translation-rule with the help of which any sentence in which 'a' occurs may be translated into a sentence in which 'a' does not occur, but 'b', 'c', ..., do, and vice versa" (Carnap 1932–33, 166–67). So psychophysical translation requires psychophysical *definitions*, definitions of psychological expressions in terms of physical expressions. And, intuitively, translation must preserve reference: that is, if P and Q are intertranslatable, they must "describe the same state of affairs" (Carnap 1932–33, 166). Thus definitions are the chief enablers of translation. What then is a definition? During this period, Carnap had a broad notion of a definition; definitions for him are not necessarily, or even chiefly, matters of conceptual analysis and need not represent logical/analytic equivalences. Rather, they could be based on nomological connections. He says, for example, "If for anger we knew a sufficient and necessary criterion to be found by a physiological analysis of the nervous system or other organs, then we could define 'angry' in terms of the biological language. ... But a physiological criterion is not yet known" (Carnap 1938).

As we saw, Hempel's notion of translation too allowed the use of empirical lawlike connections as translation rules. We now see that Carnap's notion was similarly broad, and he was more explicit about this. There is no need to debate the question whether this is an acceptable sense of "translation"; we can take "translation" as it appears in Carnap and Hempel as a term of art with a stipulated meaning. What is not a verbal issue is the question whether this overly generous notion of translation can underwrite the kind of physicalism/behaviorism that these philosophers wanted to formulate and defend. We earlier saw that Hempel's weak notion of translation undermined his behaviorist/physicalist project in "The Logical Analysis of Psychology," turning his physicalist position into something very different—in fact, something that is too weak to be called physicalism. Obviously, the same consequences follow for Carnap as well. The root difficulty lies in this fact: translatability in the sense of Carnap and Hempel does not satisfy the reasonable desideratum, explicitly stated by Carnap himself, that mutually translatable sentences "describe the same state of affairs" (Carnap 1932–33, 166). It is surely uncontroversial that a lawlike correlation between two predicates, F and G, does not sanction the claim that "x is F" and "x is G" describe the same state of affairs. To support this claim what is needed is the stronger premise that F and G express, or designate, the same property, and this is not something we can get from a mere lawlike connection between F and G. (One might even say, as I believe J. J. C. Smart once did, that correlations and identities exclude each other; for how could anything "correlate" with itself?)

However, we can bracket this issue of translation and move on. The reason is that, at least initially, Carnap is interested chiefly in a behavioristic reduction of the psychological, and if psychological concepts stand in any analytic/conceptual relations to nonpsychological concepts, the best bet is that they do so with respect to behavioral concepts, not neurobiological or microphysical concepts. So let us turn to Carnap's behavioral analysis of psychological concepts. As we will see, simple and straightforward behaviorism is only a starting point for him; he has other fish to fry.

Unlike Hempel, Carnap is very clear that it is behavioral dispositions, not actual behavior, that he considers appropriate for defining psychological properties. He says, "Every psychological property is marked out as a disposition to behave in a certain way" (Carnap 1932–33, 186), and goes on to explain what dispositions are. He writes: "by 'disposition' (or 'dispositional concept') we shall understand a property which is defined by means of an implication (a conditional relationship, an if-then sentence)" (Carnap 1932–33, 186). As an example, "Person X is excited," according to Carnap, goes into "If, now, stimuli of such and such a sort were applied to X, X would react in such and such a manner." It is assumed that both stimuli and responses are to be characterized in purely physical/behavioral terms.

If Carnap stopped here, he would simply be a behaviorist, of the Rylean sort. What led him to take an important further step was his realism about dispositions; for him, dispositions, including behavioral dispositions, were not mere conditionals, or Rylean "inference tickets." Rather, these dispositions were seen as grounded in the physical structure of the systems to which they are ascribed. A physical dispositional sentence, like "X is firm," Carnap argues, has "the same content as a sentence . . . which asserts the existence of a physical structure characterized by the disposition to react in a specific manner to specific physical stimuli" (Carnap 1932–33, 172). This means:

"X is firm"

Goes not into a simple conditional like:

"If a light load is placed on X, X will not bend"

But rather into an existential statement:

"X has a certain physical structure (or property) S such that when a light load is placed on X, S causes X to resist bending"

The Mind-Body Problem

Likewise, Carnap says that "*A* is excited" refers to "the physical structure of *A*'s body—though this structure can only be characterized by potential perceptions, impressions, dispositions to react in a specific manner, etc." (Carnap 1932–33, 175). So "*A* is excited" would go into an existential statement about an internal physical state of *A:*

> "*A* is in some physical state *P* such that when stimuli *S* are applied
> to *A*, *P* causes *A* to respond in manner *R*"

This transition is of great significance because it shows that in his work during the 1930s Carnap was anticipating functionalism, in particular the version defended by David Lewis (1966) and David Armstrong (1968). This version of functionalism has an aspect of the identity theory—the theory that identifies mental states with neural states. For Carnap, the ascription of a mental state to a person is, conceptually, the ascription of some physical state that has a certain causal role—it causes the person to respond in certain specific ways when the specified stimuli are applied to that person. It is a state that can explain why an organism responds in the way it does when certain stimulus conditions are presented to it. It turns out, under biological investigation, that the state that fills this causal role is a state of the brain. Carnap claims that as our knowledge of neurophysiology grows, enabling us to give physiological descriptions of this inner physical state, the behavioral definitions of psychological properties and states will give way to physiological definitions. He says, "the aim of science is to change the form of the definition: more accurate insight into the micro-structure of the human body should enable us to replace dispositional concepts by actual properties" (Carnap 1932–33, 187).

In current idiom, Carnap's proposal would amount to this: mental concepts are functional concepts, concepts defined in terms of causal roles connecting physical inputs and behavioral outputs. That is, to be in state *M*, where "*M*" is a mental predicate, is to be in some state *P* such that *P* is caused to instantiate by physical inputs of kind *K* and *P*'s instantiation in turn causes responses of kind *R* to be emitted. Any state that fills the specified causal role—that is, any value of "*P*" that satisfies the functional definition—is called a "realizer" of the functionally defined concept. This is combined with the physicalist thesis that all the realizers of functional mental states are physical states—that is, there are no nonphysical realizers of mental states (like states of Cartesian souls). In the last quoted sentence, Carnap is saying that the discovery of microstates and structures that realize mental properties will yield a neural reduction of mental properties. If being in pain is being some state that is caused by tissue damage and that in turn triggers avoidance behavior, and if we find that neural state *N* is just the state that, say in humans, is caused by tissue damage and triggers escape behavior, that would justify the reductive claim that, for humans, being in pain is just being in neural state N (for details, see Kim 1998).

It is clear, then, that on the mind-body problem Carnap went considerably beyond simple logical behaviorism of the Rylean kind, clearly anticipating the functionalist approach that dominated the scene during the second half of the twentieth century and whose influence is still deep and pervasive. In developing these ideas, Carnap was doing serious metaphysics; he was not merely brandishing the verifiability criterion of meaning as an all-purpose philosophical weapon. His robust good sense as a philosopher overshadowed a doctrinaire allegiance to the positivist slogans.

Let us finally turn to the physicalist phase of Schlick during his Vienna period. In their introduction to the English edition of Schlick's *General Theory of Knowledge* (1974), Herbert Feigl and Albert Blumberg refer to Schlick's position on the mind-body problem as "this beautiful solution" and consider it "the most original contribution made by Schlick to ... metaphysics." According to them, Schlick's solution was what is now called the "mind-brain" or "psychoneural" identity theory, or "type physicalism," the claim that mental events are neural events in the brain. This is the theory that had its moments in the sun during the late 1950s and 1960s but has largely been given up since then, although there have been signs for a revival (for example, Block and Stalnaker 1999). For my part, though, I do not find a reasonably clear and unambiguous statement of the mind-brain identity theory in Schlick's *General Theory of Knowledge*. We should recall that the first edition of Schlick's book was published in 1918, and the second edition in 1925, and that 1918 is only two years after the publication date of his paper "Ideality of Space, Introjection and the Psycho-Physical Problem," a paper we discussed earlier. This paper, which is steeped in the Idealist tradition, contains nothing remotely like the mind-brain identity theory. And as I noted earlier, the Vienna Circle's transition to physicalism occurred around 1930. This chronology alone makes it doubtful that Schlick could have worked the identity theory into the second edition of the book.

Be that as it may, I believe we find a much clearer and pretty unambiguous statement of mind-body physicalism, a version of the identity theory, in his paper "On the Relation between Psychological and Physical Concepts" published in 1935. He frames the mind-body problem as follows: "the essential feature of physical concepts is that they are arrived at by selecting out of the infinite variety of events a special class, namely these "coincidences" [meter readings in measurement], and describing their inter-relationships with the help of numbers. Physical magnitudes are identical with the number-combinations which are thus arrived at. The question which we are seeking to answer (in principle) can therefore be put as follows: What is the relation of these coincidences to all other events, for example to the occurrence of a pain to the change of a color, to a feeling of pleasure, to the emergence of a memory, and so forth?" (Schlick 1935, 398). His answer is this: "To the assertion of the one-sidedness and limitations of the methods of physics, there stands in sharp opposition the claim that an absolutely complete description of the world is possible by the use of physical methods; that every event in the world can be described in the language of physics, and therefore specifically, that every psychological proposition can be translated into an expression in which physical concepts alone occur. This claim—which is referred to ... as 'physicalism'—is correct, if the physical language is not only objective, which we have already seen, but is in addition the only objective language ... This seems indeed to be the case" (Schlick 1935, 399).

This is clearly a form of psychophysical identity theory. But is it a mind-brain, or psycho-neural identity theory as Feigl and Blumberg claim? This I don't think is so clear. Schlick does bring in considerations of the brain. But consider this passage: "Rather the dependence of the feeling [of grief] on the state of the subject is so obvious that everybody looks to the body of the griever himself for the coincidences which are here principally in question. Once again we do not need to consider the events in the nervous system—which are for the most part unknown —for it is sufficient to pay attention to his expression, his utterances, his whole deportment. In these processes—which are describable in terms of coincidences— we have the facts by which feelings are expressible in the physical language" (Schlick 1935, 402). This sounds more like a psycho-behavioral identity theory than a psychoneural identity theory.

To sum up this somewhat cursory examination of the logical positivist treatment of the mind-body problem, I believe we can say the following: first, although the early position advocated by philosophers like Carnap and Hempel looked and sounded like behaviorism, especially analytical/logical behaviorism, a close examination of their arguments show that the view really defended by them is a broader form of physicalism, and that if we take seriously the concept of translation used in their arguments, it isn't clear that their position amounted to a serious form of physicalism. Second, we find in Carnap—and perhaps a hint of it in later Schlick —a form of physicalism that anticipates important later developments, in particular, functionalism and psychoneural identity theory based on the functionalist approach. Reading the literature again, I was impressed by the metaphysical depth and sophistication in the positivist philosophers, especially Carnap.

References

Armstrong, D. 1968. *A Materialist Theory of the Mind.* New York, Humanities Press.
Block, N., and R. Stalnaker. 1999. "Conceptual Analysis, Dualism, and the Explanatory Gap." *Philosophical Review* 108:1–46.

Carnap, R. 1928. *Der logische Aufbau der Welt*, Berlin: Weltkreis-Verlag.

———. 1931. "Physical Language as the Universal Language of Science." In *Readings in Twentieth-Century Philosophy*, ed. W. P. Alston and G. Nakhnikian. Glencoe, Ill.: Free Press. Originally published in German as "Die physikalische Sprache als Universalsprache der Wissenschaft," *Erkenntnis* 2, 1931 (pp. 393–424 in 1964 volume).

———. 1932–33. "Psychology in Physical Language." In *Logical Positivism*, ed. A. J. Ayer, Glencoe, Ill.: Free Press. English translation of "Psychologie in physikalischer Sprache," *Erkenntnis* 3, 1932–33.

———. 1937. "Testability and Meaning." *Philosophy of Science* 3 (1936): 419–71; 4 (1937): 1–40.

———. 1938. "Logical Foundations of the Unity of Science." In *Encyclopedia of Unified Science*. Chicago: University of Chicago Press. Page references are to its reprinted version in *Readings in Philosophical Analysis*, ed. H. Feigl and W. Sellars. New York: Appleton-Century-Croft, 1949.

Hempel, C. G. [1935] 1980. "The Logical Analysis of Psychology." In *Readings in Philosophy of Psychology*, vol. 1, ed. Ned Block, 14–23. Cambridge: Harvard University Press, 1980; originally published in French.

Kim, J. 1996. *Philosophy of Mind*. Boulder, Colo.: Westview Press.

———. 1998. *Mind in a Physical World*. Cambridge: MIT Press.

Lewis, D. 1966. "An Argument for the Identity Thesis." *Journal of Philosophy* 63:17–25.

Schlick, M. 1916. "Ideality of Space, Introjection and Psycho-Physical Problem." In Schlick 1979, 190–206. The German original published in 1916.

———. 1974. *General Theory of Knowledge*. Trans. Albert E. Blumberg. New York: Springer Verlag. English translation of *Allgemeine Erkenntneslehre* (Berlin: Springer, 1918).

———. 1979. *Philosophical Papers*. Vol. 1 (1909–22). Dordrecht, Holland: Reidel.

———. 1935. "On the Relation between Psychological and Physical Concepts." In *Readings in Philosophical Analysis*, ed. H. Feigl and W. Sellars, 393–407. New York: Appleton-Century-Croft, 1949. English translation of "De la relation entre les notions psychologiques et les notions physiques," *Revue de Synthèse*, 1935.

Wittgenstein, L. 1921. *Tractatus Logico-Philosophicus (Logisch-Philosophische Abhandlung)*. German version in *Annalen der Naturphilosophie*, 1921; German and English bilingual version. London: Kegan Paul, 1922.

V I

Scientific Rationality

Kinds of Probabilism

MARIA CARLA GALAVOTTI

Department of Philosophy, University of Bologna

The first part of the article deals with the theories of probability and induction put forward by Hans Reichenbach and Rudolf Carnap. It will be argued that, despite fundamental differences, Carnap's and Reichenbach's views on probability are closely linked with the problem of meaning generated by logical empiricism and are characterized by the logico-semantical approach typical of this philosophical current. Moreover, their notions of probability are both meant to combine a logical and an empirical element. Of these, Carnap over the years put more and more emphasis on the logical aspect, while for Reichenbach the empirical aspect has always been predominant. Seen in this light, Carnap's and Reichenbach's theories of probability can be taken to represent the Viennese and Berlinese mainstreams of the common logical empiricist approach. The second part of the article contrasts the position of these authors with that of Bruno de Finetti, who is the main representative of the subjective interpretation of probability. Though the latter is sometimes associated with the position taken by Carnap in his late writings, it will be argued that the two are in many ways irreconcilable.

Reichenbach's Frequentism

Before tackling Carnap's and Reichenbach's views of probability, it is important to point out their strong concern with the problem of meaning. As is well known, this problem plays a central role in the development of logical positivism, where it is strictly connected to the confirmation of scientific hypotheses. The link between confirmation and the problem of meaning is a distinctive feature of the perspective taken by these authors, a feature that is absent from the approach of other logical positivists, such as Friedrich Waismann, who is considered a forerunner of Carnap's logicism, or Richard von Mises, who gave a most valuable contribution to frequentism. These authors address the question of the nature of probability quite independently of problems like the criteria of cognitive significance, the meaning of theoretical terms, or the confirmation of scientific hypotheses. Even Carl Gustav Hempel, who has made strenuous efforts to clarify the notion of "confirmation," does not seem to consider it strictly linked to the interpretation of probability. The latter is for Hempel a sort of side interest that he took up when writing his dissertation under Reichenbach but did not cultivate in later writings.[1]

On the one hand, Carnap's interest in probability was triggered in the thirties by the need to overcome the strictures connected with the verifiability theory of meaning that imprinted the first stage of logical positivism. On the other hand, Reichenbach, who had been working on probability in connection with the interpretation of contemporary science since 1915, claims to have been the first to recognize the need to go beyond verifiability, which exercised a profound influence on his views of probability and induction. He points out the close ties between the significance of scientific statements and their predictive character, which is a condition for their testability. At the same time, he reaffirms that "the theory of knowledge is a theory of prediction" (Reichenbach 1937, 89) and puts forward his theory of probability as a "theory of propositions about the future" (Reichenbach 1936, 159). Such a theory includes in the first place a probabilistic theory of meaning: "The theory of propositions about the future will ... be a theory in which the two truth-values, true and false, are replaced by a continuous scale of probability" (154). Reichenbach's probabilistic theory of meaning "substituted probability relations for equivalence relations and conceived of verification as a procedure in terms of probabilities rather than in terms of truth ... it abandoned the program of defining 'the meaning' of a sentence. Instead, it merely laid down two principles of meaning; the first stating the conditions under which a sentence has meaning; the second the conditions under which two sentences have the same meaning" (Reichenbach 1951, 47).

Reichenbach always regarded his own approach as a confutation of the reductionist attitude he attributed to logical positivists, including Carnap. After having been one of its chief proponents, Carnap soon became aware of the difficulties connected with the principle of strict empiricism and started working at its revision. His main contribution in this connection is the theory of partial definability contained in "Testability and Meaning." Carnap's reduction chains are bitterly criticized by Reichenbach, who says that they are "too primitive instruments for the construction of scientific language" (Reichenbach 1951, 48), because Carnap's testability criterion of meaning is based on logical implication, not on probability. As a matter of fact, Reichenbach charges Carnap with reductionism and lack of consideration for the probabilistic aspects of science already in his review of the *Aufbau*, where he says: "It is a puzzle to me just how logical neo-positivism proposes to include assertions of probability in its system, and I am under the impression that this is not possible without an essential violation of its basic principles" (Reichenbach [1933] 1978, 1:407).

Reichenbach held the conviction that it is probability, not truth, that should underlie a reconstruction of science in tune with scientific practice. "The ideal of an absolute truth," he says, is "a phantom, unrealizable; certainty is a privilege pertaining only to tautologies, namely those propositions which do not convey any knowledge" (Reichenbach 1937, 90). Moreover, "it would be illusory to imagine that the terms 'true' or 'false' ever express anything else than high or low probability values" (Reichenbach 1936, 156). Reichenbach's attitude toward truth probably exercised some influence on Hempel's thought while marking a divergence with Carnap's position.[2]

That Carnap's theory of probability is rooted in the problem of cognitive significance is testified by the fact that he put forward for the first time the idea of "degree of confirmation" toward the end of "Testability and Meaning." Carnap does not share Reichenbach's probabilism and strives to bring the notion of confirmation as close as possible to that of truth. In fact, he defines the notion of "degree of confirmation" as a semantical concept that is by definition time-independent, exactly like the notion of truth. In the thirties and forties Carnap regarded confirmation as a relationship between the meanings of two sentences, respectively describing evidence and hypothesis. Sentences expressing degrees of confirmation are analytic, and their logic, namely inductive logic, is analogous to deductive logic, the difference being "only the fact that the first [statements of inductive logic] contain the concept of degree of confirmation and are based on the definition of this concept, while the latter [statements of deductive logic] are independent of it" (Carnap1946, 591).

Reichenbach also embarks on his treatment of probability from logic, presenting his frequentist theory as a sort of "probability logic." Reichenbach points out that the attempt to combine probability with the logic of truth faces a peculiar problem, arising from the fact that when a statement about a future event is called probable, such a statement can be verified only after the event has occurred. This calls for some way of bringing together probability, which can take many values, with the two values of truth and falsehood. According to Reichenbach, this problem finds a solution in the frequency theory, which combines statements about single events and statements about frequencies in the proper way, because "the frequency interpretation derives the degree of probability from an enumeration of the truth values of individual statements" (Reichenbach [1933] 1949, 311). When interpreted as frequency, probability refers to sequences (namely, to series of statements), while truth refers to single sentences, but since the propositional sequence "can be conceived as an extension of the concept of statement" (312), probability logic can be seen as a logic of propositional sequences and "appears as a generalization of the logic of statements" (313).

While embracing frequentism, Reichenbach repeatedly emphasizes that his position should not be conflated, or seen as a continuation, of the kind of frequentism worked out by Richard von Mises. In a letter to Bertrand Russell written in 1949 and published in *Selected Writings, 1909–1953*, Reichenbach wrote the following with reference to Russell's book *Human Knowledge:*

> I was surprised to find myself hyphenated to von Mises . . . —as much surprised, presumably, as he. You even call my theory a development of that of von Mises. I do not think this is a correct statement. My first publication on probability [*Der Begriff der Wahrscheinlichkeit für die mathematische Dartstellung der Wirklichkeit*, Leipzig, 1915], which is earlier than Mises' publications, has already a frequency interpretation and a criticism of the principle of indifference, although later I abandoned the Kantian frame of this paper . . . Mises' merit is to have shown that the strict-limit interpretation does not lead to contradictions and, further, to have provided a means for the characterization of random sequences. I then could show that my earlier frequency interpretation (which was weaker than a strict-limit interpretation) in combination with Bernoulli's theorem leads to the limit interpretation and thus took over this interpretation. But my mathematical theory is more comprehensive than Mises' theory, since it is not restricted to random sequences; furthermore, Mises does not connect his theory with the logical symbolism. And Mises has never had a theory of induction or of application of his theory to physical reality. (Reichenbach 1978, 2:410)

Indeed, Reichenbach's frequentism is more flexible than von Mises's, because it allows for single-case probabilities, develops a theory of induction, and contains an argument for its justification. These features bring Reichenbach's theory closer to Carnap than to von Mises, because the latter is more influenced by operationalism than by formalism and regards statistics as more important than logic.

Reichenbach's point of departure is the conviction that degrees of probability can never be ascertained a priori, but only a posteriori. The method by which degrees of probability are attained is "induction by enumeration." This "is based on counting the relative frequency [of a certain attribute] in an initial section of the sequence, and consists in the inference that the relative frequency observed will persist approximately for the rest of the sequence; or, in other words, that the observed value represents, within certain limits of exactness, the value of the limit for the whole sequence" (Reichenbach [1949] 1971, 351). This procedure is reflected by the "rule of induction": if an initial section of n elements of a sequence x_i is given, resulting in the frequency f^n, we posit that the frequency f^i $(i > n)$ will approach a limit p within $f^n \pm \delta$ when the sequence is continued. As suggested by the formulation of the rule of induction, a probability attribution is a "posit"— namely, "a statement with which we deal as true, although the truth value is unknown" (373).

The notion of "posit" occupies a central role within Reichenbach's construction. It is introduced by analogy with the gambling behavior: "The gambler has to make a prediction before every game, although he knows that the calculated probability has a meaning only for larger numbers; and he makes his decision by betting, or as we shall say, by *positing* the more probable event . . . The frequency interpretation justifies, indeed, a *posit* on the more probable case. It is true that it cannot give us a guarantee that we shall be successful in the particular instance considered; but instead it supplies us with a principle which in repeated application leads to a greater number of successes than would obtain if we acted against it" ([1933] 1969, 314). The concept of posit supplies a bridge between the probability of a sequence and the probability of the single case. The idea here is that a posit regarding a single occurrence of an event receives a weight from the probabilities attached to the reference class to which the event in question has been assigned, which should be "the narrowest class for which reliable statistics can be compiled." In the terminology of "propositions" and "sequences," "the concept of weight replaces the untenable concept of the probability of a single statement; we cannot coordinate a probability to a single statement, but we can coordinate a weight to it, by which the probability of the corresponding propositional sequence assumes an indirect meaning for the single case" (315). Therefore, "*a weight is what a degree of probability becomes if it is applied to a single case*" (Reichenbach 1938, 314).

Posits differ depending on whether they are made in a situation of "primitive" or "advanced" knowledge. A state characterized by knowledge of probabilities is "advanced," while a state where this kind of knowledge is lacking is "primitive." In a state of primitive knowledge the rule of induction represents the only way of attaining probability values, while in a state of advanced knowledge the calculus of probabilities applies. Posits made in a state of advanced knowledge have a definite weight and are called "appraised." They conform to the principle of the greatest number of successes that makes them the best posits that can be made. Posits whose weight is unknown are called "anticipative," or "blind." Although the weight of a posit of this kind is unknown, its value can be corrected. The blind posit has an approximate character: we know that by making and correcting such posits we will eventually achieve success, in case the considered sequence has a limit. It is on this idea that Reichenbach grounds his justification of induction.

Induction is justified on pragmatical grounds, on the basis of the following consideration: "We know: if the sequences occurring in nature possess a limit of the frequency we shall eventually arrive at reliable predictions by applying the method of the approximate posit; and if there is no limit we shall never attain this goal. If anything can be achieved at all, we shall reach our aim by applying the method of the approximate posit; otherwise we shall not attain anything" (Reichenbach [1933] 1949, 321). This argument is meant to justify scientific method itself, which "is nothing but a continuous correction of posits by incorporating them into more general considerations" (318). In other words, scientific method is a self-correcting procedure that starts with blind posits and goes on to formulate appraised posits that become part of a complex system, in a continuous interplay between experience and prediction, as suggested by the title of one of Reichenbach's major works (see Reichenbach 1938). The soundness of this system is largely guaranteed by logic, induction is its only nonanalytical assumption; therefore, once induction is justified nothing more is needed.

Reichenbach's argument for justifying induction has inspired much literature. Various authors, including Ian Hacking and Wesley Salmon, have tried to supply Reichenbach's argument, which justifies a whole class of asymptotic rules, with further conditions, apt to restrict it to the rule of induction. In Salmon (1991) the author applies to the justification of induction the distinction, also due to Reichenbach, between a "context of justification" and a "context of discovery." Salmon's suggestion is based on the idea that in the context of justification the inferences and assumptions that are made in the context of discovery can be justified by means of statistical tests. Reichenbach's distinction between the two contexts, however, is not unproblematic, and the use that Salmon makes of it is questionable, first, because statistical tests are often used as tools in the context of discovery,[3] and, second, because their application requires precise assumptions, which are themselves in need of justification. It is not always obvious to which context the application of certain probabilistic or statistical methods belongs. Nevertheless, Reichenbach's pragmatic justification of induction points to the only promising direction to proceed, in order to circumvent Hume's problem, which is insoluble on logical grounds.

Reichenbach's theory of probability is highly objectivistic. First of all, his notion of probability has an objective character: correct probability values exist, but we usually do not know them; we approach them by the method of approximated posits. Reichenbach's whole idea of scientific method as a self-correcting procedure does not make sense if not grounded on the conviction that there are correct, or "true," frequencies. Equally objectivistic is Reichenbach's approach to the confirmation of hypotheses. In this connection, he embraces an objective form of Bayesianism according to which the probability of hypotheses is obtained by Bayes's rule, combined with a frequentist determination of prior probabilities.

The question might be raised whether Reichenbach's theory of probability supports some form of realism. An answer to this question, in the positive or in the negative, would require an argument that I will not attempt to develop. Salmon has argued that Reichenbach's insistence on the "overreaching character of probability inferences" (Reichenbach 1938, 127) can be taken as opening the way to a mild, non-metaphysical form of realism (see Salmon 1994). Indeed, it is very hard to make sense of an objectivistic view of probability like the one endorsed by Reichenbach, including the idea of unknown probabilities, outside a realistic framework.

Carnap's Logicism

Let us now look at Carnap's views on probability and induction. While Reichenbach's position is a monolithic monument to the frequency view of probability, Carnap not only takes a more articulated approach by admitting two concepts of probability but also has revised his position over the years. As stated above, Carnap's work on probability originated in the thirties, in connection with the problem of meaning (see Carnap 1936, 1936–37). As a first step, he elaborated a notion of confirmation that can be seen as a continuation of his work in semantics. Around the mid-forties his notion of degree of confirmation became fully developed, together with the distinction between two notions of probability: probability$_1$, or logical probability, and probability$_2$, or frequency. Carnap charged Richard von Mises and Harold Jeffreys with making the same mistake of regarding their own approach, frequentist in the case of von Mises and logicist in the case of Jeffreys, as the only right one. On the contrary, Carnap claimed that the two notions of probability are both important and useful. Probability$_1$ is a semantical concept, having to do with the degree to which a given hypothesis is confirmed by a given body of evidence. A statement of probability$_1$ "can be established by logical analysis alone . . . It is independent of the contingency of facts because it does not say anything about facts (although the two arguments do in general refer to facts)" (Carnap [1945a], 1949, 339). A statement of probability$_2$, on the contrary, "is factual and empirical, it says something about the facts of nature, and hence must be based upon empirical procedure, the observation of relevant facts" (339).

Carnap makes clear that both concepts have an objective import. As a matter of fact, around the mid-forties Carnap seemed unable to even conceive of a sub-jective notion of probability. "I believe—he says—that practically all authors re-ally have an objective concept of probability in mind, and that the appearance of subjectivist conceptions is in most cases caused only by occasional unfortunate formulations" (340). Commenting on the upholders of an epistemic notion of probability, like Laplace, Keynes, and Jeffreys, he affirms that "most and perhaps all of these authors use objectivistic rather than subjectivistic methods. . . . It ap-pears, therefore, that the psychologism in inductive logic is, just like that in de-ductive logic, merely a superficial feature of certain marginal formulations, while the core of the theories remains thoroughly objectivistic" (342). He praised Ramsey for holding a position similar to his own, thereby revealing a deep misunderstand-ing of Ramsey's ideas, because Ramsey was a subjectivist and an upholder of that kind of psychologism that Carnap condemns.[4] Moreover, Ramsey opposed fre-quentism and, as we will see in the following pages, he made an attempt to make sense of objective chance within the subjectivist interpretation of probability. More will be said on the attitude toward subjectivism taken by Carnap in his late writings.

A fundamental aspect of the distinction between probability$_1$ and probabil-ity$_2$ lies in the fact that the value of probability$_2$ can be unknown "in the sense that we do not possess sufficient factual information for its calculation" (345). Proba-bility$_1$ cannot be said to be unknown in the same sense, though "it may, of course, be unknown in the sense that a certain logico-mathematical procedure has not yet been accomplished, that is, in the same sense in which we say that the solution of a certain arithmetical problem is at present unknown to us" (345). Probability$_2$ "has only one value" that is usually not known; what is known is the observed rel-ative frequency. One can speak of the best estimate of a probability$_2$ on the basis of a certain piece of evidence, but in this case one is referring to probability$_1$. In fact, probability$_1$ can also be seen as an estimate of probability$_2$, and this interpre-tation offers a way of bridging the gap between the two notions of probability. Probability$_2$ represents a physical magnitude, therefore a statement of probabil-ity$_2$ has to be established empirically, like any other statement regarding physical properties (like temperature) it can be tested in order to be confirmed or discon-firmed. According to this interpretation, we can have a statement of probability$_1$ expressing the degree of confirmation of a statement of probability$_2$, but it does not make sense to talk about a probability$_1$ of a probability$_1$, because a statement of probability$_1$ is, "like an arithmetical statement," analytically true or false. Two things are worth noticing here. First, when defining probability$_2$ Carnap adopts a realistic language. He says that "this use does not imply acceptance of realism as a metaphysical thesis but only of what Feigl calls 'empirical realism'" (345). Second, higher order logical probabilities are not admitted by Carnap, simply because statements of probability$_1$ are analytically true or false, and cannot be the object of other probability statements.

The interpretation of probability$_2$ developed in the forties remains unchanged in Carnap's later writings. In these writings Carnap concentrates on probability$_1$, in the conviction that probability$_2$, which he used to call "Big Rudi," had been sufficiently developed by others, such as Reichenbach, while probability$_1$, or "Little Rudi," still needed a lot of care in order to grow up.[5] As a matter of fact, Carnap's interpretation of probability$_2$ is similar to Reichenbach's. Commenting on the latter's work, Carnap says: "Since Reichenbach is one of the leading representatives of the frequency conception, it might at first appear as if our views must be fundamentally opposed. However, a closer examination of Reichenbach's argumentation shows that the two points of view are actually quite close to each other" (Carnap 1950, 175). Carnap regards as a strong analogy between himself and Reichenbach the latter's admission of an inductive notion of probability in addition to the frequentist one. Obviously, this also marks their main divergency, because Carnap interprets inductive probability as logical, while Reichenbach wants to attach a frequentist interpretation also to his notion of "weight." In this connection, Carnap remarks: "It seems to me that it would be more in accord with Reichenbach's own analysis if his concept of weight were identified instead with the *estimate* of relative frequency. If Reichenbach's theory is modified in this one respect, our conceptions would agree in all fundamental points" (176). Clearly, Carnap recognizes that he and Reichenbach are treading on the same ground, they both hold the need to define a statistical and an inductive notion of probability, meant to apply to confirmation. However, Carnap makes an important distinction between "inductive logic" and "rules of application," which is absent from Reichenbach's work. Were such a distinction applied to Reichenbach's theory of "primitive" and "advanced" knowledge, the analogy pointed out by Carnap in the above passage would become very strict.[6] Unfortunately, Reichenbach never included in his theory this distinction that would have helped to clarify the nature of the two stages of knowledge acquisition he defines.

For a long time Carnap shared Reichenbach's approach to the problem of induction. This is openly admitted in Carnap (1947a, 1945b), where he says "Reichenbach was the first to raise the problem of the justification of induction in a new sense and to take the first step towards a positive solution" (Carnap [1945b] 1972, 78). Elsewhere, Carnap is not so explicit but still retains the idea that the only viable justification of induction is based on its success. As we will see, in the sixties he changed his mind radically and turned to the idea of "inductive intuition".

The ideas sketched in the forties were fully developed in Carnap's major works of the fifties: *Logical Foundations of Probability* and *The Continuum of Inductive*

Methods. In these works the interpretation of probability$_1$ and probability$_2$ remains pretty much the same with respect to Carnap's earlier writings, though their relationship is clarified in more detail. Probability$_1$ "has its place in inductive logic and hence in the methodology of science," probability$_2$ "in mathematical statistics and its applications" (Carnap 1952, 5). Within the methodology of science, probability$_1$ plays a twofold role, being used both as a method of confirmation and as a method of estimation of relative frequencies. The task accomplished in *The Continuum of Inductive Methods* is that of showing that there is a complete correspondence between the two meanings of probability$_1$, in the sense that there is a one-to-one correspondence between the confirmation functions and the estimate functions and that these functions form a continuum within the specified logical calculus (a first-order calculus with identity). Once the continuum has been constructed, the problem of justification mingles with the problem of the choice of a particular inductive method. The problem does not receive a definite answer, though Carnap's suggestion still indicates in the success of inductive methods the canon for their choice.

In the sixties the notion of probability$_1$ underwent a significant change. This is clearly expressed in the preface to the second edition of *Logical Foundations* (1962; republished with some modifications in 1963a), and in "The Aim of Inductive Logic" (1962; republished in a modified and expanded version in 1971 as "Inductive Logic and Rational Decisions"). Of the three interpretations of probability$_1$ mentioned in the first edition of *Logical Foundations,* namely: (1) the degree of inductive support given to a hypothesis *h* by an evidence *e,* (2) a fair betting quotient, and (3) an estimate of relative frequency, the first is discarded "because of . . . [its] ambiguity" (Carnap 1963a, 67). Therefore, when it is not used as a method of estimation of relative frequencies, probability$_1$ is interpreted as a fair betting quotient. This is meant to provide a tool for "a rational reconstruction of the thoughts and decisions of an investigator" that "could best be made in the framework of a probability logic" (67). In this vein Carnap's late writings regard inductive logic as a theory of decision. Contextually, such writings incorporate a justification of the basic principles of inductive logic in terms of coherence, typical of the subjectivistic approach of Ramsey and de Finetti. Abner Shimony (1992) attributes to himself the merit of having called Carnap's attention to the fact that probability$_1$, interpreted as "the fair betting quotient for bets on *h,* given *e* as the only evidence," "must satisfy the condition of coherence" (Shimony 1992, 267, 269). Shimony describes this occasion as "the one time when I made Carnap happy," because this approach solved the vexed question of the adequacy of Carnap's confirmation functions, in the sense of their ability to satisfy the probability calculus.[7]

Carnap's appeal to coherence has fostered the opinion that in his late writings he became a subjectivist. As a matter of fact, in the second part of "A Basic System of Inductive Logic" Carnap labels his own position a "(modified) *subjectivist point of view*" (Carnap 1980, 112). In other passages of his late writings, however, he says that "the use of 'subjective' for the concept of personal probability seems to me highly questionable" (Carnap 1971, 13). Between Carnap's position and that of the upholders of a truly subjectivist point of view, such as de Finetti, there are deep divergences, of which he seems to be well aware. In "The Aim of Inductive Logic," after stating that his own concept of probability concerns "*rational* credence," Carnap observes that de Finetti "says explicitly that his concept of 'subjective probability' refers not to rational, but to actual beliefs" and adds: "I find this puzzling" (Carnap [1962] 1972, 108). He claims that the other subjectivists, like Ramsey, retain the notion of rational credence rather than that of actual belief. These views —which, insofar as Ramsey is concerned, are questionable[8]—testify to Carnap's attitude toward inductive logic as a theory of decision: it has to be a theory of *rational* decisions, dealing with *rational credence*. In this connection he clarifies that "*rational credence* is to be understood as the credence function of a completely rational person X; this is, of course, not any real person, but an imaginary, idealized person" (108).

Carnap distinguishes between "credence" and "credibility." While credence reflects the beliefs of an agent at certain specified times, credibility expresses his permanent dispositions for forming and changing his beliefs in the light of information. Credibility can also be seen as the initial credence of a hypothetical human being, before experiencing empirical data. It is credibility, not credence, that can offer a good basis for rational decision theory. In order to define the reasonableness of a person's credibility function, "a sufficient number of requirements of rationality" (Carnap 1971, 22) have to be fixed, like coherence, regularity, and symmetry. In this way a system of inductive logic is obtained, which has a normative function, because it can indicate the road to rational decision making.

Decision theory and, more specifically, the approach in terms of "beliefs, actions, possible losses, and the like" gives reasons for accepting the axioms and choosing among different credibility functions. In this way, purely logical constructs are selected on the basis of considerations that are not purely logical. Indeed, in Carnap (1980) the author admits that a λ-function can be chosen on a personal basis on account of subjective and contextual elements. But he warns that his theory is "not in the field of descriptive, but of normative decision theory. Therefore, in giving my reasons, I do not refer to particular empirical results concerning particular agents or particular states of nature and the like. Rather, I refer to a *conceivable* series of observations . . . , to conceivable sets of possible acts, to possible states of nature, to possible outcomes of the acts, and the like. These features are characteristic for an analysis of *reasonableness* of a given function ... in contrast with an investigation of the *successfulness* of the (initial or later) credence function of a given person in the real world. Success depends upon the particular contingent circumstances, rationality does not" (Carnap [1962] 1972, 117; also 1971, 26). Carnap's notion of rationality and the normative character ascribed to inductive logic, reflected by the above-mentioned passage, sets him far apart from the subjectivism of Ramsey and de Finetti, upholders of a descriptive approach to decision theory and probability.

The stress on the rationality of inductive methods, as opposed to their successfulness, makes Carnap abandon the pragmatic approach to the justification of induction, to embrace the notion of "inductive intuition." This position is expressed in the section called "An Axiom System for Inductive Logic" of (1963b) and in the article "Inductive Logic and Inductive Intuition." To the question as to what reasons can be given for accepting the axioms of inductive logic, Carnap answers that "the reasons are based upon our intuitive judgments concerning inductive validity, i.e., concerning inductive rationality of practical decisions (e.g. about bets)" (Carnap 1963b, 978). The concept of inductive intuition is fundamental to Carnap's attempt to keep inductive logic entirely within the field of a priori knowledge. The reasons for accepting the axioms that are suggested by intuition "are a priori"; they "are independent both of universal synthetic principles about the world, e.g. the principle of the uniformity of the world, and of specific past experiences, e.g., the success of bets which were based on the proposed axioms" (978–79). However, inductive intuition does not seem to provide a solid ground for justifying induction. As Salmon has pointed out, Carnap's solution comes "dangerously close to the view that induction needs no justification precisely because it is incapable of being justified" (Salmon 1967, 738).

The pragmatist turn regarding the choice of probability functions, showed by Carnap's late writings, is counterbalanced by his renunciation of the pragmatic justification of induction and by the increasingly aprioristic character ascribed to the notion of rationality. This is reflected by his conviction, expressed in Carnap (1963b), that "questions of rationality are purely a priori," and by his aversion to "the widespread view that the rationality of an inductive method depends upon factual knowledge, say, its success in the past" (981). It is this attitude toward rationality that makes his logicism irreconcilable with subjectivism.[9]

The peculiar combination of empiricism and apriorism that constitutes the specificity of Carnap's position is described in the following way by Richard Jeffrey, who names Carnap's approach "rationalistic Bayesianism": "One side is a purely rational, 'logical' element: a prior probability assignment M characterizing the state of mind of a newborn Laplacean intelligence ... The other side is a purely empirical element, a comprehensive report D of all experience to date. Together, these determine the experienced Laplacean intelligence's judgmental probabilities, obtained by conditioning the 'ignorance prior' M by the *Protokollsatz D*. Thus M (*H*/*D*) is the correct probabilistic judgment about *H* for anyone whose experiential data base is *D*" (Jeffrey 1992a, 2–3).

De Finetti's Subjectivism

Working independently in the mid-twenties, Frank Plumpton Ramsey in England and Bruno de Finetti in Italy worked out the subjective interpretation of probability. In a nutshell, this interpretation says that probability is a quantitative expression of the degree of belief in the occurrence of an event, entertained by a person in a state of uncertainty. In his celebrated paper "Truth and Probability," read at the Moral Sciences Club in Cambridge in 1926 but published posthumously in 1931 in the collection *The Foundations of Mathematics and Other Logical Essays*, Ramsey put forward for the first time a definition of probability as degree of belief and argued that the overall criterion of acceptability of degrees of belief is that of coherence. This means that subjective probabilities are to be assigned so that if they are used as the basis of betting ratios, they should not lead to a sure loss. This requirement ensures that subjective probabilities obey the usual rules of probability calculus. This result was actually proved by de Finetti, though it was certainly seen also by Ramsey. Starting from 1928, de Finetti, in a series of important articles culminating with "La prévision: Ses lois logiques, ses sources subjectives," the text of a series of lectures delivered in 1935 but published in 1937, added to the definition of probability in terms of coherent degrees of belief the important notion of "exchangeability." Taken together with Bayes's method, exchangeability, which is equivalent to the probabilistic property that Carnap calls "symmetry," added a dynamical dimension to Ramsey's "static" notion of subjective probability, providing it with direct applicability to statistical inferences. This was the decisive step toward a fully developed subjective definition of probability.[10]

De Finetti not only endowed subjectivism with its most powerful tool but also developed a philosophical position that combines pragmatism and empiricism with a radical form of probabilism. "Radical probabilism" is the expression used by Jeffrey to qualify de Finetti's position (see Jeffrey 1992a and 1992b), which de Finetti himself also calls "subjective Bayesianism" to stress the fundamental role played by Bayes's rule in his view of induction. De Finetti's philosophy of probability opposes the objectivistic character shared by frequentism and logicism to embrace a view of probability as a product of human activity, vindicated by its success rather than grounded on an abstract notion of rationality.

The starting point of de Finetti's radical probabilism is a refusal of the notion of truth, together with the related notions of "immutable and necessary" laws. "No science," he says, "will permit us say: this fact will come about, it will be thus and so because it follows from a certain law, and that law is an absolute truth. Still less will it lead us to conclude skeptically: the absolute truth does not exist, and so this fact might or might not come about, it may go like this or in a totally different way, I know nothing about it. What we can say is this: *I foresee* that such a fact will come about, and that it will happen in such and such a way, because past experience and its scientific elaboration by human thought make this forecast seem reasonable to me" (de Finetti [1931] 1989, 170). If these claims remind us of Reichenbach's position with respect to truth, the conclusion drawn from such premises is completely different, since for de Finetti the instrument enabling us to face the future is subjective probability. This is so because one can show not only that subjective probability is the sole noncontradictory notion of probability but also that it covers all uses of probability, in science and everyday life. All the other interpretations: the frequentist notion of von Mises and Reichenbach, the logical notion of Keynes and Carnap, and the classical notion of Laplace, are ill-founded in some way or other.

The main advantage of the subjective notion of probability is identified with the fact that it can be amenable to an operational definition. The latter can be made in terms of fair betting quotients, or of some other device, like the method based on penalties, on which more will be said in what follows. The operational character of subjective probability also works as a criterion of admissibility, offering the best and simplest possible foundation for it. Equally operational and pragmatical is the foundation of probabilistic inference, grounded in de Finetti's main result, namely the so-called representation theorem. This is a combination of exchangeability and Bayes's rule, which shows how information about frequencies interacts with subjective factors within statistical inferences. As clearly seen also by Carnap, who regarded symmetry as an important "requirement of rationality" to be imposed on probability functions, exchangeability allows learning from experience. For de Finetti, exchangeability is a subjective assumption, in the sense that its adoption depends on a personal choice. The choice of priors is again subjective, but the representation theorem proves that under exchangeability subjective probabilities and observed frequencies approach each other, thereby allowing the best possible predictions. Within this perspective the acceptability of inductive inferences depends on their success, which rests on the behavior of exchangeable functions, embedded in a Bayesian inferential method. By giving a mathematical argument that shows the coherence of our inductive habits, de Finetti intends to offer a pragmatical solution to Hume's problem of induction, and he claims to proceed one step further along the same road that Hume indicated, namely in the direction of constructing a purely empiricist theory of induction.

Scientific Rationality

Some aspects of de Finetti's perspective are worth mentioning, to clarify the differences between his position and that of the authors we have dealt with in the preceding sections. First of all, for de Finetti probability judgments involve more than a combination of empirical evidence and aprioristic criteria: a whole array of factors can influence probability assessments, so that it is perfectly possible for two people to assign a different probability to the occurrence of an event, on the basis of the same evidence. De Finetti's perspective is "anti-rationalist," in the sense that "the subjective theory . . . does not contend that the opinions about probability are uniquely determined and justifiable. Probability does not correspond to a self proclaimed 'rational' belief, but to the effective personal belief of anyone" (de Finetti 1951, 218). As we saw, this feature of de Finetti's position is criticized by Carnap. Conversely, de Finetti, commenting on Carnap's "The Aim of Inductive Logic," claims to be in agreement with Carnap "if his 'credibility function' is interpreted simply as subjective personal probability" (de Finetti 1972, 183). Indeed, this does not seem to be the case.

According to de Finetti, when new judgments are made in the light of new information, one cannot say that the evaluating subject "changes opinion," but rather that he updates his opinion. "If we reason according to Bayes's theorem, we do *not* change opinion. We keep the same opinion and we update it to the new situation. If yesterday I said 'Today is Wednesday', today I say 'It is Thursday'. Yet, I have not changed my mind, for the day following Wednesday is indeed Thursday" (de Finetti 1995, 100). Arguing along these lines, de Finetti opposes Reichenbach's idea that probability assignments are embedded in a self-correcting procedure, approaching some "true" probability value. At the same time, de Finetti rejects the idea of "truthlikeness," entertained in the first place by the Popper's school. In this connection, he claims that "if we knew the truth we would not approach it, we would reach it. If we don't know the truth, we don't even know how far it is" (119).

Against frequentism, de Finetti argues that the linkage between probability and frequency is a matter for demonstration and therefore should not be taken for granted and postulated as frequentists do. He also criticizes the fundamental assumptions of frequentism, like the notion of randomness or the idea of an infinite sequence of experiments, which he considers "meaningless" (130). To be sure, de Finetti does not deny that frequency values can "suggest probability attributions close to such frequencies" (218), what he opposes is the pretense to *define* probability in terms of frequencies. He distinguishes between the definition of probability and its evaluation and claims that while subjectivism recognizes this distinction and keeps its components separate, the upholders of the other interpretations of probability confuse them: they look for a unique criterion and ground on it both the definition and the evaluation of probability. This mistake is also made by logicists, who put symmetry considerations at the basis of the definition of probability.

In connection with logicism, he remarks that grounding the notion of probability on the logical properties of propositions, as Carnap does, means "to build it on 'the structure of nothing', or, in other words, on the attribution of a realistic content to logical-mathematical expressions which are purely formal or symbolic" (de Finetti [1938] 1985, 86).[11] Unlike the upholders of other interpretations of probability, subjectivists do not commit themselves to a single criterion or to one particular method of estimating probability; they regard all coherent functions as admissible and leave the choice of one particular method to the evaluating subject, who operates his choice within a certain context. While opposing the other interpretations of probability, de Finetti also rejects non-Bayesian statistics, including all kinds of testing procedures and methods for randomizing.

De Finetti's subjectivism should not be thought of as a perspective where "anything goes." If the explicit recognition of the role played by subjective elements in the assessment of probability is for him a prerequisite for a correct appraisal of objective elements, a probability judgment results from "the conjunction of both objective and subjective elements at our disposal" (de Finetti 1973, 366). In fact, "subjectivism is one's *degree of belief* in an outcome, based on an evaluation making the best use of all the information available to him and his own skill . . . Subjectivists . . . believe that every evaluation of probability is based on available information, including objective data" (De Finetti 1974b, 16). Therefore, probability evaluations are here too the result of two components: "(1) the objective component, consisting of the evidence of known data and facts; and (2) the subjective component, consisting of an opinion concerning unknown facts based on known evidence" (de Finetti 1974a, 7). However, the interpretation attached to them by de Finetti is quite distant from Carnap's. First, de Finetti warns that the objective component, namely factual evidence, is in many ways context dependent: the collection of evidence must be made carefully and skillfully, and its exploitation depends on the judgment of what elements are relevant to the problem under consideration and useful to the evaluation of related probabilities. Equally context dependent and subjective is the decision on how to let objective elements influence belief; as we also saw, the adoption of exchangeability is seen as a free decision of the evaluating subject. When information on frequencies is available, probability judgments will typically depend on it, and indeed much information of this kind makes the evaluation of probability easy. Scant information, instead, raises the problem of good probability appraisers.

The issue is addressed by de Finetti in a series of articles of the sixties and seventies, where he adopted an approach based on penalty methods, such as scoring rules of the kind of the well-known "Brier's rule." Scoring rules are inspired by the idea that the resulting device should oblige those who make probability evaluations to be as accurate as they can, and, in case they have to compete with others, to be honest.[12] A rule of this kind, working as a measure of the success of predictions, is apt to improve probability evaluations. Within a subjectivistic framework, such methods of calibration have the meaning of measures of the "goodness of the evaluation" (de Finetti 1962, 360).

If devices like scoring rules or other methods of calibration can improve the evaluation of probability, the latter is and remains subjective. On this point, de Finetti's position is uncompromising. He always denied that an objective meaning can be attached to probability. As is well known, he wanted the sentence "Probability does not exist!" printed in capital letters in the preface to the English edition of his *Theory of Probability*. By this he meant to say that an objective notion of probability is meaningless, because probability can only represent the degree of belief of the person who evaluates it, and as such it is always known and definite. The notion of "unknown probability" is banished from de Finetti's subjectivism.[13] As a consequence of his radical subjectivism, de Finetti never felt the need to tackle notions like "chance" or "probability in physics." For him, science is just the continuation of everyday life, and it does not require any special notion of probability, besides the subjective one. Only the volume *Filosofia della probabilità* contains some remarks that point to the idea that in science we encounter certain evaluations of probability that, being determined by scientific theories, have a stronger character than those encountered in everyday life (see de Finetti 1995).[14] For instance, referring to the probability distributions belonging to statistical mechanics, de Finetti claims that "they provide more solid grounds for subjective opinions" (de Finetti 1995, 117).

The idea behind de Finetti's hint is developed by Ramsey, who defines "chance" and "probability in physics" within the framework of subjective probability. To accomplish this aim, Ramsey refers these notions to systems of beliefs that typically include scientific laws and theories. Since everybody agrees on scientific theories, such systems of beliefs are stronger than the actual beliefs encountered in everyday life, and this confers a peculiarly "objective" character to those probability assignments that are grounded upon them. The notion of "objectivity" involved in this view is pragmatical, like the conception of theories underlying it. Chance attributions, like all general propositions belonging to theories, are taken by Ramsey as "variable hypotheticals" or "rules for judging" that provide tools with which the users meet the future. This, in turn, combines with a pragmatical notion of truth, according to which a "true scientific system" is a system on which the opinion of everyone, supported by experimental evidence, will eventually converge.[15] Not surprisingly, Ramsey's position is far more sophisticated than that of de Finetti, a working mathematician, not trained in philosophy within a formidable milieu like that surrounding Ramsey in the Cambridge of the twenties. However, their ideas point in the same direction. In view of this, it is perfectly viable to think of a combination of de Finetti's conception of probabilistic inference with Ramsey's notion of objective chance, which would allow for a mature form of subjectivism, apt to provide a basis for an interpretation of probability in tune with a pragmatist and nonrealistic epistemology.

Conclusions

Reichenbach, Carnap, and de Finetti represent divergent conceptions of probability, in many ways irreconcilable, despite the fact that all of these authors share an empiricist approach rooted in the work of authors like E. Mach and H. Poincaré.[16] In addition to this common background, Reichenbach and Carnap share the logical empiricist matrix, while de Finetti embraces the pragmatism of C. S. Peirce and W. James, though filtered through the work of the Italian thinkers M. Calderoni, G. Vailati, and A. Aliotta. The main consequence of this philosophical choice is the different attitude taken toward rationality. Albeit based on different notions of probability, the theories of both Carnap and Reichenbach are sustained by a strong notion of rationality, while de Finetti's perspective is deeply antirationalist. In addition, de Finetti's subjectivism is inspired by what today we would call an antirealist philosophy, whereas Reichenbach and Carnap seem rather oriented toward some form of realism, at least taken as methodological realism.

The most fruitful attempt at carrying on Carnap's program of inductive logic is due to Jeffrey, whose probability kinematics opens the door to the treatment of uncertain evidence and includes such elements as interval-valued degrees of belief, higher order probabilities, and imperfect information. In order to accomplish this, Jeffrey has to distance himself from Carnap's position, to embrace a form of radical probabilism that brings him close to the subjectivism of Ramsey and de Finetti. More particularly, Jeffrey moves away from Carnap's way of interpreting the two elements of probability judgments, namely the aprioristic and empirical component. For him empirical evidence is not to be regarded as an indisputable datum, and standards of rationality represent cultural artifacts rather than aprioristic canons. Jeffrey's radical probabilism is deeply pragmatist and characterizes probability judgments as the result of a congeries of factors, of which some pertain to the evaluating subject and others are essentially social products.[17] Indeed, a move toward pragmatism looks like the inevitable price one has to pay in order to keep alive Carnap's dream to characterize the structure of the inductive process leading to acquisition of knowledge and to single out its assumptions.

Reichenbach's philosophy of science, as we noted, is less compromised than Carnap's with reductionism and apriorism. The author who has made the most strenuous efforts to carry on Reichenbach's program is Wesley Salmon (see especially Salmon 1984). Developing some of Reichenbach's ideas, like his probabilistic conception of causality, Salmon has worked out a sophisticated epistemology, which shares Reichenbach's ideal of objectivity based on frequentism, meant to suggest also a pattern of rationality. Salmon's theory of explanation elegantly combines probability with the ideal of mechanical explanation and can be considered the most outstanding development of Reichenbach's work. However, Salmon's view raises various problems that shed doubt on the idea that frequentism can do the job of providing a sound basis for epistemology.[18] Frequencies are undeniably important and should be taken into account, when available, but it is questionable whether frequentism is the right way to go.

Recent trends in epistemology are more oriented toward subjectivism, which seems to offer a richer and more flexible framework, apt to accommodate the complex procedures of knowledge acquisition. Unlike de Finetti, however, various authors, including Jeffrey, think that talking about "objective chance" is not only meaningful but also important and fruitful. Ramsey has pointed the way to go to account for this notion within the framework of subjective probability. A further elaboration of his hints in this connection constitutes a worthwhile task for future research.

Embracing subjectivism entails adopting a pragmatical view of rationality, which is neither a priori nor a posteriori but results from the concurrence of logical and empirical elements, as well as personal and social factors. Though weaker than that retained under various forms by logical empiricists, this kind of rationality is apt to embed stronger ideas, like Salmon's demand that action be guided by knowledge of causal links, an idea that has great heuristic impact.[19] The viability of this kind of rationality and the possibility of combining it with other notions, such as "objective" probability, and the adoption of various statistical methodologies, are also open to further research.

Notes

A slightly different version of this paper was read at the "Workshop at the University of Pittsburgh Dedicated to Themes from the Work of Wesley Salmon: Induction/Probability and Causation/Explanation," November 18, 2000. A few months later Wes died tragically. This paper is dedicated to his memory.

1. On Hempel's views on probability and truth, see Hempel (2000).
2. See note 1.
3. This was argued by Gigerenzer (1991).

4. A psychologistic attitude toward probability is expressed by Ramsey in the note "Induction: Keynes and Wittgenstein," written in 1922 and published in Ramsey (1991). Some remarks on this point are to be found in Galavotti (1991).

5. This story is told by Salmon (1979, 15).

6. I owe this remark to Wesley Salmon, who pointed it out to me in conversation.

7. See Shimony (1992) for details on the problem of adequacy of Carnap's confirmation functions.

8. Some remarks on Ramsey's attitude toward rationality are to be found in Galavotti (1997).

9. The difference between logicism and subjectivism is seen by Schramm (1993). However, Schramm seems to believe that Carnap in his late writings adhered to subjectivism.

10. For a comparison of Ramsey's and de Finetti's subjectivism, see Galavotti (1991).

11. A more articulated discussion of de Finetti's criticism of logicism and frequentism is to be found in Galavotti (1989).

12. For more detail on scoring rules, see de Finetti (1962, 1970, [1970] 1975, 1980, 1981). Some useful considerations on the problem of evaluation are also to be found in de Finetti (1974a, 1974b).

13. De Finetti argues that the representation theorem gets rid of the notion of "objective but unknown" probability by reducing it to subjective probability. This is explained in detail in Galavotti (1989).

14. The book contains the text of a series of lectures delivered by de Finetti in 1979 that have been transcribed from a tape recording and edited, together with the discussion following each lecture, by Alberto Mura.

15. For a more detailed analysis of Ramsey's notion of "chance," see Galavotti (1995). Some remarks on the topic are also to be found in Galavotti (1999a).

16. In de Finetti (1931) the author explicitly recognizes the influence of these thinkers upon his own ideas. On the other hand, their influence on logical empiricism is well known.

17. See Jeffrey (1992a, 1992b). Jeffrey has dedicated a number of articles to Carnap's inductive logic (1973, 1975, 1991).

18. For a discussion of Salmon's work, see Galavotti (1999b).

19. A perspective of this kind is to be found in Galavotti (1999b, 2001).

References

Carnap, R. 1936. "Wahrheit und Bewährung." *Actes du Congrès International de Philosophie Scientifique*, 18–23. Paris: Hermann.

———. 1936–37. "Testability and Meaning." *Philosophy of Science* 3:419–71, 4:1–40.

———. 1945a. "The Two Concepts of Probability." *Philosophy and Phenomenological Research* 5:513–32. Reprinted in Feigl and Sellars (1949, 330–48).

———. 1945b. "On Inductive Logic." *Philosophy of Science* 12:72–97. Reprinted in Luckenbach (1972, 51–79).

———. 1946. "Remarks on Induction and Truth." *Philosophy and Phenomenological Research* 6:590–602.

———. 1947a. "Probability as a Guide in Life." *Journal of Philosophy* 44:141–48.

———. 1947b. "On the Application of Inductive Logic." *Philosophy and Phenomenological Research* 8:133–48.

———. 1949. "Truth and Confirmation." In Feigl and Sellars, (1949, 119–27).

———. 1950. *Logical Foundations of Probability.* Chicago: Chicago University Press. Second edition with modifications 1962.

———. 1952. *The Continuum of Inductive Methods.* Chicago: Chicago University Press.

———. 1953. "Inductive Logic and Science." *Proceedings of the American Academy of Arts and Sciences* 53:189–97.

———. [1962] 1972. "The Aim of Inductive Logic." In *Logic, Methodology, and Philosophy of Science,* ed. E. Nagel, P. Suppes, and A. Tarski, 303–18. Stanford: Stanford University Press. Reprinted in Luckenbach (1972, 104–20).

———. 1963a. "Remarks on Probability." *Philosophical Studies* 14:65–75.

———. 1963b. "Replies and Systematic Expositions." In Schilpp (1963, 859–1013).

———. 1968. "Inductive Logic and Inductive Intuition." In *The Problem of Inductive Logic,* ed. I. Lakatos. Amsterdam: North Holland, 258–67.

———. 1971. "Inductive Logic and Rational Decisions." In *Studies in Inductive Logic and Probability,* vol. 1, ed. R. Carnap and R. C. Jeffrey, 7–31. Berkeley: University of California Press (modified version of 1962). This volume also contains "A Basic System of Inductive Logic, Part 1."

———. 1975. "Notes on Probability and Induction." In *Rudolf Carnap, Logical Empiricist,* ed. J. Hintikka. Dordrecht-Boston: Reidel, 293–324.

———. 1980. "A Basic System of Inductive Logic, Part 2." In *Studies in Inductive Logic and Probability,* vol. 2, ed. R. C. Jeffrey, 7–156. Berkeley: University of California Press.

de Finetti, B. [1931] 1989. "Probabilism." *Erkenntnis* 31:169–223. English translation of "Probabilismo," *Logos* (1931): 3–70.

———. [1937] 1964. "Foresight: Its Logical Laws, Its Subjective Sources." In *Studies in Subjective Probability,* ed. H. E. Kyburg and H. E. Smokler, 95–158. New York: Wiley. English translation of "La Prévison: Ses lois logiques, ses sources subjectives." *Annales de l'Institut Henri Poincaré* 7:1–68.

———. [1938] 1985. "Cambridge Probability Theorists." *Rivista di matematica per le scienze economiche e sociali* 8:79–91. English translation of "Probabilisti di Cambridge." *Supplemento statistico ai Nuovi problemi di Politica, Storia ed Economia* 4:21–37.

———. 1951. "Recent Suggestions for the Reconciliation of Theories of Probability." In *Proceedings of the Second Berkeley Symposium on Mathematical Statistics and Probability,* 217–25. Berkeley: University of California Press.

———. 1962. "Does It Make Sense to Speak of 'Good Probability Appraisers'?" In *The Scientist Speculates: An Anthology of Partly-Baked Ideas,* ed. I. J. Good, A. J. Mayne, J. M. Smith, 357–64. New York: Basic Books.

———. 1970. "Logical Foundations and Measurement of Subjective Probability." In *Subjective Probability* 36 (1920): 129–45, ed. G. de Zeeuw (*Acta Psychologica* 34).

———. [1970] 1975. *Theory of Probability.* New York: Wiley. English translation of *Teoria delle probabilità* (Torino: Einaudi).

———. 1970. "Logical Foundations and Measurement of Subjective Probability." *Acta Psychologica* 34:129–45.

———. 1972. *Probability, Induction, and Statistics.* New York: Wiley.

———. 1973. "Bayesianism: Its Unifying Role for Both the Foundations and the Applications of Statistics." *Bulletin of the International Statistical Institute, Proceedings of the Thirty-ninth Session,* 349–68.

———. 1974a. "The Value of Studying Subjective Evaluations of Probability." In *The Con-*

cept of Probability in Psychological Experiments, ed. C.-A. S. Staël von Holstein, 1–14. Dordrecht: Reidel.

———. 1974b. "The True Subjective Probability Problem." In *The Concept of Probability in Psychological Experiments,* ed. C.-A. S. Staël von Holstein, 15–23. Dordrecht: Reidel.

———. 1980. "Probabilità." In *Enciclopedia Einaudi,* vol. 10: 1146–87. Torino: Einaudi.

———. 1981. "Discussion: The Role of 'Dutch Books' and of 'Proper Scoring Rules.'" *British Journal for the Philosophy of Science* 32:55.

———. 1992. *Probabilità e induzione (Induction and Probability).* Ed. P. Monari and D. Cocchi. Bologna: CLUEB (published both in Italian and in English).

———. 1995. *Filosofia della probabilità.* Ed. A. Mura. Milan: Il Saggiatore.

Feigl, H., and W. Sellars, eds. 1949. *Readings in Philosophical Analysis.* New York: Appleton-Century-Crofts.

Galavotti, M. C. 1989. "Anti-realism in the Philosophy of Probability: Bruno de Finetti's Subjectivism." *Erkenntnis* 31:239–61.

———. 1991. "The Notion of Subjective Probability in the Work of Ramsey and de Finetti." *Theoria* 57:239–59.

———. 1995. "F. P. Ramsey and the Notion of 'Chance.'" In *The British Tradition in the Twentieth Century Philosophy. Proceedings of the Seventeenth International Wittgenstein Symposium,* ed. J. Hintikka and K. Puhl, 330–40. Vienna: Holder-Pichler-Tempsky.

———. 1997. "Probabilism and Beyond." *Erkenntnis* 45:253–65.

———. 1999a. "Some Remarks on Objective Chance (F. P. Ramsey, K. R. Popper and N. R. Campbell)." In *Language, Quantum, Music,* ed. L. Dalla Chiara, M. R. Giuntini, and F. Laudisa, 73–82. Dordrecht: Kluwer.

———. 1999b. "Wesley Salmon on Explanation, Probability, and Rationality." In *Experience, Reality, and Scientific Explanation,* ed. M. C. Galavotti and A. Pagnini. Dordrecht: Kluwer, 39–54.

———. 2001."Causality, Mechanisms, and Manipulation." In *Stochastic Causality,* ed. M. C. Galavotti, P. Suppes, and D. Costantini, 1–13. Stanford: CSLI Publications.

Gigerenzer, G. 1991. "From Tools to Theories: A Heuristics of Discovery in Cognitive Psychology." *Psychological Review* 98:254–67.

Hempel, C. G. 2000. *Selected Philosophical Writings.* Ed. R. C. Jeffrey. Cambridge: Cambridge University Press.

Jeffrey, R. C. [1965] 1983. *The Logic of Decision.* Chicago: University of Chicago Press.

———. 1973. "Carnap's Inductive Logic." *Synthèse* 25:299–306.

———. 1975. "Carnap's Empiricism." In *Induction, Probability, and Confirmation,* ed. G. Maxwell and R. M. Anderson. Minneapolis: University of Minnesota Press, 37–49.

———. 1991. "After Carnap." *Erkenntnis* 35:255–62.

———. 1992a. "Introduction: Radical Probabilism." In *Probability and the Art of Judgment,* 1–13. Cambridge: Cambridge University Press.

———. 1992b. "De Finetti's Radical Probabilism." In de Finetti (1992, 263–75).

Luckenbach, S. A., ed. 1972. *Probabilities, Problems, and Paradoxes.* Encino-Belmont, Calif.: Dickenson.

Ramsey, F. P. 1931. *The Foundations of Mathematics and Other Logical Essays.* Ed. R. B. Braithwaite. London: Routledge and Kegan Paul.

———. 1991. *Notes on Philosophy, Probability, and Mathematics.* Ed. M. C. Galavotti. Naples: Bibliopolis.

Reichenbach, H. [1933] 1949. "The Logical Foundations of the Concept of Probability." In

Feigl and Sellars (1949, 305–23). English translation with modifications of "Die logischen Grundlagen des Wahrscheinlichkeitsbegriffs," *Erkenntnis* 3:401–25.

———. [1933] 1978. "Carnap's *Logical Structure of the World.*" *Kantstudien* 38: 199–201.
Reprinted in English in Reichenbach (1978, I:405–8).

———. 1935. *Wahrscheinlichkeitslehre.* Leyden: Sijthoff.

———. 1936. "Logistic Empiricism in Germany and the Present State of Its Problems."
Journal of Philosophy 6:141–60.

———. 1937. "La philosophie scientifique: Une esquisse de ses traits principaux." In
Travaux du IX Congrès International de Philosophie, 86–91. Paris: Hermann.

———. 1938. *Experience and Prediction.* Chicago: University of Chicago Press.

———. 1949. *The Theory of Probability.* Berkeley: University of California Press. 2d ed.
1971. English translation of Reichenbach (1935) with modifications.

———. 1951. "The Verifiability Theory of Meaning." *Proceedings of the American Academy
of Arts and Sciences* 53:46–60.

———. 1978. *Selected Writings, 1909–1953.* Vols. 1–2. Ed. M. Reichenbach and R. S. Cohen.
Dordrecht: Reidel.

Salmon, W. C. 1967. "Carnap's Inductive Logic." *Journal of Philosophy* 64:725–39.

———. 1979. "The Philosophy of Hans Reichenbach." In *Hans Reichenbach: Logical Empiricist,* ed. W. C. Salmon, 1–84. Dordrecht: Reidel.

———. 1984. *Scientific Explanation and the Causal Structure of the World.* Princeton:
Princeton University Press.

———. 1991. "Hans Reichenbach's Vindication of Induction." *Erkenntnis* 35:99–122.

———. 1994. "Carnap, Hempel, and Reichenbach on Scientific Realism." In *Logic, Language, and the Structure of Scientific Theories,* ed. W. C. Salmon and G. Wolters, 237–54.
Pittsburgh and Konstanz: University of Pittsburgh Press and Universitätsverlag.

Schilpp P. A., ed. 1963. *The Philosophy of Rudolf Carnap.* La Salle, Ill.: Open Court.

Schramm, A. 1993. "Zwei Theorien der Induktion—Reichenbach und Carnap." In *Wien-
Berlin-Prag: Der Aufstieg der wissenschaftlichen Philosophie,* ed. R. Haller and F. Stadler,
538–54. Vienna: Hölder-Pichler-Tempsky.

Shimony, A. 1992. "On Carnap: Reflections of a Metaphysical Student." *Synthèse* 93:261–74.

Smooth Lines in Confirmation Theory

Carnap, Hempel, and the Moderns

MARTIN CARRIER

Bielefeld University

The past two decades have seen a reviving interest in logical empiricism. Distance in time typically engenders the equanimity characteristic of the historical attitude. The historical perspective mellows old front lines and tends to transform previous warfare into nuanced analysis and more detached consideration. After Karl Popper's attacks on Rudolf Carnap have shifted away from the philosophical combat zones and after the battle smoke of the fighting over theory change, as initiated by Thomas Kuhn's paradigm approach, has cleared, it is precisely this historical attitude that has begun to dominate the study of logical empiricism. If winning or losing is no longer the issue, the woodcut type approach of marked contrasts gives way to expounding multifaceted shades and variegation. Continuing scholarly occupation with logical empiricism has brought to light a wide variety of views among its advocates and has shown that there never was anything like a monolithic Viennese ideology. Moreover, it never was the primary objective of logical empiricists to debunk statements as meaningless gibberish. Rather, the top of the agenda was characterized by more positive challenges. For instance, the double-language model paved the way to the eventual recognition of a genuine contribution of theory to conferring meaning to scientific terms (Irzik and Grünberg 1995, 289–90). In sum, logical empiricism was more contrasting and more modern than its opponents would have it.

I wish to broaden this recent reevaluation of logical empiricism by bringing confirmation theory into the picture. Carnap, Hans Reichenbach, and Carl Hempel contributed to laying the foundations of present-day confirmation theory. So it was logical empiricists, among others, who planted the seeds that were later brought to fruition.

Reshaping the Ground in Confirmation Theory

Conventional wisdom has it that science provides us with insights that are more accurate and more trustworthy than the commonsense views we entertain in everyday life. Most of us are willing to grant a high degree of credibility to scientific tenets. The reason is that we take such tenets to be better confirmed than the usual folklore, which in turn places the issue of the nature of scientific confirmation on the agenda.

Let me prepare my sketch presentation of logical-empiricist contributions to this issue by delineating in broad strokes the most striking lines in the development of confirmation theory. This development is characterized by two peaks, separated by a period of roughly two decades in which the relevant interest dropped to a low. The first peak extends from roughly 1930 to 1960 and includes the relevant work of logical empiricists. Outstanding among them are Reichenbach's *Experience and Prediction* (1938), Hempel's "Studies in the Logic of Confirmation" (1945), and Carnap's *Logical Foundations of Probability* (1950). All these accounts placed the logical or probabilistic relationship between a hypothesis and a piece of evidence at the center of consideration. It was asked whether or to which degree a given observation report was capable of supporting a given assumption. Within this framework it was conceded at once that general theories are in need of empirical assessment as well. However, the evaluation of individual hypotheses was considered primary. Overarching theories were construed as collections of hypotheses so that the support of the former was thought to be derivable from the confirmation of the latter.

The publication of Kuhn's *Structure of Scientific Revolutions* (1962) shifted the focus to a different ground and initiated a distinct approach to the evaluation of scientific achievements. Let me call this approach the tradition of theory change, as opposed to confirmation theory proper. Kuhn regarded the issue of the viability of comprehensive theoretical traditions or paradigms as the fundamental one.

The evaluation of individual hypotheses was held to be inextricably tied up with the credibility of the associated paradigm. Thus, the theory-change tradition replaced the emphasis on individual assumptions by a holistic approach according to which large-scale theories constitute the primary object of scientific assessment. Second, the historical development of a theory was considered essential in addition. In the Kuhnian framework, the paradigmatic principles are decreed exempt from critical examination during the periods of normal science but might well be subjected to severe tests in the times of crisis or imminent revolution (Kuhn 1962, 35–37, 66–68).

Imre Lakatos's methodology represents another approach within the theory-change tradition. According to Lakatos, one of the chief measures of acceptability is whether a theory manages to anticipate relevant data or whether it merely accommodates observations after their discovery and only responds to anomalies. Although two theories might be able to account for the same set of data, their support by these data might yet be different. If one of the theories predicted the relevant findings while the other had to be adapted to new problems and fitted to known results, the former is buttressed more strongly by these data than the latter (Lakatos 1970, 32–36; Carrier 2002).

In sum, there is a twofold contrast between confirmation theory and the theory-change tradition as to the evaluation of scientific achievements. First, confirmation theory considers the appraisal of individual hypotheses in light of the relevant evidence as the core and kernel of judging scientific merit. The tradition of theory change, by contrast, places comprehensive theories at the focus and regards confirmation as a profoundly holistic endeavor. Second, confirmation theory distinguishes logical or probabilistic relations between the data and the theoretical claims as exclusively relevant for assessing these claims. The theory-change tradition, on the other hand, additionally features historical patterns of development and temporal relations between the enunciation of a theory and the discovery of the pertinent empirical effects.

The thriving of the theory-change tradition corresponded to the eclipse of confirmation theory. The latter was revived after about 1980 when interest in theory change began to fade. This period constitutes the beginning of the second peak of confirmation theory. It is characterized, first, by Clark Glymour's bootstrap model of confirmation, which takes up essential principles from Hempel's entailment approach, and, second and predominantly, by Bayesian confirmation theory, which incorporates elements of the probabilistic accounts of Carnap and Reichenbach. In each case the later theories owe much to their logical-empiricist sources. Confirmation theory can be regarded as a thoroughly cumulative endeavor. When confirmation began to re-attract attention and the second peak came into view, the work was resumed at the locus, where it was dropped decades ago. Today's theories of confirmation have smoothly grown out of their logical-empiricist predecessors. My chief example is Hempel's approach, which profoundly shaped the bootstrap model. But let me begin by offering some remarks on the joint characteristics of probabilistic theories of confirmation.

Probabilism in Confirmation Theory

Underlying all relevant approaches is the attempt to clarify confirmation by taking advantage of the probability calculus. Degrees of confirmation are intended to be captured by the probability of a hypothesis in light of the available evidence. The differences between Carnap and Reichenbach, on the one hand, and present-day Bayesianism, which largely goes back to Bruno de Finetti, on the other, predominantly concern two aspects. Bayesianism of this sort is characterized, first, by the commitment to Bayes's theorem as the measure of empirical support and, second, by the subjective interpretation of the relevant probabilities (and of prior probabilities in particular). Probabilities of hypotheses are supposed to rely on or to merely express personal belief strengths that are only constrained by the condition that their relations are in agreement with the probability calculus (Howson and Urbach 1989, 10, 67).

On both counts, Carnap and Reichenbach held dissenting views. Their approaches to confirmation differ in detail but agree on invoking the theorem of conditional probability rather than Bayes's theorem, and on trying to confer objective meaning to all relevant probabilities. Carnap's inductive logic employs language-dependent measures of the ratio of the favorable cases to the possible ones; Reichenbach's approach is based on frequentist assumptions. Adopting Bayes's theorem rather than conditional probability marks an important improvement but does not indicate an in-principle change. Bayes's theorem features the likelihood or expectedness of data, which is crucial to the reconstruction of theory choice and theory change. Still, this is a technical point that in no way indicates a switch in the fundamental orientation. In this vein, Colin Howson and Peter Urbach, in their now standard account of Bayesianism, regard Bayesian confirmation theory as the continuation of inductive logic by other means (Howson and Urbach 1989, 290; Howson 1997, 278).

The gap between the objective probabilities of the empiricist tradition and the subjective probabilities of standard Bayesianism is deeper and has wider ramifications. Still, there are factions in the present debate, Wesley Salmon prominently among them, who advocate an objective interpretation of probabilities within the

Bayesian framework (Salmon 1967, 115–31; Salmon 1990, 180–87). All in all, there is a striking continuity between the early and the modern debates. In spite of superficial dissimilarities, strong conceptual ties bind Carnap's views of confirmation to latter-day Bayesianism. This bears testimony to the fecundity of logical empiricism. In confirmation theory logical empiricism was not abandoned but gradually transformed. Probabilistic confirmation theory proceeded along smooth lines.

Hempel's Entailment Account of Hypothesis Confirmation

I want to support this claim of gradual transformation by reviewing one example more extensively and more carefully. This example concerns the relationship between Hempel's entailment account of confirmation and Glymour's bootstrap model. So I set aside the grand lines for the moment and give a more detailed account of what I take to be the salient aspects of the transition.

Hempel's account of confirmation is based on the intuition that a hypothesis is confirmed by its positive instances rather than its consequences—as the hypothetico-deductive approach has it. Hypothetico-deductive confirmation involves the derivation of the data from tentatively presumed assumptions. If these empirical consequences are found in experience, the theoretical premises of the deduction are considered confirmed. Thus, the key to hypothetico-deductivism is what Hempel calls the *"converse consequence condition."* This condition stipulates that if an observation report e confirms a hypothesis h, which is in turn implied by another hypothesis h^* (that is, $h^* => h$), then e also confirms h^*. In the hypothetico-deductive framework, it is in virtue of this condition that pieces of evidence are able to bear out general theories. Suppose the observation of a freely falling body can be accounted for by the law of free fall and thus confirms this law. Since the law is a consequence of Newtonian mechanics (drawing on appropriate initial and boundary conditions in addition), the observation supports the overarching theory as well. Hempel rejects the converse consequence condition. His objection is that the hypothetico-deductive approach is circular: the move from the correctness of the consequences to the soundness of the premises is nondeductive and for this reason in need of justification. However, any such justification presupposes the very concept of confirmation one is about to establish in the first place (Hempel 1945, 28–29, 32–33).

Hempel's objective is to give an explication of a qualitative relation of confirmation that moreover is, in contrast to hypothetico-deductivism, deductively valid. The aim is to clarify what it means that a piece of evidence e confirms a hypothesis h: $c(h,e)$. Hempel's suggestion is that confirmation involves the derivation of the hypothesis from the data rather than the other way around. The basic relation is *instantiation:* hypotheses are confirmed by their positive instances. An instance of a hypothesis is characterized by the fact that variables contained in it are replaced by individual constants. That is, $\forall x\,(Px \rightarrow Qx)$ is a hypothesis; $Pa \wedge Qa$ is a positive instance. These instances are observation statements; they ascribe certain properties to individual objects. The general assumption that all pixies are quirky is instantiated by Archibald, the quirky pixy. Observation statements are critical to confirmation; hypotheses are appraised on their basis. The key intuition is captured by Hempel's "*entailment condition:* Any sentence which is entailed by an observation report is confirmed by it" (Hempel 1945, 31; my emphasis). That is, a hypothesis is confirmed if instantiations of this hypothesis follow from the data.

However, as Hempel proceeds, this account has the awkward consequence that confirmation relations may depend on the formulation of a hypothesis. In particular, logically equivalent hypotheses may come out confirmed differently by the same data. Consider the contrapositive to the mentioned schematic hypothesis: $\forall x\,(\neg Qx \rightarrow \neg Px)$. What is nonquirky is not a pixy. This version is no longer instantiated by Archibald, the quirky pixy. The evidence which supported the hypothesis is confirmatory neutral to its contrapositive reformulation. Hempel considered this feature an unacceptable flaw of his model and suggested fixing it by adopting the "*equivalence condition*" as an additional constraint (Hempel 1945, 13, 31). It says that logically equivalent hypotheses are confirmed by the same evidence: If c(h,e) and $h \Leftrightarrow h^*$ then c(h*,e). Accepting this further condition means that instantiation is merely sufficient, not necessary, for confirmation. If a hypothesis remains noninstantiated but is equivalent to a positively instantiated one, the former receives empirical backing as well.

A further condition brings out the spirit of Hempel's account more distinctly. This "*special consequence condition*" says: "If an observation report confirms a hypothesis *h*, then it also confirms every consequence of *h*" (Hempel 1945, 31). Hempel argues that this condition should be obvious. Deriving consequences from a statement does no more than making the content of this statement explicit. It should be clear, then, that the logical consequences of a confirmed hypothesis are confirmed themselves.[1] It thus militates against the adequacy of the hypothetico-deductive approach to confirmation that the special consequence condition comes out violated within its framework. If hypothesis *h* entails evidence *e*, it is by no means guaranteed that a weaker hypothesis *h'* that follows from *h* still entails *e* (Hempel 1945, 31–32).

These considerations make it clear that the converse consequence condition, on the one hand, and the entailment and special consequence conditions, on the other, are conceptually antagonistic. They belong to divergent approaches to confirmation—namely, traditional hypothetico-deductivism and Hempel's entailment account, respectively. In particular, it is impossible simply to conjoin all three conditions. The result would be that each observation e would support a completely arbitrary hypothesis h. Here is the argument: e entails and thus confirms itself; hence, in virtue of converse consequence, it also supports $e \wedge h$ which, by dint of special consequence, implies the confirmation of h (Hempel 1945, 32; Carnap 1950, 474). Consequently, the two sets of conditions have to be kept separate.[2]

Hempel specifies one more requirement, namely, the "consistency condition." One of the clauses attached to this condition says that all the hypotheses an observation statement (which is not self-contradictory) confirms are logically compatible with one another. Given the special consequence condition, there is no way to avoid the demand for consistency. If an observation statement supported two incompatible hypotheses, it also confirmed all the consequences of this contradiction. But anything is entailed by a contradictory premise. The acceptance of a violation of the consistency condition would amount to regarding arbitrary hypotheses as confirmed. This is an unwholesome result that needs to be ruled out (Hempel 1965, 33–34).

Hempel proceeds by specifying a conception of confirmation that satisfies these three conditions. The challenge is to show that general hypotheses are subject to confirmation in this sense by observation reports. The pivot of Hempel's conception is the notion of the "development of a hypothesis." The development serves to unfold the content of the hypothesis for a finite class of objects. The development represents a list of the cases to which the hypothesis is supposed to apply. This is achieved by substituting the variables in the hypothesis by the relevant individual constants. Consider the hypothesis "all swans are white" or $\forall x \, (Sx \rightarrow Wx)$ for a finite class of objects $\{a, b, c\}$. The development of this hypothesis is: $(Sa \rightarrow Wa) \wedge (Sb \rightarrow Wb) \wedge (Sc \rightarrow Wc)$. Hempel's condition of "direct confirmation" says: An observation statement e directly confirms a hypothesis h, if e entails the development of h for those objects that are mentioned in e. Let the observation statement be: $Sa \wedge Wa$, "Albert is a swan and Albert is white." It implies the first clause of the development, which is the only one that mentions Albert.[3] It follows that the hypothesis is directly confirmed by the observation report (Hempel 1945, 36–37).

The idea underlying the entailment account is to conceive of confirmation as deduction of positive instances of a hypothesis from the data. This idea is realized using the notion of development. The point is that hypotheses are of a general nature, whereas observation statements refer to individuals. Thus, the latter cannot imply a general claim. The development serves to remove the universal quantifier from the hypothesis and to make its content explicit for each of the relevant objects. The key to confirmation is that an observation report entails the entire content of the hypothesis with respect to the objects the report refers to. The hypothesis says no more as to these objects than the report does. This is why the observations are capable of confirming general hypotheses.[4]

As yet, Hempel has delivered precisely what he had advertised: a qualitative relation of confirmation that is deductively valid. The conception of deriving the relevant parts of the development of a hypothesis makes it transparent how a piece of evidence can possibly bear on a general assumption in a logically sound fashion. In contrast, it is obvious that this procedure cannot capture all there is to confirmation in the sciences. After all, scientific evaluation draws on judgments of the kind of how well a hypothesis is doing in light of the available evidence. This goes beyond assessments such as that a given piece of evidence supports a given hypothesis. What is needed is a comparative or quantitative concept of confirmation. In order to tackle this task, Hempel takes recourse to auxiliary criteria that have a traditional ring. The overall confirmation of a hypothesis, namely, is said to depend on the amount of confirming instances, the simplicity of the hypothesis, and its coherence with other relevant assumptions (Hempel 1945, 41–42). At this juncture, the nondeductive aspects of the confirmation process resurge.

Shortcomings of Hempel's Account

Let me highlight two shortcomings of Hempel's account that turned out to be significant for its transformation into the bootstrap model. First, the notion of development requires that hypotheses are couched in observational terms. There is no development of theoretical hypotheses. The reason is that the development of a hypothesis is supposed to catalog the observational content of this hypothesis. But without assistance of further assumptions, a hypothesis such as "the wavelength of red light is 700 nm" lacks observational content. Wavelengths are not accessible to the unaided human senses so that auxiliary assumptions are needed for endowing the hypothesis with empirical indications. It follows that no observation statement is able to entail, all by itself, the relevant content of a theoretical hypothesis.

The dependence of the observational content of a hypothesis on additional assumptions makes it impossible to unambiguously develop a theoretical hypothesis into the list of its empirical instantiations. A theoretical quantity can be measured by a multiplicity of methods. Current intensity, for instance, can be evaluated using electromagnetic interaction or electrochemical effects. In both cases the observational indications are disparate. It follows that one and the same hypothesis may translate into the observation report e_1, if auxiliary h_1 is invoked, or, alternatively, into observation report e_2, if auxiliary h_2 is resorted to. This change in the empirical indications, depending on which further assumptions are adduced, vitiates considering both reports as simply bringing out the observational content of the hypothesis. Not a single, unambiguous development is attached to the hypothesis, but rather a context-dependent multitude of disparate observational indications (Earman 1992, 68; Earman and Salmon 1992, 52).

Another trouble of Hempel's lies with the consistency condition. This condition requires that a given piece of evidence can only confirm hypotheses that are compatible with one another. But consider the notorious problem of fitting a curve to measurement results. The problem is that a given set of measurement values may lead to a number of distinct, incompatible hypotheses as to the underlying law. More than one curve can sensibly be drawn through a cloud of points representing the measurement results. One might assume that all the hypotheses instantiated by these results are supported by the results. But this prima-facie plausible judgment is ruled out by the consistency condition. Hempel grants the objection (Hempel 1945, 33), which makes Carnap wonder why Hempel sticks to a condition faced with such "a clear refutation" (Carnap 1950, 476). In the 1964 postscript to his seminal 1945 paper, Hempel acknowledged his predicament. He conceded, on the one hand, that the curve-fit objection seems "to carry considerable weight," but he insisted, on the other, that giving up consistency would be tantamount to abandoning the condition of special consequence as well—which went against the grain of the instantiation approach (Hempel 1945 [1964], 49). In fact, if the special consequence condition is retained, the confirmation of two incompatible hypotheses by the same data would imply that all the consequences of these hypotheses were confirmed as well—which would mean to regard arbitrary hypotheses as confirmed. Given the special consequence condition, preservation of the consistency condition is inevitable.

This finding may raise doubts as to whether special consequence is a kosher condition. In fact, closer inspection reveals that it is not. It gives rise to two paradoxes—namely, the conjunctive and the disjunctive irrelevance paradox. For the conjunctive paradox, consider the hypothesis: "trout are gill-breathing and sparrows are warm-blooded." The observation of the gill-breathing trout Frederick counts as a positive instance of this hypothesis. Actually, an observation report to this effect implies everything the hypothesis says on the objects mentioned in the report—as demanded by Hempel. So we may conclude that the hypothesis is confirmed by the observation statement. Further, the hypothesis can be considered as a conjunction of two partial claims, one referring to trout and the other to sparrows. Each of these partial claims, and the one on sparrows, in particular, follows from the comprehensive assumption. That sparrows are warm-blooded is a special consequence of the hypothesis that trout are gill-breathing and sparrows are warm-blooded. In virtue of the special consequence condition, the observation of a gill-breathing trout should support the presumption that sparrows are warm-blooded.

Drawing a little on the machinery of formal logic, the conjunctive irrelevance paradox amounts to the following. Let the hypothesis be: $\forall x\,[(Tx \rightarrow Gx) \wedge (Sx \rightarrow Wx)]$. The observation report may be: $Tf \wedge Gf$. On Hempel's account, this statement counts as a positive instance and thus as a confirmation of the hypothesis. A special consequence of the hypothesis is: $\forall x\,(Sx \rightarrow Wx)$, which is thus supported by the evidence as well.[5]

The disjunctive paradox is similarly structured. Take the hypothesis "trout are gill-breathing" and assume it to be confirmed by the observation of trout Frederick. Among the special consequences of this empirically confirmed hypothesis is the assumption that "trout are gill-breathing or sparrows are warm-blooded." Classical logic permits us to weaken a claim by appending another clause through disjunction. The truth of an assumption entails that this assumption or some other is true. Consequently, judging by Hempel's lights, the disjunctive hypothesis should be confirmed by the observation as well—albeit the second disjunct is not supported separately.

Again speaking in terms of formal logic, let the hypothesis be represented by the expression $\forall x\,(Tx \rightarrow Gx)$, with the supporting evidence $Tf \wedge Gf$. The hypothesis implies: $\forall x\,[(Tx \rightarrow Gx) \vee (Sx \rightarrow Wx)]$, which disjunction is likewise confirmed by the evidence in virtue of the special consequence condition.

Thus, the special consequence condition issues licenses for appending irrelevant clauses to hypotheses and having them confirmed by evidence relating to the original hypothesis. Irrelevant clauses turn out to be free riders of confirmation. Actually, the conjunctive irrelevance paradox is known to haunt hypothetico-deductivism.[6] The special consequence condition serves to import the conjunctive paradox into the entailment account, for one, and to bring forth the disjunctive paradox, for another.

Basics of the Bootstrap Model

The bootstrap model, as suggested by Glymour in 1980, owes its fundamental approach to confirmation to Hempel's account but substantially alters most of the technical machinery of the latter. Glymour accepts Hempel's basic idea to place qualitative confirmation at the top and to characterize confirmed hypotheses as being entailed by the data. The most important changes concern the conception of the development and the special consequence condition. Both are replaced by what I call the "overall instantiation requirement." Let me briefly sketch these modifications.

Within the general framework of the entailment account, the confirmation of hypotheses is appraised by examining their positive or negative instances. On the one hand, Glymour's bootstrap model involves a liberalization of Hempel's demands in that it permits one to perform the derivation of such instances by drawing on additional principles. That is, the hypotheses to be confirmed need no longer be couched in observational vocabulary; they need no longer be translatable into the collection of their empirical instances. Rather, it is acknowledged that hypotheses may be expressed in theoretical terms so that their instances could only be identified using auxiliary suppositions. This means that the notion of the development of a hypothesis is dropped and that, consequently, the theory-ladenness of empirical confirmation is recognized. On the other hand, Glymour tightens the demands in requiring that the values of *all* the quantities in the hypothesis in question are fixed uniquely by the evidence and that the agreement between the data and the hypothesis was not guaranteed in advance on logical grounds.

More specifically, an empirical test of a hypothesis is performed by producing a positive or, as the case may be, negative instance of it. In such an instantiation of a hypothesis all the variables figuring in it have assumed definite values. If these values match the hypothesis, the instantiation is positive; if not, it is negative. Hypotheses are confirmed by their positive instantiations and discredited by their negative ones. As a rule, the quantities in scientific hypotheses are not directly observable but need to be inferred from the data using additional assumptions. The auxiliary assumptions serve to furnish instantiations of hypotheses couched in theoretical terms. A positive instantiation is disqualified as a confirmation if the nature of the situation ruled out the appearance of a negative one.

Bootstrap confirmation is thus characterized by the following conditions.

1. *Overall instantiation:* Each quantity in the hypothesis in question has been unambiguously evaluated on the basis of the data and by eventual recourse to auxiliary assumptions.
2. *Positive instantiation:* the values are in agreement with the hypothesis.
3. *Risk of failure:* accordance between data and hypothesis is not made sure by the logical characteristics of the test (Glymour 1980, 114–23; 130–32).

The adoption of the overall instantiation requirement marks the crossroads where Glymour parts Hempel's company. In virtue of this commitment, instantiation becomes necessary for confirmation (and is not alone sufficient as in Hempel).

The recognition of theory-ladenness calls for the adoption of an *auxiliary condition.* If arbitrary assumptions were allowed to be employed for producing positive instances of a hypothesis, the instantiation requirement could be trivially satisfied. The auxiliary condition demands that the ancillary assumptions used in instantiating a hypothesis be confirmed themselves. If this was meant to say that each bootstrap confirmation needs to rely on bootstrap-confirmed hypotheses, an infinite regress would ensue. But Glymour specifies three options for judging particular hypotheses without prior recourse to bootstrap confirmation. First, some hypotheses can be supported by applying them twice to the same situation. One can employ the ideal-gas law for evaluating the gas constant and check the invariance of this constant by drawing on another application of the same law (Glymour 1980, 111, 121–22, 140; Mitchell 1995, 244; see, however, Culler 1995). That is, a hypothesis is used as its own auxiliary hypothesis. Second, under suitable circumstances it can be decided in a noncircular fashion that a hypothesis is not confirmed. The bootstrap conditions imply that a hypothesis remains unsupported if either definite values cannot be obtained for all relevant quantities on the basis of the data or if these evaluations are not liable to risk of failure (Glymour 1980, 118–21, 134–35, 143). Third, a "concordance procedure" can be used for establishing the mutual support of auxiliary assumptions, none of which is bootstrap confirmed in advance. In this case one and the same theoretical quantity is evaluated by relying on different auxiliary hypotheses or different data sets. If the results of these calculations are in agreement with one another, the correctness of the relevant ancillary assumptions employed is thereby buttressed (Glymour 1980, 122–23; see Carrier 1994, 52–55).

I will not go into these ramifications but will focus on the bearing of the overall instantiation condition. This condition is apt to dissolve both the conjunctive and the disjunctive irrelevance paradox. The reason is that the requirement makes it impossible to regard hypotheses or their consequences as empirically supported if one of the relevant variables fails to be uniquely determined by the data. The conjunctive paradox is removed on the grounds that evidence relating only to partial claims entertained in a hypothesis is disqualified as confirmation of the entire hypothesis. Data on trout alone fail to fully instantiate the conjunctive hypothesis on trout and sparrows. The overall instantiation requirement is violated so that the comprehensive hypothesis remains without support. The disjunctive paradox is dismissed in virtue of the fact that the special consequences of an instantiated hypothesis may fail to be instantiated themselves. The hypothesis on gill-breathing trout is, in fact, confirmed by the observation of gill-breathing trout Frederick. But these data are unsuitable for instantiating the logical consequence of this hypothesis that is generated by appending a disjunctive clause. Since this clause remains uninstantiated, the entire hypothesis receives no empirical support. This shows that adopting the overall instantiation requirement is tantamount to abandoning the special consequence condition.

It follows that there is no need to retain the consistency condition whose application to problems like curve fitting appeared implausible. If consequences of confirmed hypotheses are no longer supposed to be supported automatically, consistency can be dropped without detrimental effects. Two incompatible hypotheses may be considered confirmed by the same set of data without having to admit that everything implied by an inconsistency (that is, arbitrary assumptions) are borne out as well. Hempel's consistency condition is not part of the bootstrap model.

Let me expound the characteristics of the bootstrap model by briefly turning to a famous example of confirmation by deduction from the phenomena—namely, Newton's derivation of a central inverse-square force from Kepler's third law and his own laws of motion. For reasons of simplicity let us confine to the case of the circular motion of a single planet of mass m orbiting the sun at a distance r. It follows from Newton's laws of motion that a centripetal force $F = mv^2/r$ is necessary for maintaining the circular motion. That is, this force is exerted from the sun on the planet. Assume that it takes the period T for the planet to complete one revolution around the sun so that its uniform velocity comes out as $v = 2\pi r/T$. Plugging in this quantity in the equation for the centripetal force gives $F = 4\pi^2 mr/T^2$. Kepler's third law of planetary motion says that the squared periods of revolution are proportional to the cubes of the semimajor axes of the planetary orbits. In the simplified case of circular motion under consideration, the semimajor axis coincides with the radius of the planetary orbit. Kepler's third law thus amounts to $T^2 = kr^3$ (with some constant k). Plugging in yields for the centripetal force $F = 4\pi^2 mr/kr^3 = 4\pi^2 m/kr^2$. The result is that the centripetal force of gravitation decreases with the squared distances from the sun $F \sim 1/r^2$.

This example places essential features of bootstrap confirmation at the focus. First, the basis of the derivation is Kepler's third law, which is regarded as the relevant "phenomenon." However, it is obvious that this law involves a general claim and is not to be identified with a single experience. So, what bootstrap confirmation actually amounts to is the derivation of instances of more comprehensive hypotheses from more restricted ones.

Second, the procedure essentially draws on the invocation of additional laws and on the assumption of initial and boundary conditions. The central nomic premise is the equation of centripetal force that derives from the Newtonian equation of motion. The factual premise involves the constraint to a single planet. From these premises the existence of an inverse-square attractive force from the sun to the planet can be deduced, indeed. This demonstrates the importance of loosening the demands on the derivation of instantiations of hypotheses from the data. In contrast to Hempel, Glymour licenses the intrusion of theory in this process. It is only through this liberalization that realistic examples can be brought into the scope of confirmation theory.

Third, in contrast to hypothetico-deductivism, empirical support in the bootstrap vein concerns specific hypotheses. Glymour shares Hempel's nonholistic approach to confirmation. These nonholistic features are codified by the overall instantiation requirement that says only the instantiated parts or aspects of an assumption are confirmed by the relevant data. Let us take hypothetico-deductivism as a template so as to realize more distinctly the contrasting features of bootstrap confirmation. Within the hypothetico-deductivist framework, Kepler's third law is thought to be derived from Newton's laws of motion and his law of gravitation, along with the relevant initial and boundary conditions. Thus, Kepler's third law is shown to be a consequence of this set of premises. The empirical success of the law indiscriminately supports all the premises used in the deduction. Hypothetico-deductive confirmation is directed at large-scale theoretical networks; it is holistic in kind. Kepler's third law confirms the "science-wide web" of Newtonian mechanics.

The bootstrap picture is essentially different. The overall instantiation requirement entails that confirmation is restricted to the instantiated parts or aspects of the hypotheses at issue. Conversely, what is not instantiated remains unsupported. The only part of the law of gravitation that is actually instantiated by the sketched deduction is the assumption of an inverse-square force from the sun to the planet in question. Consequently, in virtue of the overall instantiation requirement, this one assumption alone is confirmed by Kepler's third law. This means, in particular, that neither the reciprocality nor the universality of gravitation, which are both essential to the Newtonian conception of gravity, receive any support from the third law. Nothing is derived with respect to a force exerted from the planet on the sun; nothing is deduced as to the forces among various planets. So, it is only a restricted aspect of the law of gravitation, and by no means the entire network of Newtonian mechanics, that is confirmed by Kepler's third law. Bootstrap confirmation is directed at specific assumptions and is thus nonholistic.[7]

Glymour joins Hempel in considering qualitative confirmation as central. The core issue is whether or not a piece of evidence supports a given hypothesis. Just as in Hempel's original version, additional criteria have to be invoked in order to assess degrees of confirmation. The comparative or quantitative merits of a hypothesis or theory are only appraised at a later stage of the procedure. Glymour resorts to standards such as the paucity of untested hypotheses, the variety of evidence brought to bear on a theory, the uniformity of the explanations achieved, and so on (Glymour 1980, 153–54). Neither Hempel nor Glymour leave any doubt that qualitative confirmation is insufficient for capturing the whole of hypothesis appraisal in the sciences. Further criteria of a nondeductive nature have to be taken into account in any event. The purpose in this section was to reveal the strong conceptual ties between the basic principles of Hempel's and Glymour's approaches. This can be accomplished by focusing exclusively on the fundamental notion of qualitative confirmation.

Hempel's Paradox of Confirmation

The irrelevance paradoxes, as sketched earlier in this chapter, are not to be confused with what is usually called "Hempel's paradox of confirmation" or "the ravens paradox." This paradox, in contrast to the irrelevance paradoxes, is not due to the special consequence condition, but arises from the equivalence condition. This condition requires that a piece of evidence that supports a hypothesis also confirms each hypothesis that is logically equivalent with the first one. The catch is that this not implausible demand has somewhat tricky ramifications. Let the hypothesis be "all ravens are black." This statement is equivalent to its contrapositive, "all nonblack things are nonravens." A pink elephant is neither black nor a raven; its observation instantiates and thus certainly supports the latter claim. But then, in virtue of the equivalence condition, it should also confirm the original version that all ravens are black—which has a somewhat paradoxical ring (Hempel 1945, 12–15).

The entailment approach to confirmation is in no way particularly liable to Hempel's paradox. After all, the nonravenhood of nonblack things is a consequence of the presumption that all ravens are black, so that pink elephants bear out this presumption by hypothetico-deductive lights as well. Each theory of confirmation is threatened by the difficulty, and the extant literature contains a large number of attempts to cope with it (Maher 1999, 57–65). The reason I briefly want to go into the matter is that the entailment approach in general, and the bootstrap model in particular, allows for a natural treatment.

Let me begin by briefly sketching Glymour's own suggestion for handling the issue. He requires that genuine confirmation be selective. A hypothesis is borne out by only such data that do not also indiscriminately confirm alternative hypotheses, expressed in the same concepts as the first one but incompatible with it. What makes the support of the raven hypothesis by the pink elephant appear suspect is that this piece of evidence, using the same logic, likewise buttresses the claim that all ravens are green, or that they are yellow and littered with blue spots, or what have you. After all, pink elephants do not alone constitute nonblack nonravens but also nongreen nonravens or nonyellow, nonspotted nonravens. The piece of evidence at issue fails to undermine any competing hypothesis. Its confirmatory impact is highly diffuse and nonselective, and this is the reason why it is bereft of corroborating force (Glymour 1980, 156–60).

I do not deny that the selective-confirmation requirement is plausible and suitable for coping with the paradox. Still, its acceptance is less than satisfactory since it involves the introduction of a special condition of adequacy so as to defuse one particular problem. It would certainly be more appealing to resolve the difficulty by drawing on the natural virtues of the instantiation account. Each confirmation theory could invoke the selective-confirmation requirement as additional principle; nothing hinges on featuring positive instances as the key to confirmation. Adopting the condition of selective confirmation does not confer a selective distinction to the instantiation account.

So let me try to advance bootstrapping a little further. There appears to be a way out of the quandary that is paved, indeed, by the emphasis on instantiations. It deserves notice that insistence on instantiations helps avoid the emergence of the paradox. A pink elephant is no positive instance of the generalization that all ravens are black and should fail to support it for this reason. The confirmatory force of such observations wholly derives from the equivalence condition; it does not arise from the first principles of the entailment approach. This constitutes a natural advantage over hypothetico-deductive theories of confirmation. Whatever the particulars of such theories may be like, they all subscribe to the converse consequence condition. This condition embodies the spirit of hypothetico-deductivism. But each hypothesis qualifies as a premise for the derivation of its contrapositive so that each confirmation of the latter is automatically transferred to the hypothesis itself. No additional condition is needed for generating the paradoxical result.

The bootstrap model, by contrast, demands positive instances. The overall instantiation requirement stipulates that each quantity in the hypothesis in question must be subject to unique evaluation on the basis of the data. It follows that non-instantiated hypotheses receive no support. But the pink elephant is unsuitable for fixing any quantity in the hypothesis about ravens. It only instantiates the contrapositive, not the hypothesis itself. The paradox does not arise in the first place.

It is obvious that this type of approach exacts renunciation of the equivalence condition; the very pivot of this treatment is that logically equivalent hypotheses could be borne out differently. However, in all the benign cases the equivalence condition is not needed, and in the malign ones it issues in paradoxes. So the condition could or should be dropped, respectively. Consider the law of gravitation: $F = Gm_1m_2/r^2$. A relevant piece of evidence, such as a given planetary constellation, indiscriminately instantiates all the logically equivalent versions of this equation: $r^2 = Gm_1m_2/F$ or $m_1 = Fr^2/Gm_2$. If the quantities are instantiated in one case, they are automatically instantiated in the others as well. Demanding equivalence explicitly is quite superfluous. Problems only emerge with laws in the form of material conditionals. To be sure, there are real-life laws of this sort. The law of inertia can be expressed to the following effect: if a body is not subject to any external force, it performs a uniform rectilinear motion. If the equivalence condition is renounced the curvilinear motion of a body under the action of an external force is not covered by this law. But there is nothing to worry about that. Noninertial motion is captured by a different law, the equation of motion. After all, these two types of situations were treated separately in Newton's original axiomatization of mechanics; they were addressed by the first and the second law of motion, respectively. If we want to have it one way and the contrapositive way alike, we should state it both ways.

Abandoning the equivalence condition serves to resolve Hempel's paradox of confirmation without the need to appeal to a particular, tailor-made requirement. It is true that it might prove difficult to extend this treatment to cover comparative and quantitative notions of confirmation (Maher 1999, 53); but given Hempel's and Glymour's strategy of placing qualitative confirmation at the top, degrees of confirmation are not yet at issue.

Passing on the Torch: Hempel and Glymour

Glymour places his model explicitly in the tradition initiated by Hempel (Glymour 1980, 128–29). Hempel's theory of confirmation is said to possess "admirable qualities" but to suffer from its restriction to simple cases (Glymour 1980, 27). This supposedly means that Hempel's basic conception is all right but needs to overcome its limitation to toy propositions of the "all swans are white" variety. What is called for, Glymour suggests, is to enlarge the framework so as to make real-life cases tractable.

I have attempted to point out that the fundamentals of the bootstrap model are directly linked to Hempel's approach. In the rational reconstruction I have given, the bootstrap tenets are shown to grow naturally out of the deficiencies of the Hempelian views. Hempel's notion of the development of a hypothesis is abandoned and Hempel's special consequence condition is dropped (although Glymour fails to recognize the fatal properties of the latter condition) (Glymour 1980, 132–33, 155, 174). The two are replaced by the overall instantiation requirement. So, what Glymour did in 1980 was to take up, pursue, and improve ideas from the heyday of logical empiricism. The connection is tight, as I have tried to argue, and it stretches right across the "theory-change period" in which interest in confirmation theory proper had dropped off to almost nothing.

Further, none of the criticisms leveled against Hempel's account was substantially connected with the demise of logical empiricism—a change in the philosophical landscape that happened right in between the enunciation of the two approaches in question. The odd properties of the special consequence condition could have been noted at any time. And the restriction to observational terms, as expressed by the commitment to the development of hypotheses, should have appeared inadequate at any point after the acceptance of the double-language model. This model was advocated by Carnap and Hempel from the 1950s onward and involved the claim that science encompassed two linguistic levels—namely, observational and theoretical terms. Logical empiricists insisted that both levels are to be kept separate. To be sure, so-called correspondence rules were supposed to establish links between the two linguistic levels. But since one theoretical term might be tied to several observational ones and vice versa, the theoretical language was thought not to be reducible to observation predicates.

The pivotal aspect is that the double-language view is incompatible with the notion of development. And since Hempel was among the chief proponents of this view, he himself could have well conceived all the objections that were later directed against this notion. Abandoning the notion of the development of a hypothesis has nothing to do with the transition to postempiricism; the changes within logical empiricism would have suggested this move anyway. In this area, nothing hinges on the passage of philosophical time.

Consequently, the bootstrap model smoothly continues the lines Hempel had drawn in confirmation theory. To be sure, the bootstrap model significantly modifies Hempel's earlier approach. If the equivalence condition is dropped (as suggested above), nothing but the entailment condition is left from Hempel's original version. On the other hand, this condition is the key to Hempel's view; it embodies the very spirit of the entailment account. The severe changes introduced by Glymour do not militate against regarding the bootstrap model as a continuation of Hempel's conception. Rather, this is how progress in philosophy is produced. It hardly ever happens that a philosopher takes over another philosopher's views unchanged; and if it happens it is a boring episode that deserves to pass unnoticed. To take up philosophical ideas in a fruitful fashion means to change them; fertile philosophical traditions are characterized by frequent alterations. Philosophical argument thrives on continuing lines of thought by improving them. As I tried to make plausible, this is precisely what is distinctive of the relationship between Hempel and Glymour. My point is that this testifies to the lasting fecundity of Hempel's views on confirmation. To be philosophically alive is to be criticized, to be pursued, and to be modified. In this vein, Salmon took up Hume's challenge to identify causal processes by relying exclusively on conceptual means acceptable to an empiricist and advocated his process theory of causation (Salmon 1984, 136–37, 182–83). Although the result is largely at variance with Hume's own opinion on the subject, the attempt to meet Hume's challenge bears witness to the enduring force of his philosophy. The indication of philosophical fecundity is not acceptance but the power to stimulate new thoughts. I hope to have shown that Hempel's theory of confirmation passes this test easily.

It follows that it is unjustified to pass in silence over Hempel's contributions to confirmation theory when it comes to appraising the "spirit of logical empiricism" (Salmon 1999). It is true that Hempel, in the 1964 postscript to his original article, lists objections to his proposal and comes close to abandoning the entire project in favor of Carnap's inductive logic: "My general formal definition of qualitative confirmation now seems to me rather too restrictive. . . . Perhaps the problem of formulating adequate criteria of qualitative confirmation had best be tackled, after all, by means of the quantitative concept of confirmation. This has been suggested especially by Carnap" (Hempel 1945 [1964], 48–49, 50). Analogously, Hempel does not even mention his own earlier efforts in the chapter on confirmation in his introductory *Philosophy of Natural Science* of 1966. Instead, Carnap's inductive logic is presented as a promising approach (Hempel 1966, 45–46). The foregoing considerations suggest that Hempel should have stuck to his earlier views more tenaciously. The theory in its original shape could not be upheld; it was bound to collapse. But there were jewels to be found in the debris.

Notes

1. The special consequence condition implies the equivalence condition since equivalence amounts to reciprocal implication (Hempel 1945, 21). For this reason, the adoption of the special consequence condition makes it superfluous to adduce the equivalence condition as a separate constraint.

2. In view of the fact that the two sets of conditions are part of incongruous approaches to confirmation, it is a queer endeavor to probe into options for nevertheless reconciling the converse consequence condition with Hempel's conditions of entailment and special consequence—as Le Morvan (1999) does.

3. The relevant logical rule is $p \land q \Rightarrow p \to q$.

4. For this reason the restriction to a finite domain of application is inessential and could be dropped. To be sure, the development of a hypothesis cannot be stated comprehensively for an infinite class of objects, but an observation statement could still imply those parts of the development that deal with the objects mentioned in the statement.

5. It is important to realize that the argument does not invoke the converse consequence condition. Hempel used a reply to this effect in order to rebut a similar but more

schematic objection. Let an observation e confirm a hypothesis h_1; then e might be thought also to confirm $h = h_1 \wedge h_2$ where h_2 may be completely irrelevant to e. However, as Hempel replies, the move from the confirmation of h_1 to the confirmation of h tacitly appeals to the converse consequence condition that appeal is illicit in the framework of the instantiation account (Hempel 1945, 33).

But in the more detailed scenario sketched in the text, the corresponding move to the confirmation of the conjunctive hypothesis is not licensed by the converse consequence condition but by drawing more specifically on the concept of the development of a hypothesis. In the example given in the text, the piece of evidence directly confirms the conjunctive hypothesis in question. No auxiliary recourse to any general condition is necessary.

6. If a hypothesis h entails evidence e, the conjunction of this hypothesis with some irrelevant clause i likewise entails e: $h \rightarrow e \Rightarrow h \wedge i \rightarrow e$ (Glymour 1980, 31).

7. See Carrier (1994, 56–61) for another, more extensively discussed example of bootstrapping.

References

Carnap, R. 1950. *Logical Foundations of Probability*. Chicago: University of Chicago Press.

Carrier, M. 1994. *The Completeness of Scientific Theories: On the Derivation of Empirical Indicators within a Theoretical Framework: The Case of Physical Geometry*. Vol. 53, Western Ontario Series in the Philosophy of Science. Dordrecht: Kluwer.

———. 2002. "Explaining Scientific Progress: Lakatos's Methodological Account of Kuhnian Patterns of Theory Change." In *Appraising Lakatos: Mathematics, Methodology, and the Man (Vienna Circle Library)*, ed. L. Kvasz, G. Kampis, and M. Stöltzner, 53–71. Dordrecht: Kluwer.

Culler, M. 1995. "Beyond Bootstrapping: A New Account of Evidential Relevance." *Philosophy of Science* 62:561–79.

Earman, J. 1992. *Bayes or Bust? A Critical Examination of Bayesian Confirmation Theory*. Cambridge Mass.: MIT Press.

Earman, J., and W. C. Salmon. 1992. "The Confirmation of Scientific Hypotheses." In *Introduction to the Philosophy of Science*, M. H. Salmon et al., 42–103. Englewood Cliffs N.J.: Prentice Hall.

Glymour, C. 1980. *Theory and Evidence*. Princeton: Princeton University Press.

Hempel, C. G. 1945. "Studies in the Logic of Confirmation/Postscript 1964." In *Aspects of Scientific Explanation and Other Essays in the Philosophy of Science*, ed. C. G. Hempel, 3–51. New York: Free Press; London: Collier-Macmillan, 1965.

———. 1966. *Philosophy of Natural Science*. Englewood Cliffs, N.J.: Prentice Hall.

Howson, C. 1997. "A Logic of Induction." *Philosophy of Science* 64:268–90.

Howson, C., and P. Urbach 1989. *Scientific Reasoning: The Baysesian Approach*. La Salle, Ill.: Open Court.

Irzik, G., and T. Grünberg. 1995. "Carnap and Kuhn: Arch Enemies or Close Allies," *British Journal for the Philosophy of Science* 46:285–307.

Kuhn, T. S. 1962. *The Structure of Scientific Revolutions*. Chicago: University of Chicago Press.

Lakatos, I. 1970. "Falsification and the Methodology of Scientific Research Programmes." In *Imre Lakatos: The Methodology of Scientific Research Programmes (Philosophical Papers I)*, ed. J. Worrall and G. Currie, 8–101. Cambridge: Cambridge University Press.

Le Morvan, P. 1999. "The Converse Consequence Condition and Hempelian Qualitative Confirmation." *Philosophy of Science* 66:448–54.

Maher, P. 1999. "Inductive Logic and the Ravens Paradox." *Philosophy of Science* 66:50–70.

Mitchell, S. 1995. "Toward a Defensible Bootstrapping." *Philosophy of Science* 62:241–60.

Reichenbach, H. 1938. *Experience and Prediction: An Analysis of the Foundations and the Structure of Knowledge.* Chicago: University of Chicago Press.

Salmon, W. C. 1967. *The Foundations of Scientific Inference.* Pittsburgh: University of Pittsburgh Press.

———. 1984. *Scientific Explanation and the Causal Structure of the World.* Princeton: Princeton University Press.

———. 1990. "Rationality and Objectivity of Science or, Tom Kuhn Meets Tom Bayes." In *Scientific Theories,* ed. C. W. Savage, 175–204. Vol. 14, *Minnesota Studies in the Philosophy of Science.* Minneapolis: University of Minnesota Press.

———. 1999. "The Spirit of Logical Empiricism: Carl G. Hempel's Role in Twentieth-Century Philosophy of Science." *Philosophy of Science* 66:333–50.

Changing Conceptions of Rationality

From Logical Empiricism to Postpositivism

GÜROL IRZIK

Bogaziçi University

One of the central issues that came into focus in the sixties and the seventies with the transition from the logical empiricist to the postpositivist philosophy of science was scientific change. This issue was bound to arise when a historical-developmental approach to science replaced a logico-structural one adopted by the logical empiricists. It involved, among others, questions such as these: How does scientific change occur? What sort of criteria, standards, or norms are employed in preferring one scientific theory over another? Are they fixed and universal or local and historically changing? Clearly, these are questions about the rationality of science, and while they interested the logical empiricists and received considerable attention by the falsificationists, they became the center of stormy debates only after the publication of Thomas Kuhn's *The Structure of Scientific Revolutions* in 1962.

As the questions in the preceding paragraph indicate, these rationality debates were at the same time debates about scientific methodology, which was typically understood as laying out the rules that govern scientific practice—in particular, the acceptance of a theory or the choice among alternative theories. The relationship between methodology and rationality was taken to be straightforward: if the methodology involved, say, rules like "always formulate your conjectures boldly and test them severely," it was rational to prefer the hypothesis that was bolder and had been tested more severely than its rivals. Thus, rationality was conceived in terms of compliance with rules. A philosopher was a rationalist if he held that science employed a universal method, a universal and fixed set of rules for theory evaluation; he was accused of being a relativist if he denied such rules.

Roughly, this is the framework within which emerged the standard account concerning the rationality of science. According to this account, the logical empiricists and Popper are rationalists in the sense just explained. Logical empiricists, notably Carnap, advocate confirmationism as the method of science; Popper favors falsificationism, but this is a relatively minor difference when compared with Kuhn's denial of any universal method. According to Kuhn, the standard account goes, theory change can be understood in terms of neither confirmationism nor falsificationism, but only in terms of such notions as paradigm shifts, gestalt switches, conversions, and incommensurability. Furthermore, since the standard account takes Kuhn to believe that there are no paradigm-independent reasons for choosing one theory over another, Kuhn becomes a relativist, and to the extent that he claims that the choice involves personal tastes, he is a subjectivist and even an irrationalist.

Without doubt Kuhn's conception of rationality of science is different from Carnap's and Popper's, but the standard account is a caricature that I intend in this chapter to replace with a more realistic picture. I will try to explain how the conception of rationality of science changed from logical empiricism to postpositivism in Popper's, Carnap's, and Kuhn's writings, with emphasis on the latter two. I will argue that the standard account is seriously flawed with respect to Carnap and Kuhn. My claim that the charges brought against Kuhn are exaggerated if not totally unfounded is not original. Kuhn himself denied them emphatically. Nevertheless, the conception of rationality that Kuhn was attributing to science and how it differed from Carnap's and Popper's is not fully appreciated. Similarly, the evolution of his views about theory change is not well understood.

Carnap's case is more complicated. For one thing, Carnap wrote very little on the issue of scientific theory change, though he was deeply interested in questions of justification, in particular with respect to inductive rules and linguistic frameworks. In addition, the complexity of his views about justification makes it difficult to label him. Thus, while the standard account I mentioned above makes him a rationalist like Popper, Michael Friedman, who has arguably given us the most sophisticated and sobering revisionary picture of logical empiricism, turns him into a relativist like Kuhn (Friedman 1998). I argue that these conflicting interpretations all fail to recognize that in Carnap there are two distinct notions of justification, and a fortiori, two distinct kinds of rationality, and the recognition of these distinctions is the key to the understanding Carnap's position vis-à-vis the rationality-relativism issue.

The plan of this chapter is as follows. I will begin by documenting the standard account of Carnap's views on rationality and show that it misses the mark completely. I then take up Friedman's interpretation of Carnap and argue that it is only partially right; Carnap is a relativist in one respect but not in another. I then turn to Kuhn's conception and defend it against charges of relativism with respect to theory choice. Kuhn's views about theory choice evolved considerably, and my defense will make much use of this evolution. However, the rationalism vs. relativism debate is not the right axis for understanding Kuhn's originality in this context. In my opinion, one of Kuhn's most important contributions is to redefine our notion of scientific rationality. We can appreciate this contribution better if we consider it from the viewpoint of Aristotle's notions of phronesis and practical wisdom. This point becomes clear when I compare Kuhn's views with Carnap's and Popper's at the end. I conclude my essay by noting that the issue of rationality is ultimately bound up with questions of ethics and that the limits of Carnap's account of rationality arise from his failure to see this.

The Standard Account Documented

We find what I have called the standard account in many places: Kuhn (1970a, 1970b), Lakatos (1970), Brown (1977), Feyerabend (1978), and Putnam (1981). All of these philosophers attribute to the logical empiricists the view that theory choice or acceptance is simply a matter of experimentation and logic alone, which, with sufficient ingenuity and effort, could be reduced to an algorithm that would determine the degree of confirmation of a hypothesis by a given body of evidence. Rationality then is seen as a matter of accepting or choosing that hypothesis that has a greater degree of confirmation. Kuhn, for example, writes that "for Sir Karl and his school, no less than for Carnap and Reichenbach, canons of rationality thus derive exclusively from those of logical and linguistic syntax" (Kuhn 1970c, 234). He has argued repeatedly that "this issue of paradigm choice can never be unequivocally settled by logic and experimentation alone" (Kuhn 1970a, 94).

What Kuhn and others have in mind, of course, is Carnap's inductive logic of confirmation, and Brown's formulation not only makes this explicit but also links it to algorithmic computability: "Again the project is to find an algorithm on the basis of which we can evaluate scientific theories, the assumption being that ... we can produce a set of rules which will allow us to determine the degree to which it has been confirmed by the available evidence ... The attempt by logical empiricists to identify rationality with algorithmic computability is somewhat strange, since it deems rational only those human acts which could in principle be carried out without the presence of a human being" (Brown 1977, 146–48).

That logical empiricists identified rationality with scientific method appears also in Putnam: "In the past fifty years the clearest manifestation of the tendency to think of the methods of 'rational justification' as given by something like a list or canon was the movement known as Logical Positivism. Not only was the list or canon that the positivists hoped 'logicians of science' would one day succeed in writing down supposed to exhaustively describe the 'scientific method'; but, . . . according to the logical positivists, the 'scientific method' exhausts rationality itself" (Putnam 1981, 105). No name is mentioned, but it is clear that it is Carnap more than anybody else who represents logical empiricists here too.

In short, the following theses are attributed to the logical empiricists: (1) scientific rationality is secured by the scientific method; (2) the scientific method consists of a set of universally valid logico-methodological rules, epitomized by Carnap's system of inductive logic; and (3) these rules exhaust not only scientific rationality but also rationality as a whole.

It is interesting to note that this characterization makes Carnap look like a rationalist about theory choice much as Popper is, except that the latter had a falsificationist methodology and never believed in the notion of inductive support. It makes Carnap's position virtually indistinguishable from Popper's because Popper too characterizes the rationality of science in terms of the scientific method understood as a set of universal rules: other things being equal, choose that theory which has more empirical content; choose that theory which has withstood against more severe tests; avoid ad-hoc hypotheses, etc. (see also Feyerabend 1978, 179).

At any rate, once Carnap's and the logical empiricists' conception of rationality is characterized in terms of (1) through (3), it becomes an easy target. Accordingly, Kuhn can claim that theory acceptance or choice is never determined by logico-methodological rules; Feyerabend can find historical cases where every such rule is violated; Lakatos can simply point to the failure of Carnap's heroic efforts for constructing an inductive logic; Brown can draw attention to the oddity of a conception of human rationality that can be exercised by machines; and, finally, Putnam can triumphantly announce that such a conception of rationality is self-refuting because once the scientific method is claimed to exhaust rationality, we are deprived of discussing the rationality of the claim itself in a rational way (Putnam 1981, 113).

A Critique of the Standard Account

Clearly, rationality is tied up with justification, and to justify is to provide good reasons. Of course, this is almost trivial unless we explain what counts as good reason. Minimally, it is compliance with the constraints of logic. Another constraint on good reasoning is learning from experience. Scientific activity displays both: scientists try to construct theories that are free of contradiction, that entail observational consequences that can be tested against observations, and so on. Naturally, one can try to codify both aspects of good reasoning in terms of certain general rules and principles. Indeed, modern logic does form such a codified system, and many scientists as well as philosophers have devoted considerable attention to the development of scientific methodology in this sense.

Although Carnap believes in the necessity of abiding by the constraints of logic and of learning from experience, he does not think that the rules of logic and scientific methodology are fixed and unchanging. The standard account's first mistake results from a total blindness to Carnap's principle of tolerance and his distinction between internal and external questions of existence. These are the sources of the relativistic dimension in Carnap's philosophy with respect to both deductive and inductive logic, and, a fortiori, with respect to confirmation. The second mistake of the standard account is to ignore the important distinction Carnap makes between the methodology of induction and inductive logic proper. When we look at what Carnap says about the former, the criticisms that pertain to algorithmic computability simply evaporate. Finally, Carnap never held the view that rules (of deductive and inductive logic) exhaust rationality. I will take up this last point when I discuss Friedman's views. Let me now consider the first two.

In *Logical Foundations of Probability* Carnap carefully draws a distinction between what he calls "inductive logic proper" and "the methodology of induction" (Carnap [1950] 1962, 202). The task of inductive logic proper is to assign a quantitative degree of confirmation to a hypothesis on the basis of a body of evidence. This requires a set of exact rules, and most of Carnap's efforts have gone into their specification. Inductive logic tells us nothing about theory acceptance because, according to Carnap, hypotheses are never accepted but only probabilified. The methodology of induction, on the other hand, specifies no exact rules, but only advises us about the application of inductive logic proper for certain purposes such as the construction of appropriate experiments for a given hypothesis (203).

Consequently, it is simply a myth that Carnap's inductive logic aims to produce an algorithm for theory or hypothesis acceptance. Indeed, Carnap has answered such misconceptions as follows:

> Those who are sceptical with respect to quantitative inductive logic point to the fact—and here they are certainly right—that in the practice of science factors of very different kinds influence the choice of a hypothesis. Some seem to think that to determine this choice by a simple calculatory schema would be just as preposterous as to propose rules of calculation which are to determine for every man which of the available women is the best for him to marry. In judging objections of the kind described, it is important to be clearly aware of what is and

what is not the nature and task of inductive logic and especially its distinction from the methodology of induction. Inductive logic alone does not and cannot determine the best hypothesis on a given evidence, if the best hypothesis means that which scientists would prefer. This preference is determined by factors of many different kinds, among them logical, methodological, and purely subjective factors. (221)

Obviously, Carnap was aware that factors other than the degree of confirmation influence the acceptance of hypotheses in actual practice, but his aim was to determine in a precise way the role of evidence in science from the viewpoint of empiricist philosophy. On the one hand, as an *empiricist* he believed that no choice can be rational unless it takes empirical evidence into account, so it was necessary to determine, above all, the degree of support that a given body of evidence bestows upon a hypothesis. On the other hand, as a *philosopher* he also firmly believed that it was of utmost importance to distinguish between the psychology and the logic of science. Although he granted that psychological and sociological approaches to science were valuable, strictly speaking they had nothing to do with philosophy; for him philosophy was a normative, analytical discipline concerned with the logic of science whose aim was to promote the best available science (see Carnap [1934] 1967).

Let me now take up the issue of acceptance in the form of comparison. It may come as a surprise to most readers that Carnap's new "Basic System of Inductive Logic" in *Studies in Inductive Logic and Probability* leaves little room for comparison of two hypotheses based on the degree of confirmation or, in other words, logical probability. Here is why. Carnap's new system is constructed for a very simple observational language with monadic predicates, in which the absolute probability of a hypothesis H and the conditional probability H relative to evidence E depend, among other things, on judgments of similarity relations (that is, the distances) between the properties picked up by these predicates. As an example, Carnap considers the family of colors. Since the basic system consists of only observational predicates, the similarity relations between colors will depend on perceived similarity. But, Carnap says, if we were to construct an inductive logic for a theoretical language, those relations would be determined by theoretical considerations such as frequency of light waves, which would give us totally different values for the probabilities. Consequently, Carnap accepts the possibility that different hypotheses cannot be compared on the basis of inductive (logical) probabilities (see Hilpinen 1973). We might call the realization of this possibility "confirmational incommensurability."

Consequently, the degree of confirmation is not only relative to a framework but also to a theory; scientific theory enters into the way the degree of confirmation is calculated. This is to be expected since, according to Carnap, every scientific theory is formulated in some linguistic framework. By a theory he means the conjunction of theoretical postulates and correspondence rules. Given what he says about them, we have every reason to believe that he considered the theoretical postulates of a theory as belonging to the linguistic framework of that theory. To begin with, as I have shown elsewhere in detail (Irzik and Grünberg 1995), Carnap distinguishes between "normal" activity within a scientific theory cum its linguistic framework and "revolutionary" activity outside it. The latter is characterized in terms of a radical change in the framework rules or a change in the theoretical postulates (see Carnap 1963b, 921). These postulates have both a factual and a semantic function. They not only serve as laws but also introduce primitive theoretical terms and partially interpret them. Thus, though they are synthetic, they do not change their truth-value during "normal" science: "To be sure, this status [analyticity in a language] has certain consequences in case of changes of the second kind [changes in the truth values of indeterminate statements], namely, that analytic sentences cannot change their truth-value. But this characteristic is not restricted to analytic sentences; it holds also for certain synthetic sentences, e.g., physical postulates and their logical consequences" (Carnap 1963b, 921).

This suggests that the theoretical postulates behave just like meaning postulates, forming a part of the linguistic framework. There is further textual evidence for this. For example, consider the following passages from *Studies in Inductive Logic and Probability*: "The basic laws of the theories are incorporated into the language Lt as theoretical postulates" (Carnap and Jeffrey 1971, 51); "laws of nature (which may sometimes be taken as B-postulates)" (94), where a B-postulate is a basic assumption accepted generally for any language L, not just for a special investigation (80). Finally, look at this passage in Carnap's reply to Putnam in the Schilpp volume: "It seems to me that a basic role should be assigned both to the primitive magnitudes of the theoretical language and to the [theoretical] postulates and correspondence rules. Since I regard d.c. [degree of confirmation] as always relative to a language, the influence of both factors on the values of d.c. is in accord with my conception" (Carnap 1963b, 988). In these passages Carnap explicitly takes the theoretical postulates of a scientific theory as part of the linguistic framework of that theory, and the last passage indicates clearly that the degree of confirmation is relative to both a given theory and its linguistic framework.

So much, then, for the myth that Carnapian confirmation theory attempts to provide an algorithm for theory choice and that it is a cumulative enterprise. Now let me turn to the other mistake of the standard account according to which Carnap believed in fixed and universally valid logico-mathematical rules. The trouble with this is that Carnap is a relativist about logic and mathematics, as pointed out by several philosophers (see, for example, Earman 1993 and Friedman 1998). This relativism results from his famous principle of tolerance: "*In logic, there are no morals.* Everyone is at liberty to build up his own logic, i.e. his own form of language, as he wishes" (Carnap [1934] 1937, 52). Thus, there is a classical logic and mathematics, an intuitionistic logic and mathematics, etc., and no one logic or mathematics can be said to be *the* correct one. And since every sufficiently rich scientific theory is formulated in some logico-mathematical framework, its evaluation will be relative to it.

A similar relativism also follows from Carnap's distinction between internal and external questions of existence. The significance of this distinction for the issue of rationality cannot be overestimated and is striking when we compare Carnap to Popper. Appearances notwithstanding, there are fundamental differences between Carnap's and Popper's conceptions of rationality. One of them is well known: according to Popper, hypotheses are never confirmed or supported by evidence but only corroborated. Carnap and Popper intensely disagree on this issue. But a deeper difference results from Carnap's distinction between internal and external questions of existence relative to a linguistic framework. Both the principle and the distinction are anathema to Popper's entire philosophy. When we ask, for example, whether we should believe in the existence of electrons, Popper says that the answer depends on whether our theory of electrons has been severely tested and corroborated. If so, Popper concludes, it is rational to believe in the existence of electrons. Carnap the homophonist, too, agrees that there are electrons, that it is rational to believe in their existence, provided that we interpret the question of existence as an internal one, internal, that is, to the linguistic framework within which the claim is made (Carnap [1950] 1956, 218). If the framework has rules for formulating claims about electrons and testing, confirming and disconfirming them, then the question whether electrons exist can be justifiably answered on the basis of those rules and the results of observation. Otherwise, Carnap believes, the question does not make cognitive sense. Hence, the answers to existence questions, whether of electrons, numbers, or sets, cannot be given independently of the rules of a linguistic framework. Since, by the principle of tolerance, there is a multiplicity of such rules, an answer that is justifiable by one set of rules can be unjustifiable relative to another. This is nothing short of a relativism with respect to existence questions, a relativism that the standard account misses badly.

Internal vs. External Justification; Validation vs. Vindication

By interpreting Carnap as a relativist, Michael Friedman corrects the standard account significantly. His mistake, however, is the opposite of the standard account; while the latter makes Carnap look like an arch-rationalist, Friedman turns him into more of a relativist than he really is. Let us see how. Friedman characterizes Carnap's relativism as the view that there are no universal standards of rationality, objectivity, and truth and argues that it follows from his principle of tolerance as well as his distinction between internal and external questions. Recall that according to the principle of tolerance, we are free to choose our framework rules and thereby our logic in any way we wish. Since what counts as rational and correct depends on these rules, rationality and truth become relative to them. But there is a multiplicity of such rules and therefore a plurality of logics, so it follows that "there is *no* over-arching notion of universal validity or 'correctness' independent of the particular and diverse rules of the particular, equally possible and legitimate, formally specifiable calculi. For this reason, the very notions of 'rationality', 'objectivity', and 'correctness' must be *relativized* to the choice of one or another formal language or linguistic framework" (Friedman 1998, 249).

So far, our interpretations agree; but then Friedman adds that the choice between different systems of rules and logic is purely conventional, governed by "pragmatic—*as opposed to rational*—criteria," meaning that a pragmatic choice cannot be rational (249). Friedman makes the same point when he discusses Carnap's internal-external distinction. He says that while internal questions arise within an already agreed upon linguistic framework and therefore can be rationally and objectively answered relative to it, external questions "*cannot be rationally and objectively answerable,* because the over-arching rules that could define these notions, are necessarily missing" (250; emphasis mine).

Ironically, if Friedman is right, then Carnap's view this time becomes virtually indistinguishable from Kuhn's, as Friedman himself notes. At any rate, it seems to me that Friedman overstates his case because (1) he opposes pragmatic criteria to rational ones, and (2) according to his reading of Carnap, external questions "cannot be rationally and objectively answerable." I argue that for Carnap there is no such opposition and that the choice among competing linguistic frameworks can be perfectly rational in a sense I will explain below.

In "Empiricism, Semantics, and Ontology" Carnap identifies "internal" questions of existence with "theoretical" ones and "external" questions of existence with "practical" ones (Carnap [1950] 1956, 206–8, 214). The decision to accept the thing language, for example, "is not a theoretical question" (208), but a practical one. To say that a question is a theoretical question of existence is to say that it has an answer that can be judged as being either true or false, so that one can provide confirming or disconfirming evidence for it. By contrast, the answer to a practical—that is, external—question "cannot be judged as being either true or false." External questions, therefore, do "not need any *theoretical* justification" in contrast to internal, that is, theoretical questions (214; emphasis mine).

Notice that Carnap does not say that external questions need no justification at all; he only says that they cannot be justified theoretically, meaning that the answers to them cannot be confirmed or disconfirmed. Carnap also believes that "the decision of accepting the thing language . . . will nevertheless usually be influenced by *theoretical knowledge,* just like any other deliberate decision concerning the acceptance of linguistic or other rules. The purposes for which the language is intended to be used . . . will determine which factors are relevant for the decision. The efficiency, fruitfulness, and simplicity of the use of the thing language may be among the decisive factors. And the questions concerning these qualities are indeed of a *theoretical nature*" (208; my emphasis).

While the decision to accept a certain linguistic framework is not itself theoretical, it will be influenced by theoretical factors such as efficiency, fruitfulness, and simplicity. This means that there is a fact of the matter as to whether framework *A* or *B* is simpler, more efficient, and more fruitful, a fact that can be discovered empirically (hence, theoretically). Indeed, this is precisely the case, Carnap believes, for the thing language: "The thing language in the customary form works indeed with a high degree of efficiency for most purposes of everyday life. This is a matter of fact, based upon the content of our experience" (208). However, the fact that the thing language functions efficiently for certain purposes does not provide any confirming evidence for the reality of the thing world, so the decision to adopt the thing language remains practical despite the fact that it is influenced by theoretical factors.

What emerges from our discussion is that Carnap does not simply equate "the rational" with "the theoretically justifiable," as Friedman seems implicitly to assume. A scientist chooses a framework of entities on the basis criteria like efficiency, fruitfulness, and simplicity, which are "of a theoretical nature." In contrast to Friedman's claim, these criteria themselves are not practical (or pragmatic). Carnap argues that a physicalist framework, for example, meets such criteria for doing physics better than a sense-datum framework does. In this way, he supposes that there is a theoretical answer to the question whether framework *A* is simpler, more efficient, and more fruitful than framework *B* relative to the goal *G*. This is not to say that *A* is confirmed more highly than *B*, so the choice of *A* remains practical and not theoretical. But in that way the practical choice of *A* can be said to be instrumentally rational. It is certainly neither arbitrary nor irrational. Therefore, Friedman is mistaken when he says that Carnap's external questions do not have any rational answers.

Carnap distinguishes *internal justification* from *external justification,* and, in effect, between *validation* from *vindication,* to use Feigl's better-known terminology (Feigl [1950] 1971). Internal justification or justification in the sense of validation means compliance with an already accepted system of rules (logical, methodological, or whatever, depending on the nature of inquiry). External justification or justification in the sense of vindication involves grounding of compliance with a set of rules relative to certain purposes. To give a simple example, with the rule of disjunction elimination, we can validate the inference from "AvB" and "¬A" to "B," but the rule itself cannot be validated in the same way on pain of circularity. It can be vindicated, however, if our purpose is to preserve truth in making inferences. In a similar vein, whereas internal questions of existence can be validated, external questions of existence can only be vindicated. Justification in the sense of vindication can also be called instrumental rationality because it is a means-ends rationality; the adoption of a linguistic framework is instrumentally rational if it serves a predetermined goal efficiently.

Thus, Putnam, Kuhn, and others are incorrect when they claim that according to logical empiricism the scientific method exhausts rationality, at least in so far as Carnap's philosophy is concerned. Conformity to a given set of rules that comprise the scientific method (regardless of whether the rules are local or universal, fixed or changing) is merely one aspect of rationality according to Carnap and does not exhaust his conception of rationality, which also has an instrumental component.

Once this is recognized, Putnam's charge of self-refutation also drops. Recall the charge: if scientific rationality is exhausted by scientific method, then this claim itself cannot be discussed rationally since it is not itself a scientific claim amenable to scientific scrutiny. Such a conception of rationality, Putnam contends, "leaves no room for a rational activity of philosophy" (Putnam 1981, 113). But, as we have seen, Carnap does not claim that scientific method in the sense of compliance with certain logical and methodological rules is all there is to rationality. The compliance itself can be vindicated—that is, rationally justified with respect to given goals and purposes.

Inductive Logic Revisited

It is not sufficiently noticed (if it is noticed at all) that Carnap extends his internal-external distinction about existence questions to cover inductive logic as well. Indeed, when Arthur Burks suggests in his contribution to the Schilpp volume that the internal-distinction can also be applied to the questions of induction, Carnap immediately agrees (see Carnap 1963b, 982). He points out that given a certain inductive system of axioms the computation of the degree of confirmation of a hypothesis relative to available evidence is an internal question, but that the question of the acceptance of the system itself as a whole is an external one. Carnap then writes: "I agree with Burks that important questions of the philosophy of induction belong to the external questions of induction. I regard the external questions themselves ... as practical questions. But still I would agree with Burks, first, that it is an important task of the philosophical clarification of induction to specify the factors which are relevant for a *rational decision of a practical external question;* and, second, that important *theoretical* questions are involved in the relevant deliberations leading to a practical decision of this kind" (Carnap 1963b, 982; the first emphasis is mine, the second one is Carnap's). Here, Carnap reiterates a point he has earlier made in "Semantics, Ontology, and Empiricism": the practical decision to accept a system of logic or linguistic framework is not itself theoretical; nevertheless, it does involve theoretical considerations, suggesting that there is a fact of the matter as to whether one framework is simpler, more fruitful or efficient than another relative to a given goal. Moreover, Carnap explicitly refers to *the rationality of a practical external decision,* leaving no room for doubt that such decisions can be rational—rational, that is, in the instrumental sense.

If one identifies Carnap's conception of rationality with his notion of scientific method and then this notion of method with his system of inductive logic, as Putnam does, or if one identifies Carnap's notion of the rational with the theoretical, as Friedman seems to do, then one is of course left with no choice but to conclude that, according to Carnap, the adoption of the inductive system itself cannot be rational. We have already seen what is wrong with this reasoning. The passage quoted above from Carnap provides further evidence that for Carnap (1) there is a domain of rational discourse outside the boundaries of the inductive logic or the scientific method, and (2) external questions can be answered rationally. Ultimately, while Putnam's mistake is to ignore Carnap's internal-external distinction (and the related principle of tolerance), Friedman's is to equate the rational with the theoretical.

There is, however, a peculiar asymmetry between the way Carnap justifies the axioms of inductive logic and the way he justifies the rules of a linguistic framework in general. As we saw, the latter are justifiable on the basis of criteria such as efficiency, simplicity, and fruitfulness. Although one would expect a similar justification regarding the axioms of inductive logic, Carnap appeals to something totally different—namely, intuition. Here is what he says:

> It seems to me that the reasons to be given for accepting any axiom of inductive logic have the following characteristics:
> (a) The reasons are based upon our intuitive judgments concerning inductive validity, i.e., concerning inductive rationality of practical decisions (e.g., about bets). Therefore,
> (b) It is impossible to give a purely deductive justification of induction.
> (c) The reasons are purely a priori.

By (c) I mean that the reasons are independent both of universal synthetic principles about the world, e.g., the principle of the uniformity of the world, and of specific past experiences, e.g., the success of bets which were based on the proposed axioms. (Carnap 1963b, 978–79)

Carnap makes clear that the intuitions appealed to are those of the experts who work on inductive logic and confesses that at times his intuitions change too (994). It is ironical that Carnap, after fighting against Kantian intuition throughout most of his career, falls back on intuition in the sense of preanalytic judgment. Frankly, I do not know why Carnap abstains from employing the usual criteria of efficiency, simplicity, and so on, why he insists that the reasons for the choice are a priori.[1] Perhaps he thinks that if the reasons are a posteriori, then his system of inductive logic would stop being a normative construction. At any rate, he acknowledges that there must be some sort of an equilibrium between a system of inductive logic and the scientists' decisions in their day-to-day scientific activities: "It seems to me that they show, in their behavior, implicit use of these numerical values [of logical probability]. For example, a physicist may sometimes bet on the result of a planned experiment; and, more important, his practical decisions with respect to his investment of money and effort in a research project show implicitly certain features not only of his utility function but also of his credibility function and thus of the corresponding c-function. If sufficient data about decisions of this kind made by scientists were known, then it would be possible to determine whether a proposed system of inductive logic is in agreement with these decisions" (990). The crucial question is what happens when the two do not agree. Unfortunately, Carnap does not say, and I suppose the answer depends on the scope of his normativism.

The Rebuttal of an Objection

Let me summarize what I hope I have achieved so far. I argued against the standard account that makes Carnap an arch-rationalist. My argument was based, among other things, on Carnap's principle of tolerance and his distinction between internal and external questions. I also criticized Friedman's overly relativist interpretation of Carnap. I tried to show that Carnap has a way of restricting his relativism by appealing to instrumental rationality, according to which external questions that involve the choice of linguistic frameworks can be rational relative to certain purposes.

One might object that Carnap's internal-external distinction and the related principle of tolerance are intended to apply only to philosophical questions, notably, existence and reality questions, not to scientific theories. Scientific theories, the objection goes, are not part of the linguistic framework, so for Carnap there is no such thing as one kind of justification within a scientific theory and another outside it, no distinction, that is, between theoretical and instrumental justification as far as scientific theories are concerned. If Carnap were to make such a distinction with respect to scientific theories, my objector claims, he would have no way of claiming the cognitive superiority of Einstein's theory over Newton's, no way of preferring one to the other except on practical grounds. But, presumably, Carnap does believe that Einstein's theory is cognitively better than Newton's, so my account cannot be right.[2]

I am not aware of any passage in Carnap's writings where he says that relativity theory is cognitively better than classical mechanics. More important, what exactly is the meaning of "better" here? It certainly does not mean that one theory is closer to truth than another, because Carnap believed that the notion of truth makes sense only relative to a language or linguistic framework, that the only clear and exact sense we can give to it is truth-in-a-language. In "Truth and Confirmation," for instance, he writes: "Furthermore, the formulation in term of "comparison," in speaking of 'facts' or 'realities', easily tempts one into the absolutistic view according to which we are said to search for an absolute reality whose nature is assumed as fixed independently of the language chosen for its description. The answer to a question concerning reality however depends not only upon that 'reality', or upon the facts but also upon the structure (and the set of concepts) of the language used for the description" (Carnap [1936] 1949, 126). For this reason, Carnap rejects any progress of science in terms of verisimilitude. For him progress means only instrumental progress. Indeed, that is what Carnap finds most appealing in Kuhn's *Structure*. He expresses this enthusiastically in a letter to Kuhn: "you emphasize that the development of theories is not directed toward the perfect true theory, but is a process of improvement of an instrument. In my own work on inductive logic in recent years I have come to a similar idea" (April 28, 1962; qtd. from Reisch 1991, 267).

Perhaps, then, "better" means "better confirmed"? That will not do, either. The notion "better confirmed" presupposes a unique or, at the very least, a shared system of inductive logic for two theories, but where is that system? As we saw, the degree of confirmation of a hypothesis depends on the linguistic framework in which the hypothesis is formulated and is therefore a relative measure; there is no absolute sense in which one hypothesis can be said to be better confirmed than another. That sense is barred by what I have dubbed as the confirmational incommensurability; the degree of confirmation is not only framework relative but also theory relative.[3]

But if there is no way in Carnap's philosophy to say that one theory is more true or better confirmed than another, in what sense can it be said to be better? The only sense I can think of is an instrumental one, as Carnap's letter to Kuhn attests. For Carnap a theory can be a simpler, more efficient, and more fruitful instrument than another relative to whatever goal a scientist has, but never better in the sense of giving us a more correct description of reality. This is in perfect agreement with his antirealism: to say, as Carnap does, that existence claims concerning electrons or elephants (Carnap [1950] 1956, 218) make cognitive sense only internal to a linguistic framework is to admit a form of antirealism, no matter how vehemently he rejects that label as "metaphysical" (see Salmon 1994a and 1994b).

Thus, relativizing truth and confirmation to theory language is indeed costly: Carnap cannot claim that one scientific theory is better than another on cognitive-epistemic grounds. This opens up his philosophy to more troublesome charges. Consider two people, a scientist and a witch doctor. The scientist prefers the physicalist language not because it describes reality more correctly, but because it serves his goal of "doing physics" better; it is more convenient, simpler, and so on for that purpose. On the other hand, the witch doctor prefers the language of witchcraft to that of physics because it is better suited to his purpose of pleasing the gods. Can Carnap offer any rational justification to the effect that one is a more noble goal than another? Worse yet, the witch doctor can try to "do physics" with the language of witchcraft. Carnap can reject this only on the grounds that it is not simple, efficient, and so on, not on the grounds that it is less likely to give a true description of reality. But then the witch doctor can say that witchcraft theory is damned hard to work with, but after all no one ever said that it would be easy to theorize about reality![4]

Carnap's internal judgments always have the form "Given the framework F, evidence E confirms theory T." The antecedent clause can never be detached, so T never enjoys confirmation simpliciter. The witch doctor can duplicate the serious work of scientists with his pseudoscientific framework R, and given that framework he can report with just as much validity, "given R, evidence E confirms T^*." I do not think Carnap has any resources to argue that in all likelihood there are forces, fields, or subatomic particles rather than witches.

Is Kuhn a Relativist?

I now want to turn to Kuhn's conception of the rationality of science. When the *Structure* was published, the reaction of most philosophers was to accuse Kuhn of subjectivism, relativism, and even irrationalism (see Lakatos 1970; Scheffler 1982;

Shapere 1964; and Suppe 1977). Despite Kuhn's later clarifications, many philosophers continue to call Kuhn a relativist about theory choice (see, for example, Chalmers 1982; McGuire 1992; Putnam 1981) because of his thesis of incommensurability with respect to standards. Friedman too thinks that Kuhn relativizes rationality to paradigms because there are no paradigm-transcending criteria to judge the merits of rival paradigms, so that what is rational to believe in one paradigm may not be rational in another: "For it is precisely in this case—where we are faced with the choice between competing scientific paradigms—that overarching standards of rationality and validity are entirely missing. Hence, the traditional ideal of scientific rationality must here evidently give way to non-rational factors in explaining the emerging consensus" (Friedman 1998, 250).

Indeed, many of Kuhn's bold claims and cunning rhetoric in the *Structure* leave hardly any room for a different interpretation. In that book Kuhn emphatically talks about "the incommensurability of standards," "gestalt switches," and "political and religious conversions" and writes that "their [the proponents of competing paradigms] standards or their definitions of science are not the same" (Kuhn 1970a, 148), and that "the choice between competing paradigms regularly raises questions that cannot be resolved by the criteria of normal science. To the extent . . . that two scientific schools disagree about what is a problem and what a solution, they will inevitably talk through each other when debating the relative merits of their respective paradigms" (109).

However, already in the "Postscript" Kuhn tones down his rhetoric and acknowledges the existence of criteria (such as quantitative accuracy, simplicity, consistency, and so on) for paradigm comparison (184). He there emphasizes that two scientists working under rival paradigms may share the same criteria but apply them differently to concrete cases. This requires deliberation and is learned mostly not by being taught abstract rules but by being exposed to concrete exemplars. Yet this is a kind of knowledge that is almost impossible to detach from the cases from which it was acquired. Therefore, when the scientists are confronted with a new puzzle, they may disagree, for instance, whether paradigm *A* or *B* provides a simpler solution, or they may attach different weights to the shared criteria. This is a perfectly rational disagreement, and the only way to resolve it is through the techniques of persuasion, through reasoned argument to make the other see the superiority of one's choice.

It is for these reasons that in his paper "Objectivity, Value Judgment, and Theory Choice" Kuhn classifies the criteria for theory choice as values, not as rules. They are quantitative accuracy, simplicity, consistency, broad scope, and fruitfulness, known as "the five values." While "they [do] provide the shared basis for theory choice" (Kuhn 1977, 322), "they function not as rules, which determine theory choice, but as values, which influence it" (331; see also Kuhn 1970c, 262). This shared basis is hardly sufficient to determine the choice of individual scientists so that "every individual choice between competing theories depends on a mixture of objective and subjective factors, or of shared and individual criteria" (Kuhn 1977, 325).

Do the values themselves change historically? If so, Kuhn opens himself to the charge of relativism. In "Objectivity, Value Judgment, and Theory Choice" Kuhn gives a qualified answer: "Throughout this paper I have implicitly assumed that, whatever their initial source, the criteria or values deployed in theory choice are fixed once and for all ... Roughly speaking, but only very roughly, I take that to be the case. If the list of relevant values is kept short and if their specification is left vague, then such values as accuracy, scope, and fruitfulness are permanent attributes of science" (325). But after his linguistic turn in the early 1980s, Kuhn dropped the qualification. Incommensurability as varying standards disappeared altogether, and what survived was an innocuous local semantic incommensurability understood as untranslatability, which does not imply incomparability (see Irzik and Grünberg 1998). Indeed, in his latest work, "Afterwords," the five values become permanent features of science: "As the developmental process continues, the examples from which practitioners learn to recognize accuracy, scope, simplicity, and so on, change both within and between fields. But the criteria that these examples illustrate are themselves permanent, for abandoning them would be abandoning science together with the knowledge which scientific development brings" (Kuhn 1993, 338).

I suggest that we understand Kuhn's position concerning the stability/instability of criteria for theory choice in the following way. As a historian, Kuhn sees that some criteria do change, albeit very slowly. For example, he notes that in ancient times very few disciplines, notably astronomy, require quantitative accuracy. One major result of the scientific revolution in the sixteenth and seventeenth centuries was the mathematization of many scientific disciplines, routinely requiring quantitatively accurate predictions from physics, chemistry, and so on. Kuhn thinks that the five values became permanent features of science after the scientific revolution. They are operative during both normal and revolutionary science. New features may be added in the future, but Kuhn thinks that the old ones are unlikely to disappear and that "[their] rejection would be irrational" (338); they become constitutive of science and scientific rationality. I conclude, therefore, that Kuhn should not be seen as a relativist with respect to criteria for theory comparison.

Conclusion

The popular picture of a momentous shift from rationalism to relativism when postpositivism replaced logical empiricism and falsificationism must be revised drastically. As far as the rationality of science is concerned, Carnap's views are much closer to Kuhn's than to Popper's. Popper conceives of scientific rationality as a matter of abiding by universal rules. For him there is absolutely no difference between internal and external questions or between internal and external justification. The same rules apply across the board. Carnap, on the other hand, believes that such a distinction is crucial because the criteria employed in each case are radically different. Internal questions are rationally answered relative to the framework rules. But the external question of how to justify the framework itself is answered on different grounds. To the extent that the framework involves axioms of inductive logic, Carnap claims that their justification is based on intuition and is therefore a priori; otherwise, the framework is justified on the basis of efficiency, simplicity, and fruitfulness. The crucial point according to Carnap is that external questions are just as justifiable as the internal ones, although his insistence on intuition for the justification of axioms of inductive logic is far from satisfactory. However, his relativization of truth and confirmation to language creates distressing problems. It is also worth noting that for Carnap neither the criteria for internal questions nor the criteria for external questions unambiguously dictate a unique choice of theory. There is always room for other (including personal) factors.

In this, Carnap and Kuhn fully agree. Moreover, as I have argued elsewhere (Irzik and Grünberg 1995), there is a striking similarity between scientific activity within Kuhnian normal science and scientific activity within Carnapian theory-cum-linguistic framework, on the one hand, and between Kuhn's characterization of scientific revolutions and Carnap's, on the other. Carnap and Kuhn also agree on the rationality of both types of activities. However, unlike Carnap, Kuhn does not believe that two different kinds of criteria are operative during normal and revolutionary science. Beginning with the "Postscript," Kuhn comes to believe that the same criteria apply to both kinds of activity. Since they are the sole criteria for puzzle-solving, which, according to Kuhn, is the main way knowledge of nature is gained, he takes them to be epistemic (Kuhn 1993, 338). Hence, for Kuhn there is no distinction between internal and external justification; thus, if you like, there is no distinction between "theoretical" and "instrumental" rationality. Carnap, on the other hand, denies that criteria such as simplicity, efficiency, and fruitfulness provide any knowledge, for there is no connection between the satisfaction of these criteria and truth or confirmation.

A second difference between Kuhn and all other philosophers before him including Carnap and Popper is this: For Kuhn, rationality is not just compliance with rules; it also involves judgment and deliberation. Rationality in this sense is conspicuous when one tries to apply the five values to theory choice. Since Kuhn deems none of them to be ultimate, no general rule or principle tells the scientist what to do if the criteria point to different directions; each scientist must use his lifelong experience, his "practical wisdom," to make the best possible choice.[5]

This process is akin to what Aristotle calls phronesis, a kind of deliberative reasoning essential for ethical judgments. Phronesis mediates between the general and the particular and proceeds without the employment of any rules. Typically, it is required in difficult cases where the bearing of general rules or principles on particular facts are not clear. It involves deliberation not only about how best to achieve an end, but also about the ends themselves and how to weigh them if none is ultimate. Richard Bernstein summarizes Kuhn's contribution succinctly: "The shift from a model of rationality that searches for determinate rules which can serve as necessary and sufficient conditions, to a model of practical rationality which emphasizes the role of exemplars and judgmental interpretation, is not only characteristic of theory choice but is a leitmotif that pervades all of Kuhn's thinking about science" (Bernstein 1983, 57). As we saw, one nice consequence of this model is that it explains how rational disagreement can arise in science. It seems to me that this shift is the most important contribution that Kuhn has made to our conception of scientific rationality, a point that has not been appreciated sufficiently (see also Brown 1977, 149–50).

In this way Kuhn narrows the gap between science and ethics, neither of which can do without practical wisdom. In fact, the relation between the two fields goes deeper than this. To see this, go back to Carnap's conception of rationality: a move can be justified (that is, validated) by the relevant rules, and the rules themselves can be justified (that is, vindicated) with respect to certain goals on the basis of efficiency and the like. But how are we to justify the goals themselves? More bluntly, why do physics rather than engage in witchcraft or religion? The natural, and in my view the correct, response is that it serves for the general benefit of humanity better, and, indeed, this is Carnap's answer, too. As a scientific humanist, he sees science as the most valuable instrument for improving the good life, and for him the good life essentially consists of creating conditions for developing the individual's potential abilities and removing all the obstacles against social participation (Carnap 1963a, 83). So, Carnap's answer to why we should engage in science rather than witchcraft is another instance of instrumental reasoning: science is justified on the grounds that it serves the good life. That is the deeper connection I had in mind between science and ethics.

But the trouble with this for Carnap is that he is an emotivist: value statements lack cognitive content; they do not have any truth-value. But then the justification of science on the basis of ethics is not really a justification at all since the claim about the good life merely expresses Carnap's emotive state or disposition toward life. At best, it "justifies" Carnap's engagement in science, but not that of other people such as our own children. Carnap was a true heir of the Enlightenment ideal of social progress, and he firmly believed that scientific education could contribute to this significantly. Yet, his emotivism in ethics severed the crucial link between science and the good life as he envisioned it. Perhaps this is the ultimate aporia of logical empiricism.[6]

Notes

1. Wesley Salmon suggested to me that the reason is that in all likelihood Carnap embraced the notion of synthetic a priori truth toward the end of his life. Maybe this is so, but we need other evidence; the passage I have quoted explicitly denies that the reasons are dependent on synthetic principles.

2. I am grateful to Michael Friedman for raising this objection.

3. Following common usage, I have talked about confirming a theory. But I am not sure if that notion makes any sense within the Carnapian system of inductive logic. It seems to me that for Carnap, hypotheses formulated within a theory can be probabilified, but not the theory itself. If this is right, there is further reason to believe that Carnap's internal-external distinction applies to scientific theories as well. While I cannot defend this particular point here, my colleague Teo Grünberg and I are about to complete a paper discussing this issue.

4. I owe this point to Stephen Voss. I thank him also for allowing me to use some of his formulations.

5. Kuhn's reference to Polanyi's notion of tacit knowledge should be understood in this context. This kind of scientific wisdom cannot be formulated explicitly in rules and sometimes not even in propositions, but is acquired mostly in practice by being exposed to exemplars. Alasdair MacIntyre expresses this aspect of rationality eloquently:

> Objective rationality is therefore to be found not in rule-following, but in rule transcending, in knowing how and when to put rules and principles to work and when not to. Consider how practical reasoning of this kind is taught, whether it is the practical reasoning of generals, of judges in a common law tradition, of surgeons or of natural scientists. Because there is no set of rules specifying necessary and sufficient conditions for large areas of such practices, the skills of practical reasoning are communicated only partly by precepts but much more by case-histories and precedents. Moreover the precepts cannot be understood except in terms of their application in the case-histories; and the development of the precepts cannot be understood except in terms of the history of both precepts and case-histories. (Qtd. in Bernstein 1983, 57)

My (and, I would conjecture, Kuhn's) only disagreement with MacIntyre concerns the first part of his first sentence; I take rule-following also to be part of our rationality. Unfortunately, I cannot give the original source for this passage because Bernstein quotes it from an unpublished manuscript of MacIntyre's, for which he himself provides no reference.

6. I thank Stephen Voss and Berna Kilinc for their helpful comments and suggestions.

References

Bernstein, R. 1983. *Beyond Objectivism and Relativism: Science, Hermeneutics, and Praxis.* Oxford: Basil Blackwell.

Brown, H. 1977. *Perception, Theory, and Commitment.* Chicago: University of Chicago Press.

Carnap, R. [1934] 1937. *The Logical Syntax of Language.* Trans. A. Smeaton, Countess von Zeppelin. London: Kegan Paul Trench, Trubner and Co. Originally published as *Logische Syntax der Sprache, Schriften zur Wissenschaftlichen Weltauffassung.* Vol. 8. Vienna: Verlag von Julius Springer.

———. [1934] 1967. "On the Character of Philosophical Problems." *Philosophy of Science* 1:5–19. Reprinted in *The Linguistic Turn,* ed. R. Rorty, 54–62 (Chicago: University of Chicago Press, 1967).

———. [1936] 1949. "Truth and Confirmation." In *Readings in Philosophical Analysis,* ed. H. Feigl and W. Sellars, 119–27. New York: Appleton-Century-Crofts.

———. [1950] 1956. "Empiricism, Semantics, and Ontology." In *Meaning and Necessity,* 205–21. 2d ed. Chicago: University of Chicago Press.

———. [1950] 1962. *Logical Foundations of Probability.* Chicago: University of Chicago Press.

———. 1963a. "Intellectual Autobiography." In *The Philosophy of Rudolf Carnap,* ed. P. A. Schilpp, 3–84. La Salle, Ill.: Open Court.

———. 1963b. "Replies and Systematic Expositions." In *The Philosophy of Rudolf Carnap,* ed. P. A. Schilpp, 859–1013. La Salle, Ill.: Open Court.

Carnap, R., and R. Jeffrey. 1971. *Studies in Inductive Logic and Probability.* Berkeley: University of California Press.

Chalmers, A. 1982. *What Is This Thing Called Science?* 2d ed. Queensland: University of Queensland Press.

Earman, J. 1993. "Carnap, Kuhn, and the Philosophy of Scientific Methodology." In *World Changes,* ed. P. Horwich, 9–36. Cambridge, Mass.: MIT Press.

Feigl, H. [1950] 1971. "De Principiis Non Disputandum . . . ?" In *Philosophical Analysis,* ed. M. Black. Freeport, N.Y.: Books for Libraries Press.

Feyerabend, P. 1978: *Against Method.* London: Verso.

Friedman, M. 1998. "On the Sociology of Scientific Knowledge and Its Philosophical Agenda." *Studies in History and Philosophy of Science* 29:239–71.

Hilpinen, R. 1973. "Carnap's New System of Inductive Logic." *Synthese* 25:307–33.

Irzik, G., and T. Grünberg. 1995. "Carnap and Kuhn: Arch Enemies or Close Allies?" *British Journal for the Philosophy of Science* 46:285–307.

———. 1998. "Whorfian Variations on Kantian Themes: Kuhn's Linguistic Turn." *Studies in History and Philosophy of Science* 29:207–21.

Kuhn, T. 1970a. *The Structure of Scientific Revolutions.* 2d ed. Chicago: University of Chicago Press.

———. 1970b. "Logic of Discovery or Psychology of Research?" In *Criticism and the Growth of Knowledge,* ed. I. Lakatos and A. Musgrave, 1–23. Cambridge: Cambridge University Press.

———. 1970c. "Reflections on My Critics." In *Criticism and the Growth of Knowledge,* ed. I. Lakatos and A. Musgrave, 231–78. Cambridge: Cambridge University Press.

———. 1977. "Objectivity, Value Judgment, and Theory Choice." In *The Essential Tension,* 320–39. Chicago: University of Chicago Press.

———. 1993. "Afterwords." In *World Changes,* ed. P. Horwich, 311–41. Cambridge, Mass.: MIT Press.

Lakatos, I. 1970. "Falsification and the Methodology of Scientific Research Programmes." In *Criticism and the Growth of Knowledge,* ed. I. Lakatos and A. Musgrave, 91–196. Cambridge: Cambridge University Press.

McGuire, J. 1992. "Scientific Change: Perspectives and Proposals." In *Introduction to the Philosophy of Science,* M. Salmon et al., 132–78. Englewood Cliffs, N.J.: Prentice Hall.

Putnam, H. 1981. *Reason, Truth, and History.* Cambridge: Cambridge University Press.

Reisch, G. 1991. "Did Kuhn Kill Logical Empiricism?" *Philosophy of Science* 58:264–77.

Salmon, W. 1994a. "Carnap, Hempel, and Reichenbach on Scientific Realism." In *Logic, Language, and the Structure of Scientific Theories,* ed. W. Salmon and G. Wolters, 237–54. Pittsburgh: University of Pittsburgh Press and Universitatsverlag Konstanz.

———. 1994b. "Comment: Carnap on Realism." In *Logic, Language, and the Structure of Scientific Theories,* ed. W. Salmon and G. Wolters, 279–86. Pittsburgh: University of Pittsburgh Press and Universitatsverlag Konstanz.

Scheffler, I. 1982. *Science and Subjectivity.* 2d ed. Indianapolis: Hackett.

Shapere, D. 1964. "The Structure of Scientific Revolutions." *Philosophical Review* 73:383–94.

Suppe, F. 1977. Afterword to *The Structure of Scientific Theories.* 2d ed. Ed. F. Suppe, 617–730. Urbana: University of Illinois Press.

VII

Nonlinguistic Empiricism

Reason and Perception

In Defense of a Nonlinguistic Version of Empiricism

PAOLO PARRINI

University of Florence

Empiricism and the Negation of the Synthetic a Priori

The notion of "empiricism" I would like to discuss is the one that has been adopted by the major exponents of logical empiricism and, in particular, by Moritz Schlick and the authors of the Vienna Circle's manifesto, namely Rudolf Carnap, Hans Hahn, and Otto Neurath ([1929] 1973, 308). In Carnap's last years of activity, he mentioned the observation that Schlick had made in the *Allgemeine Erkenntnislehre*, according to which "there are no synthetic judgments a priori" ([1918, 1925] 1974, §40, 384). Immediately after, Carnap added: "If the whole empiricism is to be compressed into a nutshell, this is one way of doing it" ([1966] 1974, 180).

This characterization of empiricism calls to mind the well-known criticism brought against Kant's conception of synthetic a priori judgments that logical empiricists, in particular Schlick and Hans Reichenbach, developed around the 1920s when discussing the philosophical meaning of the theory of relativity. From the beginning of the 1970s this criticism has forcefully attracted scholars' attention. This has contributed to a critical reexamination of the traditional way of conceiving logical empiricism and in some cases also of the attacks that were led against it, especially in the 1950s and 1960s, first by Willard Van Orman Quine and then by other philosophers such as Norwood Russell Hanson, Thomas S. Kuhn, and Paul K. Feyerabend.

One of the most interesting aspects brought to light by these researchers is that in the 1920s Schlick and Reichenbach did not fully agree in their way of criticizing Kant's conception and in formulating an alternative one. They both tried to find an empiricist conception that was different from both a phenomenalistic empiricism that "neglects the problems of conceptualisation" (Reichenbach [1922] 1978, 37) and a kind of Kantian formal idealism based on the idea of forms of reason that are at the same time synthetic and apodictically certain (that is, universally and necessarily valid of the objects of experience). Schlick and Reichenbach had different views regarding the way in which it was possible to positively characterize the indispensable conceptual structures (coordinative assumptions) necessary to fill that gap between the complex of our beliefs on one side and the data of experience on the other. Reichenbach ([1920] 1965, 48; [1922] 1978, 38) was in favor of a weak negation of the synthetic a priori and saw these assumptions not as definitions or conventions but as "constitutive principles" analogous to Kant's synthetic a priori judgments in all their aspects except their apodictic character. Schlick, on the contrary, was in favor of a strong negation of the synthetic a priori that had no relation whatsoever to Kantism. According to him (Schlick [1921] 1979, 333, 324), scientific judgments can be included in only two categories: hypothesis and "conventions in Poincaré's sense": "in the first case they are not a priori (since they lack apodeicticity), and in the second they are not synthetic," because they are assumptions of a purely linguistic nature devoid of any empirical content.

Here I will not consider in detail the contrast between Schlick and Reichenbach (Parrini [1995] 1998, 2:59). It will be enough to call to the mind the well-known fact that later on Reichenbach moved closer and closer to Schlick's position and the way of formulating the negation of the synthetic a priori that will prevail within the Vienna Circle after the influence exercised by Bertrand Russell's and Ludwig Wittgenstein's ideas on the formulation of the verification principle and the thesis of the tautological nature of logical and mathematical truths. In the Philosophie der Raum-Zeit-Lehre, Reichenbach conforms to the terminology used by Schlick in *Allgemeine Erkenntnislehre* and describes as "coordinative definition" the assumption of congruence necessary to confer an empirical content to the hypothesis regarding the geometrical structure of physical space (Reichenbach [1928] 1958, §4, 14). In the essay "L'empirisme logistique et la désagrégation de l'a priori," presented in 1935 in Paris during the conference on scientific philosophy, he formulates the problem of the a priori in a way that perfectly conforms to the conceptions that prevailed after the "linguistic turn": "After Kant, the development of science can be considered a steady decomposition of the foundations of rationalism; it really means the breakdown of the a priori . . . Today science no longer believes in the legislative capabilities of pure reason. All that we know about the world comes from experience, and the transformations of what is empirically given are merely tautological or analytical."[1] Personally, I think that studying the reasons that led Reichenbach to abandon the position held during the early 1920s and to adhere to a reductionist conception—in Quine's sense of the term ([1953] 1961, 40–43)—of the theory/experience relationship can be extremely relevant in order to understand the historical process that led to the formation of logical empiricism, an interest heightened by the fact that the material contained in the Archives of Scientific Philosophy in the Twentieth Century presents some puzzling aspects.[2] But this is not the question I intend to deal with in this chapter. Here I am interested in two theoretical aspects related to the characterization of empiricism intended as the negation of the synthetic a priori.

The first of these aspects regards the clarification of the sense in which we can compress empiricism in the negation of the synthetic a priori, whether we want to intend this negation either in the strong sense or in the weak one. The second aspect regards the reasons in support of the validity of an empiricist position that had been so characterized. In this respect my aim is to advance three theses: (1) The reasons lying at the basis of an empiricism intended as a negation of the synthetic a priori are strictly connected to the claim that perception is independent of reason; (2) logical empiricists have not analyzed this relationship in enough depth: they have taken for granted perception's independence of reason and discussed the relationship between theory and experience by resorting to the dichotomy between "basic" statements (variously conceived) and theoretical statements; (3) an adequate treatment of the possibility of an empiricism based on the negation of the synthetic a priori demands that we consider also the subjective-psychological components that are at the basis of the formation of a symbolism with extralinguistic references.

Such components are considered, for example, in the network model proposed by Mary Hesse in the 1970s. A brief reference to some aspects of this model will allow me to show how it is possible to consider the psychological-subjective aspect by using arguments of a strictly epistemological kind so as to avoid any contamination between epistemology and psychology. In the conclusion to this chapter I will advance the hypothesis that the fear of falling into psychologism is the reason, or one of the reasons, that kept logical empiricists from dealing with the relationship between perception and reason in a way that was adequate to their aims. On the contrary, taking into account—again from an epistemological point of view—the psychological-subjective dimension of the problem is, in my opinion, fundamental in order to elaborate a valid defense of empiricism against the objections that could be raised against it under the influence of the ideas advanced by philosophers and historians of science such as Hanson, Kuhn, and Feyerabend.

The Contingent Nature of the Cognitive Synthesis and Perception's Independence of Reason

As far as the first theoretical aspect is concerned, I think that we can answer that logical empiricists ended up summarizing the essential content of logical empiricism in the negation of synthetic a priori judgments because to their eyes this negation seemed to be tantamount to stating the "contingent character of the cognitive synthesis." I will try to explain what I mean by briefly examining both the published and unpublished material regarding the debate that at the beginning of the 1920s Schlick, Reichenbach, and Ernst Cassirer had about the philosophical meaning of the theory of relativity.

In a famous essay published in 1921, "Zur Einstein'schen Relativitätstheorie," Cassirer had tried to bridge the contrast between relativistic physics and the transcendental theory of knowledge. In order to achieve this aim, he admitted that Kant might have committed a mistake when identifying synthetic a priori judgments (in particular the one regarding Euclidean geometry); at the same time, though, he continued to support the validity of Kant's conception according to which we cannot attain knowledge unless we admit the existence of these judgments, in an appropriately generalized way, for example in the form of the causality principle, or induction or the presupposition of experience's conformity to laws ("unity of nature"). The need to admit presuppositions of this kind would prove clearly the fact that science is not made only of hypothesis or conventions: science must also avail itself of principles that cannot be justified either on the logical or on the empirical level. In a letter written to Schlick on October 29, 1921, Cassirer suggested that the general idea of the "univocal nature of coordination" must present itself in some functional form in all theories of nature. Consequently, also in Schlick's characterizations of knowledge as coordination and of truth as univocal coordination we could see a synthetic a priori principle that is extremely general and which cannot be undermined by any development that occurs in scientific knowledge.

Schlick replies to this observation in his 1921 essay "Kritizistische oder empiristische Deutung der neuen Physik?" and there is an interesting analogy between his position on knowledge and Hermann von Helmholtz's position on geometry's epistemological status. In the conclusion to his essay on the origin and meaning of geometrical axioms, the German scientist stated that it is impossible to save Kant's conception by making "the concept of the fixed geometrical structures necessary for measurement" (Helmholtz, [1870, 1921] 1977, 25) depend on Euclid's postulates, because in this case these postulates cease to be synthetic a priori judgments and become analytic judgments. In an analogous way, Schlick states that it is impossible to save transcendental epistemology by resorting to the characterizations of knowledge as coordination and of truth as univocal coordination because these traits would have to be considered as analytical components of the concepts involved.[3] The fact that so far we have seen the development of scientific theories that exemplify this type of notion of knowledge and truth is an experience datum that cannot be denied. Its acceptance is not in conflict with the empiricist credo, provided that the empiricist philosopher, unlike the Kantian one, does not claim to guarantee a priori that this realization will be perpetuated also in the future.

In this way, the thesis of the impossibility of going beyond Hume's point of view, which had already been defended in §41 of the *Allgemeine Erkenntnislehre* (Schlick, [1918, 1925] 1974, 398) is implicitly restated. To put it briefly, the answer given by the empiricist to the "rationalistic" objection moved by Cassirer is not the negation of the existence of presuppositions of knowledge, but the negation of the possibility of showing their "validity and objective necessity . . . by a transcendental deduction or in any other way. Here the critical philosopher can appeal to no physical theory, for each of them proves by its verification in experience only the factual, not the necessary, validity of the principle of the unity of nature" (Schlick [1921] 1979, 327). Thus it is from this very reply that we can see emerging the deeply empiricist meaning of the negation of the synthetic a priori—that is, of what I called the claim of the contingent nature of the cognitive synthesis.

On which basis, on the ground of which reason, can we state the contingent character of the cognitive synthesis? In other words, where does the contingent validity of all the components of our knowledge derive from? Around the 1920s the answer to this question was also clearly developed, in relation, again, to Schlick's and Reichenbach's reflection on the philosophical empiricist and antitranscendentalistic meaning of relativistic philosophy. The clearest and most synthetic formulation of this answer can be found, in my opinion, in Reichenbach's 1922 essay "Der gegenwärtige Stand der Relativitätsdiskussion." Here he develops a critical examination of the main philosophical interpretations of the theory of relativity in order to restate (with a few concessions made to Schlick) the point of view he had defended in his 1920 essay "Relativitätstheorie und Erkenntnis Apriori." At one point in the essay, where he summarizes the main thesis expressed in 1920, he writes:

> Kant had the idea that reason prescribes a certain system of principles by means of which our knowledge of the physical world is established. According to Kant, this system can never be falsified by experience because experience is possible only by means of these principles; therefore these principles can be called synthetic judgments a priori. But Kant uses an assumption for which he gives no evidence (and for which he can give no evidence); he assumes that the system of principles is not overdetermined (hypothesis of arbitrariness of co-oordination). Is it possible for a contradiction to arise between the totality of experience and a system of principles if that system provides more rules for experience than experience will admit. The ultimate instrument in the judgment of empirical truths, perception, is independent of reason; although perception always permits different interpretations, the combination of the interpretations is no longer arbitrary. It is the significance of the theory of relativity to have discovered the limits of arbitrariness. According to the theory of relativity, the choice of a geometry is arbitrary; but it is no longer arbitrary once congruence has been defined by means of rigid bodies. The combination of the principles of Euclidean geometry and the definition of congruence is excluded by experience. Thus we can show that there are overdetermined combinations; it is not possible to prove that the system offered by reason is not such an overdetermined combination. (Reichenbach [1922] 1978, 37, first emphasis mine).

This passage by Reichenbach seems to me particularly important because it allows us to establish a connection between the main reason at the basis of neo-empiricists' negation of the synthetic a priori and consequently of their acceptance of an empiricist position, and what Kant had tried to show in his *Critique of Pure Reason* in order to defend his transcendental conception. What clearly emerges from the above quotation is that the possibility of holding an empiricist position—which qualifies itself as such in relation to the negation of the synthetic a priori and thus of the statement of the contingent nature of the cognitive synthesis—is linked to the condition that perception is independent of reason. Only the fulfillment of this condition can guarantee the possibility of the insurgence of a contrast between the principles elaborated by reason and incorporated in the complex of our various hypotheses and theories and what is given in perception.

In order to bring to light the connection between this aspect of Kantian philosophy and neo-empiricist criticism to the synthetic a priori, it will be enough to look at the way in which Kant posits and solves the question of categories' transcendental deduction in the *Critique of Pure Reason*. After showing the transcendental subjectivity of the forms of space and time in the transcendental aesthetic, in paragraph 13 Kant introduces transcendental analytical philosophy's fundamental problem—that is, the problem of justification or transcendental deduction of the objective validity of intellect's forms. And he explains why the need to have such a type of deduction presents itself for the categories of understanding and not for the forms of sensibility: "The categories of understanding, [differently from the forms of space and time], do not represent the conditions under which objects are given in intuition. Objects may, therefore, appear to us without their being under the necessity of being related to the functions of understanding; and understanding need not, therefore, contain their a priori conditions" (Kant [1781, 1787] 1965, A89=B122, 123). As is well known, the interpretation of this passage has been problematic and many interpreters have been of the opinion that it reflects Kantian thought's precritical phase because the nature of the conclusion that the transcendental deduction intends to reach is opposite to what had been stated by Kant when positing the problem. This emerges clearly when Kant summarizes deduction's result in the section entitled "The Highest Principle of All Synthetic Judgments":

> The highest principle of all synthetic judgments is therefore this: every object stands under the necessary conditions of synthetic unity of the manifold of intuition in a possible experience.
> Synthetic a priori judgments are thus possible when we relate the formal conditions of a priori intuition, the synthesis of imagination and the necessary unity of this synthesis in a transcendental apperception, to a possible empirical knowledge in general. We then assert that the conditions of the possibility of experience in general are likewise conditions of the possibility of the objects of experience, and that for this reason they have objective validity in a synthetic a priori judgment. (Kant [1781, 1787] 1965, A158=B197, 194)

The connection between this aspect of Kant's philosophy and neo-empiricist criticism to the synthetic a priori principles is, thus, evident. The argument developed by Kant seems to take as its starting point the idea that perception can be considered as independent of reason in order to arrive to the opposite conclusion —that is, to state that perception cannot be independent (completely or in part) of the principles posited by reason. Consequently, if we had to recognize as valid Kant's argumentation, we would have to conclude that that very condition that, according to Reichenbach, lies at the basis of the neo-empiricist negation of the synthetic a priori is not fulfilled. It seems to me, thus, that we cannot consider as satisfactory a position that qualifies itself as empiricist with reference to such a negation if its defendants do not support, in one way or another, the claim that perception is independent of reason.

The Distinction between Theoretical and Observational Language and the "Strong Thesis" of the Theoretical Character of Observation

Putting aside the question of the theoretical validity of the statement of perception's independence of reason, on the historical level we cannot avoid recognizing that logical empiricists never paid particular attention to the illustration and justification of the idea that some contents of perception do not depend on the conceptual and theoretical structures of reference. The validity of this thesis has been considered in some ways obvious and thus not such as to need particular explanations. The question of the theory/experience relationship has been dealt with mainly by the mediation of two collateral interconnected questions: (1) the problem of the formulation of an empiricist criterion of cognitive meaningfulness (it is the question of the verification principle); and (2) the problem of elaborating an adequate characterization of the notion of "scientific theory."

The reflection on both questions led to the formulation of what is known as the "standard conception" or "received view" of scientific theories, according to which these theories can be assimilated to axiomatic-deductive systems that are only partially interpreted on the basis of experience. The distinction between observational and theoretical language is an integral part of this conception, which belongs to the general "linguistic turn" taken by philosophy. We have, thus, a vision of scientific discourse on "two levels": on the one level we have statements formulated with theoretical terms and, on the other, statements formulated with observational terms; the connection between the two levels is provided by some connective statements (variously conceived) that have been formulated ad hoc using a mixed vocabulary that includes both theoretical and observational terms.

This particular vision of scientific discourse was the result of a fairly complex thinking process that started with the phenomenal reductionism defended in the *Aufbau* and that concluded with the extreme liberalization of the criterion of cognitive meaningfulness present (in different ways) in Carnap's and Carl Gustav Hempel's works in the 1950s. And it was mainly the identification of the neo-empiricist movement with this reductionism that favored the consolidation and the spreading of the image of this movement as a fairly monolithic line of thought characterized by a conception of scientific knowledge that is foundationalistic in the empiricist sense of the term. This has led to attributing to logical empiricists the idea—which later on was criticized by Karl Popper, N. R. Hanson, T. S. Kuhn, and P. K. Feyerabend—that there is a complex of protocol or observational statements, the validity of which depends somehow only on experience. The validity of these statements would enjoy some kind of absolute empirical foundation and would not depend in any way on reason's conceptual and theoretical constructions. They would constitute, thus, a neutral basis on the ground of which we can assess the validity, or, at least, the empirical adequacy of our hypotheses and theories.

As we know, the studies conducted in the past twenty-five years on logical empiricism have shown the historiographic inadequacy of the above-described image of logical empiricism. Despite the substantial correctness of the credits that are more and more frequently given to logical empiricists,[4] though, we cannot deny the fact that they have never adequately questioned the statement regarding perception's independence of reason that lies at the basis of their empiricist stand —that is, of their negation of the synthetic a priori. When facing the relationship between theory and experience, they overlooked this philosophical question, probably because they thought this independence was an obvious fact that does not need any clarification. As I have already said, in parallel to philosophy's linguistic turn, they limited themselves to speaking about an empirical basis of knowledge that can be expressed by means of a more or less wide complex of variously construed protocol sentences, and in discussing the best way of conceiving this they ended up introducing the distinction between theoretical and observational language.

The conclusion of this story is well known to everyone and can be summarized as the "strong thesis" of the theoretical character of observation (where observation has to be construed as a kind of focused perception). This thesis has been supported mainly by Feyerabend but is present in various ways also in the ideas advanced by Hanson and Kuhn. It seems to me that such a thesis presents us again, in a linguistic and particularly drastic disguise, Kant's claim of perception's dependence on reason. It is particularly drastic because observation's dependence on the theory that is in question is linked to the whole complex of our hypotheses and theories, also the most specific ones. It is such as to deny the possibility of having a genuine empirical test of our cognitive claims that is not vacuously circular. If we accept this thesis, in other words, it becomes difficult to understand how our hypotheses and theories can clash with adverse data. It shows fully the philosophical problematicity of the idea that perception's results, since they are embedded in a complex of observational sentences, can be something independent of reason's theoretical-conceptual constructions and are capable of telling us something about which of those sentences is the empirically acceptable one.[5]

The Implausibility of the "Strong Thesis," the Paradox of Categorization, and the "Weak Thesis" of the Theoretical Nature of Observation

Independently of the cogency of the arguments that support the "strong thesis," we have to admit that one of the main consequences it leads to—in other words, the impossibility of having a genuine empirical test—seems to be sharply in contrast with what cognitive processes usually show. This thesis, in other words, seems to contrast with the well-proved fact that we held verifiable empirical expectations that are often frustrated (the well-known "stubbornness" of data). Such a conclusion seems to be endowed with such little plausibility that it has not been fully adopted even by authors such as Kuhn, Lakatos, and Feyerabend. With respect to staunch empiricists, these historians and philosophers of science have certainly aimed at attenuating the import of experience in the evaluation of theories; but they were not able or willing to deny it completely. They have opted for a position that on the whole is in line with the outcome of Duhem's criticism of crucial experiments, which endorses the empirical falsifiability and confirmability of theoretical systems globally considered.[6]

But recognizing the fact that experiential data can clash with our hypotheses and theories is not tantamount to showing how this is possible. In brief, the "strong thesis" of the theoretical nature of observation, which originated in the criticism of the neo-empiricist dichotomy between theoretical and observational language, merely reposits Kant's problem of perception's independence of reason. We can illustrate the general nature of the problem and the spectrum of the main possible solutions by taking into consideration the so-called paradox of categorization. Drawing inspiration from Clarence Irving Lewis's position on the observational basis (a position already criticized by Reichenbach because of a certain "absolutism"),[7] Scheffler formulated this paradox with the following words: "[If] my categories of thought determine what I observe, then what I observe provides no independent control over my thought. On the other hand, if my categories of thought do not determine what I observe, then what I observe must be uncategorized, that is to say, formless and nondescript—hence again incapable of providing any test of my thought. So in neither case is it possible for observation, be it what it may, to provide any independent control over thought. . . . Observation contaminated by thought yields circular tests; observation uncontaminated by thought yields no tests at all" (Scheffler [1967] 1976, 13).

It is not easy to overcome the paradox, because if, on the one hand, we state that the meaning of observational terms depends only on empirical conditions to which their application is subjected, we meet with the difficulties that the philosophers of science raised against the so-called semantic conception of observation, difficulties that had partly emerged also in the course of the controversy about protocols. If, on the other hand, we follow Feyerabend in stating a strong version of the thesis of the theoretical nature of observation (the meaning of observational terms is entirely determined by all our reference conceptions), it becomes impossible to understand how our observations can contrast with the complex of our beliefs.

After formulating the paradox of categorization, Scheffler proposed a solution that we can call the "weak thesis" of the theoretical nature of observation. What distinguishes this proposal (which in some ways is already present in Popper) from the "strong thesis" is the fact that it tries to assert the theoretical contamination of the observational vocabulary in such a way as to avoid the danger that such contamination may depend not simply upon a generic theory but also upon the very specific theoretical hypotheses that we want to test. To put it briefly, Scheffler's version of the "weak thesis" tries to overcome the objection of circularity by introducing a distinction between the two types of theoretical structures, the ones that incorporate the more general categories of our discourse (they are similar to Kant's forms of knowledge) and the ones that incorporate our beliefs and hypotheses on more specific questions: "Simply to set up an alphabetical filing system for correspondence is not yet to determine how tomorrow's correspondence will need to be filed. Conversely, to guess that the next letter will need to be filed under 'E' or 'L' is a prediction that may be made whether or not we have a place for 'X' in our system. . . . Without a vocabulary and grammar, we can describe nothing; having a vocabulary and grammar, our descriptions are not thereby determined. . . . Categorization provides the pigeonholes; hypothesis makes assignments to them (Scheffler [1967] 1976, 38). According to Scheffler's proposal, only interpretative or categorial theories play a role in fixing the meaning of observational vocabulary; theoretical hypotheses, by contrast, do not contribute to determining those meanings and can vary relatively to the categorial theory without their variation influencing the meaning of the observational terms. Therefore, theoretical hypotheses can be subjected to a nonvacuously circular empirical test. There are no purely observational terms, devoid of theoretical contamination. But this does not rule out the possibility of empirical tests, which are to some extent objective, even if relative to the background interpretative theory.

As it is well known, Scheffler's proposal has raised many objections. Many philosophers of science have tried to show that, similarly to analogous solutions, it cannot be considered as an adequate answer to the problems that, according to the supporters of the "strong thesis" and incommensurability, are posited by scientific changes or, at least, from the revolutionary aspects of these changes. Among these criticisms there is one that deserves mentioning since it raises again a connection between this problem and the criticism that neo-empiricists moved to the synthetic a priori.

Some philosophers of science pointed out the fact that, as in the case of some distinctions, such as the one between analytic and synthetic sentences or the one between theoretical and observational vocabulary, the distinction between categories and hypotheses cannot be made on a logical or naturalistic basis. None of these distinctions has an absolute value, and can be considered invariant with respect to scientific change. But, if we confer a merely contextual value on Scheffler's dichotomy, too, we must consider the possibility that the change of hypotheses will bring with it a change in the meaning of categories—in other words, that, in the passage from one theoretical context to another, the meaning of the categorial scheme may also change. After all, something analogous happened in the passage from Newton's physics to the relativistic one, and it is on this aspect that logical empiricists have insisted when criticizing Kant's conception of synthetic a priori judgments: "Philosophy"—as Reichenbach had already underlined in the early 1920s within the context of this critique ([1922] 1978, 34)—"is confronted with the fact that physics creates new categories which cannot be found in traditional dictionaries."

What emerges from Reichenbach's formulation is that in neo-empiricists' criticism of the synthetic a priori we can already find the idea of an interconnection between the changing of hypothesis and the changing of categories.[8] But if we accept this interconnection it becomes problematic to understand how it is possible to take for granted Reichenbach's statement that perception is independent of reason, or, in other words, the idea that there are some relatively neutral perceptual contents on the basis of which we can asses the empirical validity of the theory.

Mary Hesse's Network Model and the Primary Recognitions of Similarities

It is precisely in the 1970s, when the debate on the theoretical character of observation and incommensurability became more heated, that a fourth answer to the problem of how experience can exercise some empirical test on theories was advanced. This answer does not coincide either with the semantic conception of observation or with either of the two theses (the "strong" one and the "weak" one) of the theory-laden nature of observational sentences. What I am referring to is the "network model" proposed by Mary Hesse (1974, 1:9–44) under the influence of some ideas developed by Quine and Duhem. As she explicitly declares, the aim of this model is to provide the justification of a "new empiricism" (Hesse [1970] 1976). With it, in fact, Hesse intends to show how at the same time it is possible to reconcile the thesis of the theoretical nature of observation with the possibility of having a genuine empirical test of our hypotheses and theories. Using the terms I have chosen for the title of this chapter we can summarize all this, indulging ourselves in an intentionally paradoxical statement, by saying that the aim of the network model consists in showing how a certain dependency of perception on reason, which has been asserted in various ways both by Kant and the supporters of the "strong thesis," is to a degree compatible with the idea of perception's independence of reason, the idea that Reichenbach had put at the basis of neo-empiricists' criticism to Kant's synthetic a priori.

The possibility of this reconciliation is connected to the two fundamental principles that lie at the basis of the network model. According to the first one, it is impossible to explain the meaning and the functioning of those descriptive expressions that have an extralinguistic reference—also the ones that are considered as cases that represent paradigmatically some observational statements ("red," "hard," etc.)—by making reference solely to the associations that connect these expressions to experience. According to the second one, also the expressions that are introduced and learned by means of associations of an exclusively linguistic nature (this group includes also the so-called theoretical expressions, such as "atom," "electron," "gene," etc.) are in principle susceptible to empirical applications. In the network model it thus becomes possible to preserve a pragmatic and contextual value to the distinction between theoretical and observational vocabulary without expecting it to have an absolute value, since this distinction is not based on logical or naturalistic reasons. This, in its turn, allows us to reconcile the thesis of the theoretical nature of observation (a thesis that makes the thesis of perception's independence of reason problematic) with the possibility of having an empirical test of our hypotheses and theories that is not vacuously circular. This possibility, on the contrary, presupposes that perception has a certain degree of independence of reason.

To put things in a forcefully schematic way, the essential point is that the conception of scientific language that emerges from Hesse's network model is extremely different from the one we gather from neo-empiricists' two-level conception of theories. According to Hesse's "new empiricism," scientific language is "a dynamic system which constantly grows by metaphorical extension of natural language" (Hesse 1974, 4); and alternative theoretical constructions extend metaphorically in different ways to the network of functional relations between the terms of the language. In principle, this is totally compatible with the persistence of common points and areas of intersection capable of allowing an empirical comparison between competing hypotheses and theories. It is true that we cannot talk about absolutely observational sentences, whose meaning and truth value are logically determined by experience (semantical conception of observation). Nor can we talk about expressions bound to a categorial scheme completely insensitive to the changes of hypotheses and theories ("weak thesis"). But we cannot exclude a priori the independence of certain empirical applications of some words with respect to alternative hypotheses and theories.

In light of my present purposes, I deem necessary to underline mainly the reasons that according to Hesse can guarantee the existence of areas of empirical intersections between alternative theories. These reasons are strictly connected to the individuation of the epistemological conditions that allow the formation, development, and mainly the functioning of a symbolism endowed with extralinguistic reference. The starting point of this discourse is the obvious observation that it does not seem possible to understand the process of the formation and learning of such kind of symbolism if we do not start by acknowledging that "some predicates are initially learned in empirical situations in which an association is established between some aspects of the situation and a certain word" (Hesse 1974, 11). Even leaving aside the psychological and linguistic dimensions of association, we are left with the fact that we cannot understand the process of association and with it the process of formation of language without taking into consideration the following two epistemological conditions: "(1) Since every physical situation is indefinitely complex, the fact that the particular aspect to be associated with the word is identified out of a multiplicity of other aspects implies that degrees of physical similarity and difference can be recognized between different situations. (2) Since every situation is in detail different from every other, the fact that the word can be correctly reused in a situation in which it was not learned has the same implication" (Hesse 1974, 11). These epistemological conditions state our capacity of recognizing degrees of physical similarity and difference between different situations. And since, according to the well-known objection to the empiricist theory of the primacy of repetitions frequently proposed by Popper, we cannot explain this capacity by recurring to repetitions themselves (Popper [1934, 1959] 1977, 422), we have to admit that such a capacity rests on the fact that in the end there are points of view that are not explicitly verbalizable in order to have some primitive or primary recognitions of similarities:[9]

> A different sort of objections to the appeal to similarity is made by Popper who argues that the notion of repetition of instances which is implied by 1 and 2 is essentially vacuous, because similarity is always similarity in certain respects, and "with a little ingenuity" we could always find similarities in some same respects between all members of any finite set of situations. That is to say, "anything can

be said to be a repetition of anything else, if only we adopt the appropriate point of view". But if this were true, it would make the learning process in empirical situations impossible. . . . However, Popper's admission that "a little ingenuity" may be required allows a less extreme interpretation of his argument, namely that the physics and physiology of situations already give us some "point of view" with respect to which some pairs of situations are similar in more obvious respect to another than it is in the same respect to a third. This is all that is required by the assertions 1 and 2. Popper has needlessly obscured the importance of these implications of the learning process by speaking as though, before any repetition can be recognized, we have to take thought and explicitly adopt a point of view. If this were so, a regressive problem would arise about how we ever learn to apply the predicates in which we explicitly express that point of view. An immediate consequence of this is that there must be a stock of predicates in any descriptive language for which it is impossible to specify necessary and sufficient conditions of correct application. For if any such specification could be given for a particular predicate, it would introduce further predicates requiring to be learned in empirical situations for which there was no specification. Indeed, such unspecified predicates would be expected to be in the majority, for those for which necessary and sufficient conditions can be given are dispensable except as a shorthand and hence essentially uninteresting. We must therefore conclude that the primary process of recognition of similarities and differences is necessarily unverbalizable. The emphasis here is of course on primary, because it may be perfectly possible to give empirical descriptions of the conditions, both psychological and physical, under which similarities are recognized, but such descriptions will themselves depend on further undescribable primary recognitions. (Hesse 1974, 13)

The Primary Recognitions of Similarity, Kuhn's Exemplars, and the Possibility of a Genuine Empirical Test

Another fundamental aspect that characterizes the network model—and on which the possibility of showing a certain relative independence of perception of reason's constructions depends—is linked to the logic of the relation of similarity. When we consider the logical characteristics of this relation it clearly emerges that the primary recognitions of similarity are not completely insensitive to the changes of theory. These characteristics, in fact, preclude the possibility of introducing expressions whose meaning is totally based on experience and which is in no way subject to any change with respect to theoretical changes:

It may be thought that the primary process of classifying objects according to recognizable similarities and differences will provide us with exactly the independent observation predicates required by the traditional view. This however, is to overlook a logical feature of relations of similarity and difference, namely that they are not transitive. Two objects a and b may be judged to be similar to some degree in respect to predicate P, and may be placed in the class of objects to which P is applicable. But object c which is judged similar to b to the same degree may not be similar to a to the same degree or indeed to any degree. Think of judgements of similarity of three shades of color. This leads to the conception of some objects as being more "central" to the P-class than others, and also implies that the process of classifying objects by recognition of similarities and differences is necessarily accompanied by some loss of (unverbalizable) information. For if P is a predicate whose conditions of applicability are dependent on the process just described, it is impossible to specify the degree to which an object satisfies P without introducing more predicates about which the same story would have to be told. Somewhere this potential regress must be stopped by some predicates whose application involves loss of information which is present to recognition but not verbalizable. However, as we shall see shortly, the primary recognition process, though necessary, is not sufficient for classification of objects as P, and the loss of information involved in classifying leaves room for changes in classification to take place under some circumstances. Hence primary recognitions do not provide a stable and independent list of primitive observational predicates. (Hesse 1974, 13)

Thus, on the basis of the network model, there are no exclusively observational predicates the empirical application's conditions of which are completely stable and cannot change with the changing of circumstances. Among the reasons that can produce this change we can certainly find the ones underlined by the supporters of the "strong thesis" when they point to observation's dependence on theory. At the same time, though, nothing excludes the possibility that primary recognitions of similarity give a relative stability with respect to some theoretical changes—a stability such that under certain circumstances they can work as a relatively neutral common empirical basis for the test of alternative theories. In other words, nothing excludes that they can offer an area of intersection on the basis of which we can assess the empirical adequacy of alternative theories.

Along with this conclusion concerning perception's relative autonomy from reason, it seems to me necessary to mention also Hesse's general observation that the possibility of having an interconnected complex of laws and predicates such as the one described by her network model "is not a convention but a fact of the empirical world" (Hesse 1974, 16). In fact, once we have understood it in enough depth, this observation has the merit of taking us back to that thesis of contingency of cognitive synthesis on the basis of which Schlick had criticized in the 1920s Cassirer's attempt to defend Kant's philosophy. In a way that is analogous to what Schlick had stated on the contingent nature of the coordinability of the concepts to experience, we can state that from an empiricist point of view, the existence of the complex of laws and predicates that constitute the network model is neither a certainty that can be guaranteed a priori, by means of some form of transcendental deduction, nor a convention that can be maintained or rejected as we like. It is something which has so far been an undeniable fact of experience, but that nothing can guarantee us also for the future.

I would also like to underline that, however strange this might seem, the possibility of having a genuine empirical test connected to the network model is suggested, for analogous reasons, also by the considerations with which Kuhn had tried to establish the theory of incommensurability. Among the concepts that form Kuhn's notion of "paradigm" (understood as "disciplinary matrix"), there is that of "paradigmatic example" or "exemplar." This concept has a central importance.[10] Kuhn clarifies the nature of exemplars using Wittgenstein's ideas of family resemblance and linguistic game. Exemplars are "concrete problems with their solutions" (Kuhn [1974] 1977, 306), which, in the exposition and in the teaching of a theory or a discipline, constitute models to be followed for the solution of the remaining problems of "normal science." Their main characteristic is to perform the same function as general explicit rules, whilst maintaining themselves on the level of particular and concrete exemplification.

There is a precise affinity with the network model, because Kuhn, too, wanted to show how the formation and the learning of language and meanings inextricably involve both (1) the capacity of using explicit definitional criteria that can be formulated linguistically and (2) the disposition to tacitly pick out networks of similarities and dissimilarities between empirical situations. Reference to exemplars should clarify the way language "adheres" to experience and the ways in which it would be possible to learn the language and the theories "learning language and nature together by ostension":

> One of my claims is . . . that we have too long ignored the manner in which knowledge of nature can be tacitly embodied in whole experiences without intervening abstraction of criteria and generalisations. . . . One of the things upon which the practice of normal science depends is a learned ability to group objects and situations into similarity classes which are primitive in the sense that the grouping is done without an answer to the question, "similar with respect to what?" One aspect of every revolution is, then, that some of the similarity relations change. Objects which were grouped in the same set before are grouped in different sets afterwards and vice versa. Think of the sun, moon, Mars, and earth before and after Copernicus; of free fall, pendular, and planetary motion before and after Galileo; or of salts, alloys, and a sulphur-iron filing mix before and

after Dalton. Since most objects within even the altered sets continue to be grouped together, the names of the sets are generally preserved. Nevertheless, the transfer of a subset can crucially affect the network of interrelations among sets. . . . When such a redistribution of objects among similarity sets occurs, two men whose discourse had proceeded for some time with apparently full understanding may suddenly find themselves responding to the same stimulus with incompatible descriptions or generalizations. Just because neither can then say, "I use the word element (or mixture, or planet, or unconstrained motion) in ways governed by such and such criteria", the source of the breakdown in their communication may be extraordinarily difficult to isolate and by-pass. (Kuhn [1970] 1987, 275)

So, Kuhn uses the analysis of the function of exemplars in order to show how, as a consequence of important scientific changes, "breakdowns" may occur in the communication between researchers adhering to different research paradigms. This is certainly a possible outcome of scientific revolutions, and it cannot be ignored. But, considering the question from the point of view of the network model, we can understand how some primitive groupings of similarity, embodied in certain paradigmatic examples, may remain unchanged even through relevant changes, thus providing a relatively neutral basis of empirical comparison between alternative theories and hypotheses. We can prove a priori neither that there will always be the prospect of total empirical incomparability nor that there will always remain some areas of intersection at the level of verbal connections and primary recognitions of similarity. What is proved, instead, is that it is possible that one or the other case may occur; it will be the concrete discussion between scientists, or historical reconstruction, what will produce a certain interpretation of what is going on, or of what has happened.

7. Conclusion: Epistemology and Psychology

For those who, like myself, want to retain a good dose of empiricism, it is pleasant to note that there are no philosophical conceptions powerful enough to ratify a priori or to exclude a priori the empirical testability of our cognitive claims: all depends upon what we say, how we say it, and our ability in planning and carrying out the appropriate experiments (as all good experimenters have always known). What is of utmost importance for me is not so much to restate and defend this thesis, something which I have already done on other occasions (Parrini [1995] 1998, 3:4–5). Here I would like to suggest a hypothesis regarding the reasons that might have induced logical empiricists to limit themselves to stating perception's independence of reason and give a linguistic disguise to their empiricist stand that later on was largely criticized. In my opinion, we can give the most plausible answer to this question if we take into consideration one of the works that lie at the origins of the neo-empiricist movement. I am referring to Moritz Schlick's *Allgemeine Erkenntnislehre.*

As I have already said, in this work Schlick develops a conception of knowledge intended as coordination and of truth as univocal coordination. In so doing, he keeps in mind mainly the eminently abstract and theoretical developments of nineteenth-century and early-twentieth-century mathematical and physical sciences. In order to elaborate a theory of knowledge suitable for these developments he is mainly concerned to individuate a method that allows the formation of "absolutely precise concepts" (Schlick [1918, 1925] 1974, 29), such as the ones that, in his opinion, are necessary in order to obtain "the greatest possible rigor and the highest degree of certitude" (Schlick [1918, 1925] 1974, 19; emphasis mine). But how is it possible to find concepts with these characteristics? It does not seem possible to form them by employing explicit or common definitions by which a concept, or meaning, is defined by resorting to concepts or meanings that are already in use. If we do not want to incur in a vicious circularity, it will not be possible to give an explicit definition of all concepts. Some of them have to be introduced in a different way, a way that in many cases is constituted by returning to some concepts that are defined already or that can be defined by means of ostensive or concrete definitions (called also "psychological") (Schlick [1918, 1925] 1974, 30).

Definitions of this kind, though, present one defect: they contain some imprecisions and obscurities. Consequently, it will be possible to have concepts that are absolutely precise, such as the ones required by the development of exact sciences only if we use another method, along with the ostensive one. This second method used for the formation of concepts is constituted by implicit definitions. Since with this method we can obtain concepts that are devoid of any empirical content, we will have to introduce a fourth type of definitions—coordinative conventions—in order to reestablish a connection between implicitly defined concepts and concrete concepts that can be characterized by means of explicit and/or ostensive definitions.

As we can easily understand, it is from this way of formulating the problem of the formation of concepts that the two-level vision that characterizes the standard conception of scientific theories develops. It is not by chance that *Allgemeine Erkenntnislehre* is the work in which we can find the enunciation of the metaphor of the net that became famous in the 1950s thanks to Carl Gustav Hempel (Schlick [1918, 1925] 1974, 37; Hempel [1952] 1970, 688). What we need to notice here is how Schlick proposes his ideas by making some statements about the relationship between epistemology and psychology that aim to exclude any consideration regarding the psychological-subjective processes. The reason for this exclusion clearly depends on the fear of falling into forms of psychologism analogous to the ones that had characterized in the past other forms of empiricism and positivism and that had seemed to block the possibility of having a satisfactory philosophical conception of physical, logical, and mathematical sciences: "one might suppose that just as physiology seeks to analyze innervation processes, so epistemology studies the psychological processes by which scientific thinking occurs. Taken in this way, however, the analogy is altogether false. For such a study would of course be purely a task for psychology. While the carrying out of this task might to a certain extent be important for the epistemologist, it could never constitute his real goal—if for no other reason than that psychological knowledge itself is a problem for him" (Schlick [1918, 1925] 1974, 3).[11] This aspect clearly emerges when Schlick has to elaborate on what recognizing an object means—that is, on the nature of our recognizing skills. This question is particularly important because according to the conception developed in the *Allgemeine Erkenntnislehre* "to know" ("erkennen") does not mean "to be acquainted" ("kennen") but "to recognize" ("wiedererkennenn") (Schlick, Moritz [1918, 1925] 1974, §§2, 3). According to Schlick, a subject knows something when he recognizes something as something else.[12] In the case of scientific theories, for example, saying that I know that water is H$_2$O equals saying that I have recognized in water a particular compound of oxygen and hydrogen (Schlick [1932] 1979, 315; Schlick [1918, 1925] 1974, 9). In the case of everyday life and perception, for example, saying that I know that that dark thing coming toward me is an animal (for example, a dog) equals saying that in "the perception of that brown thing, I have rediscovered the mental image or idea that corresponds to the name 'animal'. The object has become something familiar, and I can call it by its right name" (Schlick [1918, 1925] 1974, 7).

But, despite the centrality of the role played by the process of recognizing in his conception, Schlick excludes the analysis of this operation from the field of epistemology and he relegates it completely to the one of psychology:

> Furthermore, from the standpoint of psychology, even so plain a process as knowing or recognizing a dog is by no means a simple and obvious matter. Indeed, it is a mystery how we can claim that any image is one with which we are already acquainted. How do we know that the same perceptual image was present once before in consciousness? As a matter of fact, what was present previously was not the exact same image but at most a similar one. Psychologists have argued a great deal about how we should conceive the process of recognition, and the question is still open. But this psychological question is none of our affair and we may leave it aside entirely. At the same time, we do have here a clear example

of the difference between the psychological and the epistemological approach, of which we spoke in the preceding section. The epistemologist is not concerned with the psychological laws that govern the process of recognition and render it intelligible. What is of moment to him is only the fact that under certain circumstances recognition does occur. And this fact stands no matter how psychology may eventually resolve questions about the mental process through which recognition occurs (Schlick [1918, 1925] 1974, 8; see also 17, 18).

This exclusion of certain questions from the field of epistemology on the ground that they are of an exclusively psychological interest, and that this involves the risk of falling into psychologism, is underlined again a few pages later, precisely where Schlick deals with the way in which we come to compare what is given in the perception of a particular object (for example, a certain dog) to what despite still being an individual representation is taken as the representative of a whole species (the canine species):

> In general, I quite correctly recognize a dog as a dog because the perceptual image agrees closely enough with the ideas or images of animals I have already seen and learned to designate as dogs. But doubtful cases may also occur. Thus some dogs resemble wolves so closely that under certain circumstances the two can be confused. In other cases it may be quite impossible for an experienced observer to compare images with certainty, as when he is called upon to tell whether two pieces of writing are by the same hand.
>
> These considerations indicate that the identification and re-cognition of mere images or ideas is generally satisfactory for the cognitive processes of everyday life (and of large areas of science). But they also prove beyond contradiction that it is impossible, in this manner, to set up a rigorous and exact concept of knowledge, one that is fully serviceable from the scientific standpoint. (Schlick [1918, 1925] 1974, 19)

If we take into consideration the problems that were brought to light by logical empiricism because it limited itself to taking for granted perception's independence of reason, by introducing the dichotomy between theoretical and observational sentences, and if we consider the arguments at the basis of the empiricist treatment of the question, such as the one embedded in the network model, it seems possible to conclude that in order to develop an empiricist conception of knowledge we cannot keep ourselves exclusively on the level of linguistic distinctions and of the logical relationships between the sentences that form scientific discourse. Even if we formulate the problem in linguistic terms, when speaking about observational sentences that can constitute a relatively neutral basis for the empirical test of our theories, in order to understand this we need to consider the psychological process that leads to the formation and development of a symbolism that has extralinguistic reference. In order to develop an empiricist episte-mology it seems relevant to take into consideration also the psychological processes that allow the formation, acquisition, and development of language and the margins of obscurity and imprecision that these processes involve in the formation of concepts.

In this respect, though, we have to be careful not to see in this conclusion a concession to psychologism. In other words, we must not think that stating the importance of taking into consideration certain psychological processes for the development of an epistemological conception is tantamount to compromising ourselves with a form of psychologism. We can easily see that the case is different if we consider the network model. As we have seen, in fact, the principles that allow us to understand the process that leads to the formation of a symbolism that has extralinguistic reference regards the epistemological presuppositions or conditions of this process and do not have anything to do with the speculations about the psychological dimensions of words' association with the properties, the objects and the processes that can be found in experience.

We can summarize everything which has been said so far in the following way. To justify an empiricist answer to certain questions and, more specifically, to the one regarding perception's independence of reason, it seems relevant (1) not to forget that they are connected to subjective and psychological facts and (2) that—different from what Schlick had imagined—taking into consideration these facts does not mean falling automatically into psychologism, because what is relevant for the epistemologist are the epistemological presuppositions at the basis of certain psychological processes.

Translated by Alessandra Parrini

Notes

1. "[Le] développement de la science, depuis Kant, peut être considéré comme une dé-composition constante des fondements du rationalisme; elle signifie, vraiment, la désagré-gation de l'a priori. . . . La science de nos jours ne croit plus aux capacités législatives d'une raison pure. Tout ce que nous savons du monde est tiré de l'expérience, et les transformations des données empiriques sont purement tautologiques, analitiques (Reichenbach 1936, 31).

2. The point I intend to make is that in the correspondence between Reichenbach and Carnap during the years that led to *Philosophie der Raum-Zeit-Lehre* and *Der logische Auf-bau der Welt*, as well as in the diary where Carnap reported a conversation he had with Reichenbach, we find no traces of the comments on Reichenbach's approach to the problem of geometry, even though this approach involves (1) the use of coordinative definitions and a way of looking at the theory/experience relationship different from the one of the *Aufbau;* and (2) Reichenbach's change of opinion about what he had stated in his discussion with Schlick in the early 1920s. I intend to deal with these questions on another occasion. I dis-

cuss this topic in "On the Formation of Logical Empiricism" (forthcoming, *Vienna Circle Institute Yearbook*). In this paper I also underline the relevance of Schlick's criticism of Schuppe's immanentism.

3. See Helmholtz ([1870, 1921] 1977, 24 [and the note by Schlick, 37 n. 55]); Schlick ([1921] 1979, 326–27). In Schlick ([1925] 1974, 384 n. 48), we find a similar remark about a similar objection by Reichenbach ([1920] 1965, 115 n. 27): "Hans Reichenbach . . . has expressed the opinion (which he must surely no longer hold) that my theory of the uniqueness of correlation in knowing is basically also a synthetic judgment *a priori* and that I have thus unwittingly taken over the erroneous portion of the Kantian philosophy. This view is of course quite wrong, since my account of knowledge and truth by means of the concept of correlation is simply a *definition* and thus most certainly a purely analytic judgment." Also in Reichenbach's correspondence with Arnold Berliner we find this incorrect interpretation of Schlick's conception (see Parrini 1993, 124–27). See the introductory notes of this work also for the relationships between Schlick, Reichenbach, and Cassirer. Here I would like to recall only that Reichenbach invited Cassirer to adhere to the Gesellschaft für empirische Philosophie and that Cassirer accepted (see Cassirer's letter to Reichenbach dated June 11, 1931 [HR-025-11-02], and the formula of adhesion signed by Cassirer [HR-025-11-01]).

4. First of all, these studies have shown the necessity to reconsider the history of the movement taking into account what Schlick, Reichenbach, and Carnap had stated in the years that preceded the foundation of the Vienna and Berlin Circles. This reconsideration has clearly shown that the empiricist stand held by the major exponents of the movement could not be assimilated to a form of naive empiricism that "neglects the problems of conceptualisation" (Reichenbach [1922] 1978, 37). Moreover, many scholars have underlined the incompatibility of this image with a well-known event that occurred in the history of the Vienna Circle—that is, the famous controversy on the protocols. So they repeatedly showed how during this discussion Neurath in particular supported theses that were very close to the ones that later on will be proposed by Hanson, Kuhn, and Feyerabend. To all this we can add the fact—which today is not very well-known and that has not been underlined enough—that also Reichenbach, who did not participate in the controversy on protocols, held an extremely complex view that cannot be assimilated to the acritical faith in an absolutely stable observational language that constitutes the unalterable basis of the empirical interpretation of theories (see note 6 below).

5. For the discussion on the thesis of the theoretical nature of observation and the distinction between the "strong" and the "weak" thesis, see Suppe ([1973] 1977; 1989, 38–77; Lanfredini, 1988; Parrini [1995] 1998, II/1 and III).

6. Kuhn's position constitutes a particularly instructive example. He insists on the idea that the cognitive activity presupposes the interpretation of nature by means of a set of "conceptual boxes." Despite this, he also maintains that "nature cannot be forced into an arbitrary set of conceptual boxes. On the contrary, the history of protoscience shows that normal science is possible only with very special boxes, and the history of developed science shows that nature will not indefinitely be confined in any set which scientists have constructed so far" (Kuhn 1987, 263). If this were not the case, Kuhn would not have been able to talk about the empirical "anomalies" that affect paradigms (understood as "disciplinary matrixes"). The Duhemian spirit of this aspect of the question is well condensed in a neglected remark by Lakatos: "it is not that we propose a theory and Nature may shout NO; rather, we propose a maze of theories, and Nature may shout INCONSISTENT." Nature can, that is, induce us to uphold "a 'factual' statement couched in the light of one of the theories

involved, which we claim Nature had uttered and which, if added to our proposed theories, yields an *inconsistent system*" (Lakatos 1987, 130). Feyerabend himself has explicitly recognized the possibility of *refuting* "incommensurable theories . . . by reference to their own respective kinds of experience"—though he specifies parenthetically that "in the absence of commensurable alternatives these refutations are quite weak" (Feyerabend 1987, 227).

7. Reichenbach was a convinced supporter of the idea that there are no unappealable "conclusive judgments" at any level of scientific discourse. This emerges clearly, as well as from some unpublished texts, also from his contribution to the controversy between Arthur Eddington and Norman R. Campbell, dating from the early thirties. He certainly did not go as far as to say (as it was to be said subsequently) that facts are "small size" theories, and theories are "large size" facts; nevertheless, he maintained firmly that there is no "clear *frontière*" between facts, laws, and theories. There are only different degrees of probability between experimental facts that are well established on the basis of solid theories, and experimental facts that are less well established on the basis of more or less problematic hypotheses (see Reichenbach 1931). Reichenbach referred to probability already in an essay published in 1929, in order to break the circularity that threatens empirical testability when we affirm the theoretical character of observation (see Reichenbach [1929] 1978, esp. sec. 9, "Physical Fact"). The unpublished documents concern the relationship between Reichenbach and Lewis. See Reichenbach's letter to Lewis of May 4, 1938 [HR-13-49-37], where he expounds the idea that even low-level assertions are more or less probable, not true or false; see also Reichenbach's letter to Richard Brandt of March 10, 1951 [HR-037-03-43], in which he says, "I agree with him [Lewis] in a fundamentally empiricist attitude, I differ from him in that I cannot accept any basic statements as absolutely certain, or as terminating judgements, as he puts it" (qtd. in Parrini 1993, 131).

8. M. Hesse has illustrated this by comparing what had been stated by the special relativity theory with Campbell's conviction "that it is possible to obtain absolutely universal agreement for judgements such as, The event A happened at the same time as B, or A happened between B and C" (Campbell [1920] 1957, 29). "This example illustrates well the impossibility of even talking sensibly about 'levels of more direct observation' and 'degrees of theory-ladenness' except in the context of some framework of accepted laws. That such talk is dependent on this context is enough to refute the thesis that the contrast between 'direct observation' and 'theory-ladenness' is itself theory-independent. The example also illustrates the fact that at any given stage of science it is never possible to know which of the currently entrenched predicates and laws may have to give way in the future" (Hesse 1974, 20).

9. See also Quine's conception of the prelinguistic quality space (Quine 1960, 83).

10. See Lanfredini (1988, chap. 3; Hoyningen-Huene 1993).

11. As far as Schlick's general position on psychologism is concerned, see the concluding remarks contained in his letter to Reichenbach dated November 26, 1920 (Parrini 1993, 135).

12. Schlick likes quoting William Stanley Jevons's saying: "Science arises from the discovery of Identity amidst Diversity" (Schlick [1918, 1925] 1974, 91). This conception of knowing as recognizing will remain, even though in a different context, also in the Schlick of the Vienna Circle years (see Schlick [1932] 1979, 312–15).

References

Campbell, N. R. [1920] 1957. *Foundations of Science: The Philosophy of Theory and Experiment.* New York: Dover Publications. [*Physics: The Elements.* Cambridge: University of Cambridge Press].

Nonlinguistic Empiricism

Carnap, R. [1966] 1974. *An Introduction to the Philosophy of Science*. Ed. Martin Gardner. New York: Basic Books. [*Philosophical Foundations of Physics: An Introduction to the Philosophy of Science.*]

Carnap, R., H. Hahn, and O. Neurath. [1929] 1973. *Wissenschaftliche Weltauffassung: Der Wiener Kreis* (Wien: Artur Wolf). [*The Scientific Conception of the World: The Vienna Circle*], in O. Neurath, Empiricism and Sociology, edited by Marie Neurath and Robert S. Cohen, Dordrecht: Reidel, "Vienna Circle Collection", pp. 299-318.

Cassirer, E. [1921] 1953. *Einstein's Theory of Relativity Considered from the Epistemological Standpoint* [*Zur Einstein'schen Relativitätstheorie*. Berlin: Bruno Cassirer Verlag]. English translation by William Curtis Swabey and Marie Collins Swabey, in E. Cassirer, *Substance and Function: Einstein's Theory of Relativity*, 347-456. New York: Dover.

Feyerabend, P. K. [1970] 1987. "Consolations for the Specialist." In Lakatos and Musgrave (1987, 197-230).

Helmholtz, H. von. [1870, 1921] 1977. "Über den Ursprung und die Bedeutung der geometrischen Axiome" (1870). English translation in Hermann von Helmholtz, *Epistemological Writings* [The Paul Hertz / Moritz Schlick Centenary Edition of 1921 with Notes and Commentary by the Editors]. Trans. Malcolm F. Lowe and ed. Robert S. Cohen and Yehuda Elkana. Dordrecht: Reidel.

Hempel, C. G. [1952] 1970. "Fundamentals of Concept Formation in Empirical Science." In *Foundations of the Unity of Science: Toward an International Encyclopedia of Unified Science*, ed. Otto Neurath, Rudolf Carnap, Charles Morris, 2:651-745. Chicago: University of Chicago Press.

Hesse, M. [1970] 1976. "Duhem, Quine and a New Empiricism." In *Can Theories Be Refuted? Essays on the Duhem-Quine Thesis*, ed. Sarah G. Hardig, 184-204. Dordrecht: Reidel.

———. 1974. *The Structure of Scientific Inference*. Berkeley: University of California Press.

Hoyningen-Huene, P. [1989] 1993. *Reconstructing Scientific Revolutions: Thomas S. Kuhn's Philosophy of Science* [*Die Wissenschaftsphilosophie Thomas S. Kuhns: Rekonstruktion und Grundlagenprobleme*. Braunschweig: Vieweg]. English translation by Alexander T. Levine, with a foreword by T. S. Kuhn. Chicago: University of Chicago Press.

Kant, I. [1781, 1787] 1965. *Critique of Pure Reason* [*Kritik der reinen Vernuft, Akademie-Textausgabe*]. English translation by Norman Kemp Smith. New York: St. Martin's Press.

Kuhn, T. S. [1970] 1987. "Reflections on My Critics." In Lakatos and Musgrave (1987, 231-78).

———. [1974] 1977. "Second Thoughts on Paradigms." In T. S. Kuhn, *The Essential Tension. Selected Studies in Scientific Tradition and Change*. Chicago: University of Chicago Press, 293-319.

Lakatos, I. [1970] 1987. "Falsification and the Methodology of Scientific Research Programmes." In Lakatos and Musgrave (1987, 91-196).

Lakatos, I., and A. Musgrave, eds. [1970] 1987. *Criticism and Growth of Knowledge: Proceedings of the International Colloquium in the Philosophy of Science, London 1965*. Vol. 4. Cambridge: Cambridge University Press.

Lanfredini, R. 1988. *Oggetti e paradigmi: Per una concezione interattiva della conoscenza scientifica*. Rome: Theoria.

Parrini, P. 1993. "Origini e sviluppi dell'empirismo logico nei suoi rapporti con la filosofia continentale: Alcuni testi inediti." *Rivista [critica] di storia della filosofia* 48:121-46, 377-93. A modified version appears in Parrini (forthcoming), *L'empirismo logico, Aspetti storici e prospettive teoriche* (Rome: Carocci editore).

———. [1995] 1998. *Knowledge and Reality: An Essay in Positive Philosophy* [*Conoscenza e realtà: Saggio di filosofia positiva*. Rome: Laterza]. English translation by Paolo Baracchi. Dordrecht: Kluwer Academic Publishers.

Popper, K. R. [1934, 1959] 1977. *The Logic of Scientific Discovery* [*Logik der Forschung*. Wien: Verlag von Jiulius Springer]. English edition with new appendices. London: Hutchinson & Co.

Quine, W. V. O. [1953] 1961. *From a Logical Point of View*. New York: Harper & Row.

———. 1960. *Word and Object*. Cambridge, Mass.: MIT Press.

Reichenbach, H. [1920] 1965. *The Theory of Relativity and a Priori Knowledge* [*Relativitätstheorie und Erkenntnis Apriori*. Berlin: Verlag von Julius Springer]. Translated and edited with an introduction by Maria Reichenbach. Berkeley: University of California Press.

———. [1922] 1978. "The Present State of the Discussion on Relativity" ["Der gegenwärtige Stand der Relativitätsdiskussion." *Logos* 10:316–78]. English translation in H. Reichenbach (1978, 3–47).

———. [1928] 1958. *The Philosophy of Space and Time* [*Philosophie der Raum-Zeit-Lehre*. Berlin: Walter de Gruyter]. Translated by Maria Reichenbach and John Freund, with introduction by Rudolf Carnap. New York: Dover Publications.

———. [1929] 1978. "The Aims and Methods of Physical Knowledge" ["Ziele und Wege der physikalischen Erkenntnis." In *Handbuch der Physik*, 1–80. Vol. 4, *Allgemeine Grundlagen der Physik*. Berlin: Julius Springer Verlag]. English translation in Reichenbach, (1978, 120–225).

———. 1931. *Letters published in Philosophy of Science* 6:398–99, 525–26.

———. 1936. "L'empirisme logistique et la désagrégation de l'a priori." In *Actes du Congrès International de Philosophie Scientifique*, 28–35. Sorbonne, Paris, 1935. Vol. 1, *Philosophie Scientifique et Empirisme Logique*. Paris: Hermann & Cie, éditeurs.

———. 1978. *Selected Writings, 1909–1953*. Vol. 2. Edited by Maria Reichenbach and Robert S. Cohen. Vienna Circle Collection. Dordrecht: Reidel.

Scheffler, I. [1967] 1976. *Science and Subjectivity*. Indianapolis: Bobbs-Merrill.

Schlick, M. [1918, 1925] 1974. *General Theory of Knowledge* [*Allgemeine Erkenntnislehre*. Berlin: Verlag von Julius Springer]. Translated by Albert E. Blumberg, with an introduction by A. E. Blumberg and Herbert Feigl. New York: Springer-Verlag.

———. [1921] 1979. "Critical or Empiricist Interpretation of Modern Physics?" ["Kritizistische oder empiristische Deutung der neuen Physik?" *Kant-Studien* 26:96–111]. In M. Schlick, *Philosophical Papers*, vol. 1, *1909–1922*, edited by H. L. Mulder and B. F. B. Van De Velde-Schlick and translated by Peter Heath, 322–34. Vienna Circle Collection. Dordrecht: Reidel.

———. [1932] 1979. *Form and Content: An Introduction to Philosophical Thinking*. In M. Schlick, *Philosophical Papers*, vol. 2, *1925–1936*, edited by H. L. Mulder and B. F. B. Van De Velde-Schlick, 285–369. Vienna Circle Collection. Dordrecht: Reidel.

Suppe, F. [1973] 1977. "The Search for Philosophical Understanding of Scientific Theories" and "Afterword—1977." In *The Structure of Scientific Theories*, edited by F. Suppe, 1–241, 615–730. Urbana: University of Illinois Press.

———. 1989. *The Semantic Conception of Theories and Scientific Realism*. Urbana: University of Illinois Press.

Commit It Then to the Flames . . .

WESLEY C. SALMON

University of Pittsburgh

In the final paragraph of his *Inquiry Concerning Human Understanding,* David Hume writes:

> When we run over our libraries, persuaded of these principles, what havoc must we make? If we take in our hand any volume—of divinity or school metaphysics, for instance—let us ask, *Does it contain any abstract reasoning concerning quantity or number?* No. *Does it contain any experimental reasoning concerning matter of fact or existence?* No. Commit it then to the flames, for it can contain nothing but sophistry and illusion. ([1748] 1955, 173)

Let me state clearly from the outset that I do not favor book burning or book banning. People should be free to read, write, and publish sophistry and illusion to their hearts' content. I see no problem here; there seems to be enough sophistry and illusion around to fulfill anybody's needs. Nevertheless, we reserve the right to identify it as such when it appears. But Hume's prose is compelling; I could not resist the temptation to quote him. I take it that he was writing metaphorically; in any case, that is the way I construe him. Moreover, we should explicitly note that he does not condemn poetry, drama, novels, and other forms of literature. He is offering a version of the criterion of *cognitive* significance. That is the topic of this essay; his is the most elegant and colorful formulation to be found anywhere in the philosophical literature. Let me state further that my aim is *not* to resuscitate any such criterion; rather, I want to examine historically the grounds on which it has been rejected by various authors at various times in the twentieth century. Although I agree that we should abandon the criterion, I will argue that its critics have often done so for defective reasons.

Let me begin by reiterating Hume's distinction between "abstract reasoning concerning quantity or number" and "experimental reasoning concerning matters of fact or existence." Applying this distinction to statements, we might distinguish two types of cognitively meaningful statements. In the first subcategory, we can place statements of logic or mathematics, as well as analytic and self-contradictory statements, if there are such things. Let us call these statements *formally meaningful*. In the second subcategory, we can place statements of fact; let us call these *empirically meaningful*. The two subcategories exhaust the class of *cognitively meaningful* statements according to the verifiability criterion.

Next, let me clear away certain early versions of the criterion that are not serious contenders. First, I reject out of hand any version that classifies all sentences that do not meet the criterion as utter nonsense; I am interested (as was Hume) in *cognitive* meaningfulness. A. J. Ayer had a similar point in mind when he referred to *literally* meaningful statements ([1946] 1952, 15). Other sorts of meaning exist and are important. The notion that defenders of the verifiability criterion condemned all sentences that did not conform to their criterion as utter nonsense is a gross distortion. Next, I will set aside any criterion that invokes conclusive verifiability or falsifiability as the standard to be met. This approach has been adequately refuted by Carl G. Hempel (1965, 102–7) and many other authors. Furthermore, I will exclude versions of the criterion—for example, the meaning of a statement *is* the means of its verification—that hypostatize *meaning* as entity. The issue concerns a property, *meaningfulness*, that can be attributed to statements. These exclusions imply that the discussion will be aimed at criteria that attempt to characterize statements as cognitively meaningful if and only if they are confirmable or disconfirmable to some degree by empirical evidence, or they are formally meaningful statements of logic or mathematics. In the end, I will suggest a different sort of approach to the problem the meaning criterion was designed to handle.

The Ayer-Church Episode

As we all know, Ayer's formulation of a meaning criterion in the first edition of *Language, Truth, and Logic* (1936) met with immediate disaster. In the second edition (1946) he offered a more sophisticated version. It did not meet with *immediate* catastrophe; Alonzo Church's devastating review came out after an interval of three years. Even though this critique is widely known, it has been almost universally misinterpreted. For this reason, I want to review the case.

Ayer was well aware of the undesirability of requiring conclusive verifiability as the basis for cognitive meaning; he therefore introduced the notion of "weak" verifiability—that is, confirmability to some degree. This is how he does it:

> I propose to say that a statement is directly verifiable if it is either itself an observation-statement, or is such that in conjunction with one or more observation-statements it entails at least one observation-statement which is not deducible from these other premises alone; and I propose to say that a statement is indirectly verifiable if it satisfies the following conditions: first, that in conjunction with certain other premises it entails one or more directly verifiable statements which are not deducible from these other premises alone; and secondly, that these other premises do not include any statement that is not either analytic, or directly verifiable, or capable of being independently established as indirectly verifiable. ([1946] 1952, 13.)

The inadequacy of this explication of empirical verifiability was shown by Church as follows:

> [L]et O_1, O_2, O_3 be three "observation-statements" . . . such that no one of the three taken alone entails any of the others. Then, using these we may show of any statement S whatever that either it or its negation is verifiable, as follows. . . . [U]nder Ayer's definition, $\neg O_1\ O_2 \lor O_3 \neg S$ is directly verifiable, because with O_1 it entails O_3. Moreover, S and $\neg O_1\ O_2 \lor O_3 \neg S$ together entail O_2. Therefore (under Ayer's definition) S is indirectly verifiable—unless it happens that $\neg O_1\ O_2 \lor O_3 \neg S$ alone entails O_2, in which case $\neg S$ and O_3 together entail O_2, so that $\neg S$ is directly verifiable. (1949, 52–53.)

When we examine Ayer's formulation and Church's critique, an extraordinary feature emerges—namely, in neither of them is there any mention of meaningfulness. In fact, Ayer has offered an analysis of weak confirmation, and Church has shown that, under Ayer's definition, any hypothesis is confirmable. The issue of the relationship between meaning and weak verifiability (or confirmation) has not been addressed in any way.

Considering all of the difficulties to be found in the literature of confirmation theory, it is hardly surprising that Ayer's twelve-line characterization of confirmation should turn out to be faulty. One might try to patch it up by placing some restrictions on the auxiliary hypothesis, but that strategy does not seem promising. However we try to fix it up, Ayer's account is basically none other than the hypothetico-deductive method. The problems associated with this schema for confirmation are well known. Noticeably lacking in Ayer's formulation is any reference to probability or induction.

In his review of Ayer's book, Church showed conclusively that Ayer's characterization of weak verifiability is inadequate; he did not claim to have done more. He did nothing to undermine the notion that statements are cognitively meaningful if and only if they are *genuinely* empirically verifiable to some degree.

The Carnap-Kaplan Episode

Immediately following the publication of Church's critique of Ayer, Hempel published two papers, one in 1950, the other in 1951, in which he cited Church's argument and expressed skepticism about the possibility of formulating a precise criterion of cognitive meaningfulness. Immediately after his mention of that episode, he writes, "I think it is useless to continue the search for an adequate criterion of testability in terms of *deductive* relationships to observation sentences. The past development of this search . . . . seems to warrant the expectation that as long as we try to set up a criterion of testability for individual sentences in a natural language, in terms of logical relationship to observation sentences, the result will be either too restrictive or too inclusive, or both" (1950, 50–51). I shall return to Hempel's views in the next section.

In a 1956 paper, Rudolf Carnap sought to overcome Hempel's skepticism regarding the criterion of cognitive significance. Maintaining a sharp division between the observation language and the theoretical language, he adopted the strategy of first dealing with the meaningfulness of terms and then moving on to sentences. His formulation is far more precise and sophisticated than Ayer's; it is too long and detailed to reproduce here. It is my impression that this attempt by Carnap did not receive much attention. Stephen Barker (1957) offered a putative counterexample, but I showed in 1959 that it was not a genuine counterexample. However, around 1958, David Kaplan wrote a paper, "On Significance," in which he severely criticized Carnap's formulation. Although this paper circulated in dittoed form for some time, it was not published until 1975 (under the title, "Significance and Analyticity"). In the appended "Postscript 1975," Kaplan cites a few works by other authors pertaining to Carnap's 1956 paper. The dittoed paper is included in the collection of Carnap's papers in the Pittsburgh Archive of Scientific Philosophy in the Twentieth Century. In a memoir ([1971] 1975a), Kaplan recounts his interaction with Carnap; he also mentions that his work was inspired by Church's critique of Ayer.

The dittoed paper, which was, of course, in Carnap's possession, is quite different in some respects from the one Kaplan published in 1975; for example, it includes proofs of theorems only stated in the 1975 paper. Apparently he wrote a different version for oral presentation at the 1959 meeting of the Pacific Division of the American Philosophical Association, and that version was published. Kaplan's argument is long and complex, so there is no possibility of displaying it here, but it will be useful to quote from Kaplan's abstract:

Carnap's approach is to distinguish between observation terms and theoretical terms. He then proposes a method of distinguishing "significant" theoretical terms from "non-significant" theoretical terms by means of their connection as given by some theory T with certain observation terms. The present paper reports two consequences of that proposal.

Given almost any theory T, there is a definitional extension T^* of T such that every theoretical term of T^* (including those of T) is significant (according to Carnap's proposal) with respect to the theory T^*; and secondly there is a "deoccamization" T^{**} of T such that no theoretical term of T^{**} is significant (according to Carnap's proposal) with respect to the theory T^{**}. The interest in these two results lies in the fact that definitions, though ordinarily thought of as adding no empirical content to a theory, seem to have the power (according to Carnap's proposal) of transforming non-significant terms into significant ones; and the process of deoccamization (which consists of "splitting" a theoretical term into a conjunction or disjunction of two new theoretical terms) which would ordinarily be thought of as subtracting no empirical content from the theory, seems to have the power (according to Carnap's proposal) of transforming a significant theory into a non-significant one. The possibility of attaining these two results is thought to constitute an inadequacy in Carnap's proposal. (Kaplan 1975, 87)

It seems that Carnap accepted Kaplan's critique as valid. In his memoir, Kaplan gives a moving account of how he, a graduate student in 1958, received Carnap's undiluted praise for this work that undermined not only one paper but also a longstanding research program: "The emotional impact upon a second year graduate student of seeing *Rudolf Carnap* respond to a student's argument with an enthusiasm completely unmitigated by his own 30-year investment on the other side has stayed strongly with me. It was a rare and cherished experience; Carnap taught me much more than logic" (1975, XLIX). Kaplan attributes the common problem of Ayer and Carnap to the fact that each formulation involves an inductive clause —inductive, that is, in the sense of mathematical induction. I have a different comment on Carnap's approach. Although it is much more detailed and sophisticated than Ayer's formulation, they share one basic feature. The meaningfulness of terms, for Carnap as for Ayer, depends upon the *deductive* consequences of statements that contain them. Deductive chauvinism has manifested itself even in the work of perhaps the greatest inductive logician of the twentieth century. Let us recall Hempel's statement, "it is useless to continue the search for an adequate criterion of testability in terms of *deductive* relationships to observation sentences" (emphasis mine).

Hempel's Criterion of Adequacy

In 1965, Hempel melded his two earlier articles on the concept of cognitive significance into a single essay, to which he added "Postscript (1964) on Cognitive Significance," for his book, *Aspects of Scientific Explanation*. In the 1951 article, he offered the following necessary "condition of adequacy" (which he reiterated in 1965): "If under a given criterion of cognitive significance, a sentence N is non-significant, then so must be all truth-functional compound sentences in which N occurs non-vacuously as a component. For if N cannot be significantly assigned a truth value, then it is impossible to assign truth values to the compound sentences containing N; hence, they should be qualified as non-significant as well" (1951, 62). These essays by Hempel, I believe, strongly abetted the demise of the criterion of cognitive meaningfulness.

Although the technical point about the inadmissibility of nonsignificant components of (two-valued) truth-functional logic is clearly valid, we need to look carefully at the way in which statements in natural language are translated into logical formulas. The conjunctions of English (and other natural languages, I imagine) are not subject to the same strictures as the operations of truth-functional logic. For example, "Shut up *and* eat your dinner," "Alas *and* alack," and "Be sure to lock the door *if* you leave," are all perfectly grammatical. Nevertheless, "Shut up," "Eat your dinner," "Alas," "Alack," and "Be sure to lock the door" are all sentences that are neither true nor false; consequently, none of them can be substituted for a sentential variable in a truth-functional formula or argument schema. What bearing has this point on the verifiability criterion?

A second criterion, contributed by Hans Reichenbach, will be helpful in sorting this out. He formulated a criterion of *equisignificance* along the following lines: "Two statements have the same cognitive meaning if and only if they are equally supported or undermined to the same degree by every item of empirical evidence" (Reichenbach 1938, 54). Consider these two sentences: "Jack is lazy" and "The Absolute is lazy." The first is a straightforward factual claim that is "weakly" verifiable or falsifiable by empirical evidence. The second is taken to represent an empirically vacuous metaphysical pronouncement. (If any metaphysicians among you feel that this is a caricature of the discipline, you might keep in mind that it is simply a less pretentious version of Josiah Royce's declaration, "The Absolute rests in eternal repose.") What happens if—still in the natural language—we conjoin the two: "Jack is lazy and the Absolute is lazy"? Given that "Jack is lazy" has clear empirical content, it seems to me that in conjoining to it the vacuous statement "The Absolute is lazy" we neither add to, nor take away from, the empirical content of "Jack is lazy." According to the equisignificance criterion, "Jack is lazy and the

Absolute is lazy" has exactly the same empirical significance as "Jack is lazy." If we choose to substitute either of them into truth-functional forms, each should be symbolized as an *atomic* sentence. If both occur in the same formula or argument form, both should be symbolized by the same letter.

In a 1966 paper, I offered the following guidelines for translations from English into the propositional calculus:

> Let *S* be a cognitive sentence and *N* a noncognitive sentence; then:
> a. *Not-N* is noncognitive.
> b. *S and N* has the same cognitive meaning as *S*.
> c. *S or N* has the same cognitive meaning as *S*.
> d. *If S then N* has the same cognitive meaning as *not-S*.

These guidelines are not intended as rigid rules; context may suggest different treatment in different cases.

The foregoing approach seems quite natural. If we take Newton's mechanics (as given in his three laws of motion and his law of universal gravitation) and add the statement that space and time are the sensorium of God, it seems to me that we neither add to, nor detract from, Newtonian mechanics. At the same time, if we choose to construe "The Absolute is lazy" as nothing more than an inspiring formulation of the law of least action, it obviously becomes a cognitively meaningful statement.

Incidentally, the criterion of cognitive equisignificance can be used as a criterion of significance, thus eliminating the need for the criterion of cognitive significance, as follows: A sentence is cognitively meaningless if it has the same meaning as its negation. It seems to me that "The Absolute is lazy" and "The Absolute is not lazy" have exactly the same empirical content—namely, none. No empirical evidence is relevant to either of them.

Turning the Criterion on Itself

As Paolo Parrini has reminded us in his 1998 book, *Knowledge and Reality: An Essay in Positive Philosophy,* "The most famous philosophical objection [to the verifiability criterion] sprung from the application of the verifiability principle to itself. This was intended to question the empirical verifiability, and therefore the meaningfulness, of the verifiability principle. This criticism cast doubt on the very possibility of an empiricist philosophy, based on verifiability as a criterion of meaning" (19). He goes on to point out, however, that for Carnap, as well as other philosophers, the principle is meant as a criterion of *cognitive* meaningfulness;

thus, the application of the principle to itself establishes, at most, that it is not an empirical statement of fact. It is not necessarily nonsense. Perhaps it has a different kind of meaning.

This is, I believe, the correct conclusion to draw. The verifiability criterion should be construed as an imperative or a rule. Hempel (1950, 60) characterizes it as an "explication" in the sense of Carnap (1950, chap. 1). Reichenbach (1951, 48) calls it a "volitional decision." In this case, its justification lies, not in empirical (or any other kind of) evidence that supports a claim for its truth. Instead, any justification must lie in the value of the results that accrue from its use. In the famous terms of Herbert Feigl (1950), rules can be *validated* or *vindicated*. One validates a rule by showing that it is a consequence of some other rule that we have already accepted. For example, the rule of conditional proof in first order logic can be *validated* by the deduction theorem, which shows that any conclusions drawn by conditional proof can be derived by use of the basic rules of inference of the system alone. Derivations using conditional proof are often much simpler than derivations of the same conclusions without conditional proof. Among the basic rules of the system we may find *modus ponens*. We vindicate *modus ponens* by somehow convincing ourselves that it is truth preserving; that is our basic desideratum for deductive logic.

I cannot think of any rule more fundamental than the criterion of cognitive meaningfulness from which it can be derived; this suggests that, if the meaning criterion can be justified in any way, it must be by showing that it serves useful or desirable ends. I recall attending a public exhibition, many years ago, at the Kitt Peak National Observatory, which is just a short distance southwest of Tucson, Arizona. Our leader related that before the construction of the observatory, the local Papago Indians had objected because they held Kitt Peak to be the home of their gods. In the course of discussions, he said, the astronomers objected that there was no conceivable evidence to support the Papago's claim that the gods inhabited the mountaintop. The Papago leaders replied that the astronomers could not show that the gods did not live there. The interlocutor seemed to think this put the Papagos one up on the astronomers. I don't. Given two apparently contradictory statements, neither of which can be supported by empirical evidence, both must be to be empirically vacuous. I don't recall the details, but somehow an accommodation was reached so that the Papago's objection was withdrawn. I believe that the astronomers carefully explained their reasons for wanting to locate their telescopes there and the kind of knowledge of the heavens they hoped to achieve. The Papago leaders evidently conferred with their gods, and the gods agreed to share the mountaintop.

It seems to me that the crucial aspect of the resolution is that both sides—

instead of engaging in futile "Yes they do," "No they don't" confrontation—shared their values with one another. Perhaps the astronomers convinced the Papagos that there are many wonderful things to be seen in the heavens that are not visible to the naked eye. The Papagos allegedly began to call the astronomers "the men with the long eyes." At the same time, presumably, the astronomers treated the social and cultural values of the Papagos with respect, so that a compromise could be worked out that would satisfy both sides. The astronomers did not say, "We are right and you are wrong; there are no gods on Kitt Peak." Turning the dispute into a disagreement about facts, instead of a consideration of values, would have been counterproductive.

Several decades later, when astronomers planned to build a large telescope on Mount Graham, a few miles northeast of Tucson, some ecologists objected on the ground that it would have an adverse effect on the habitat of a particular species of squirrel. In this case there were questions of fact that were, in principle, amenable to scientific investigation. If, indeed, it had turned out that this species would be threatened, the question of whether the results to be produced by the telescope outweighed the viability of a species of squirrel would have been a value issue. In this case the recognition of the distinction between the questions of fact and questions of values strikes me as salutary. In my present Humean mode, I recall Hume's famous skepticism regarding the derivation of "ought" from "is" ([1739] 1888, bk. III, pt. I, sec. I). In the end, the telescope was built, and, I believe, the squirrel survived.

The fundamental issue regarding the verifiability criterion is one of intellectual responsibility. The practice of asserting as a matter of fact a sentence or statement that is in principle unsupportable (or undefeatable) is arbitrary and, in many cases, immoral. When, as all-too-frequently happens, opposing sides make apparently contradictory pronouncements of this dogmatic sort, wholesale tragedy often transpires. We see it on the evening TV news reports and read about it in the daily newspapers. I realize fully that I am offering my own value judgments here—why not? That is what the issue concerning the verifiability criterion is all about. Hume's words come back to me: "Reason is, and ought only to be, the slave of the passions, and can never pretend to any other office than to serve and obey them" ([1739] 1888, bk. II, pt. III, sec. III).

At the same time, I am prepared to condone hypotheses that are offered for consideration with the explicit understanding that there can never be any evidence for or against. There may be a kind of pleasure in contemplating such things. For example, I find a certain satisfaction in the idea, advanced in *Before the Beginning* by Martin Rees (1997), the Astronomer Royal of the United Kingdom, that there is a plethora of universes of which ours is simply the one we happen to inhabit. Rees tries to argue the case, but I fail to see how *empirical* evidence can be brought to bear on it. Nevertheless, I find it an enticing speculation, mainly perhaps because it is a useful antidote to the so-called *cosmological anthropic principle* (the *strong version*, or SAP, Barrow and Tipler 1986, 21), which strikes me as an egregious contemporary example of sophistry and illusion. Evidently, some kinds of sophistry and illusion are worse than others. When we recall the close connection between SAP and the design argument for the existence of God, we must ask, as Hume did in his *Dialogues Concerning Natural Religion,* "What peculiar privilege has this little agitation of the brain that we call thought, that we must make it a model for the whole universe?" ([1799] 1970, 28).

The Antilinguistic Turn

Bas van Fraassen has written, "My own view is that empiricism is correct, but could not live in the linguistic form the positivists gave it" (1980, 3). I am inclined to agree. Consider the problems that concerned both Ayer and Carnap. Ayer wanted to rule out of cognitive discourse a great deal, if not all, of speculative metaphysics; Carnap expressed the same aim on many occasions. Carnap also dealt at length with the question of how the terms and statements of physics and other empirical sciences could acquire cognitive significance. At the very beginning of his 1956 article, Carnap writes, "In discussions of the methodology of science, it is customary and useful to divide the language of science into two parts, the observation language and the theoretical language. . . . The theoretical language . . . contains terms that may refer to unobservable events, unobservable aspects or features of events, e.g., to micro-particles like electrons or atoms, to the electromagnetic field or the gravitational field in physics, to drives and potentials of various kinds in psychology, etc. In this article I shall try to clarify the nature of the theoretical language and its relation to the observation language" (38). The problem is how to rule out metaphysics without ruling out physics and psychology as well. We should note, however, that Carnap brings in the issue of reference of theoretical terms. This means that questions of existence of unobservable entities is at stake. We must recall, of course, that for Carnap, questions of existence and linguistic issues can never be separated from one another (see his 1950, where he articulates a position he never subsequently abandoned).

In my article "Empiricism: The Key Question," I suggested that the fundamental issue is "whether inductive [that is, nondemonstrative] reasoning contains the resources to enable us to have observational evidence for or against statements about unobservable entities and/or properties" (1985, 5). One of my main points was to emphasize the infrequency with which this question has even been posed. In this chapter, I argue for an affirmative answer to the question. I maintain that analogical arguments and causal arguments together provide the nondemonstrative methods we seek. Even the analogies to which I appeal involve comparisons among causal relationships. Both Ayer and Carnap, to the best of my knowledge, accept a simple regularity view of causality according to which "A causes B" is equivalent to "Whenever A, then B," and Hempel is extremely reluctant to admit causality into his thinking on scientific explanation. It is fairly obvious, then, that none of these three philosophers will look for a causal form of nondemonstrative reasoning as the answer to what I have taken to be the "key question." Late in his work on inductive logic, Carnap attempted to deal with arguments from analogy, but this was long after his work on the meaningfulness of theoretical terms, and I seriously doubt that his work on analogies would have been helpful for purposes of the present discussion.

It would be inappropriate, in this context, to try to spell out the details of the analogical and causal arguments to which I have just referred . But let me merely mention the main steps. First I perform an experiment with the *Compact Edition of the Oxford English Dictionary* (1971), which includes (as an indispensable aid) a magnifying glass. On the basis of simple observations, I find that the magnifying glass makes small things look bigger. I then use this instrument to locate punctuation marks that are too small for me to see with my natural corrected vision. These marks appear exactly where they should to make grammatical sense. Second, I take Ian Hacking's argument in "Do We See through a Microscope?" (1981) to make a compelling case for the reality of microscopic entities. He recounts his own laboratory experience working with many types of microscopes, utilizing completely different physical principles, to identify and characterize a microscopic grid (a tool used regularly by microscopists). This surely qualifies as a common-cause argument. Third, Jean Perrin's work on Brownian motion and Avogadro's number makes a compelling case for the reality of submicroscopic entities. Again, the main point is the variety of kinds of evidence. As Perrin himself says after listing thirteen distinct ways in which Avogadro's number has been ascertained, "Our wonder is aroused at the very remarkable agreement found between values derived from the consideration of such widely different phenomena. Seeing that not only is the same magnitude obtained by each method when the conditions under which it is applied are varied as much as possible, but that the numbers thus established also agree among themselves, without discrepancy, for all the methods employed, the real existence of the molecule is given a probability bordering on certainty" (Perrin [1913] 1923, 215–16). This too, it seems to me, can be construed as an extremely powerful argument to a common cause.

Salmon / Commit It Then to the Flames . . .

Near the beginning of section IV of the *Inquiry* ("Skeptical Doubts Concerning the Operations of the Understanding"), Hume remarks, "All reasonings concerning matter of fact seem to be founded on the relation between *cause* and *effect*" (41). I think he was right—almost. Hume had well-known problems with causality; it seems to me that we can now surmount his chief difficulties. Causal reasoning is not demonstrative, but it seems to play a crucial role in scientific reasoning about unobservable entities and their properties and relations. No wonder the early logical positivists and logical empiricists had troubles with the meanings of theories and theoretical terms. My conclusion is that deductive chauvinism has led us down the wrong path. Empiricism is a viable philosophy that can stand on its own outside of the purely linguistic framework in which it was notoriously placed, as long as the critical role of nondemonstrative reasoning is clearly recognized.

Reference

Ayer, A. J. [1946] 1952. *Language, Truth, and Logic.* 2d ed. New York: Dover Publications.

Barker, S. F. 1957. *Induction and Hypothesis.* Ithaca, N.Y.: Cornell University Press.

Barrow, J. B., and F. J. Tipler. 1986. *The Anthropic Cosmological Principle.* Oxford: Clarendon Press.

Carnap, R. 1950. "Empiricism, Semantics, and Ontology." *Revue internationale de Philosophie* 11:20–40.

———. 1956. "The Methodological Character of Theoretical Concepts." In *Minnesota Studies in the Philosophy of Science,* vol. 1, *The Foundations of Science and the Concepts of Psychology and Psychoanalysis,* ed. Herbert Feigl and Michael Scriven, 38–76. Minneapolis: University of Minnesota Press.

Church, A. 1949. "Review of Ayer's *Language, Truth, and Logic.*" *Journal of Symbolic Logic* 14:52–53.

Feigl, H. 1950. "De Principiis Non Disputandum . . . " In *Philosophical Analysis,* ed. Max Black, 119–56. Ithaca, N.Y.: Cornell University Press.

Hacking, I. 1981. "Do We See through a Microscope?" *Pacific Philosophical Quarterly* 62:305–22.

Hempel, C. G. [1950] 1952. "Problems and Changes in the Empiricist Criterion of Meaning." In *Semantics and the Philosophy of Language,* ed. Leonard Linsky, 163–85. Urbana: University of Illinois Press.

———. 1951. "The Concept of Cognitive Significance: A Reconsideration." *Proceedings of the American Academy of Arts and Sciences,* vol. 80, *Contributions to the Analysis and Synthesis of Knowledge,* 61–77. Boston: American Academy of Arts and Sciences.

———. 1965. *Aspects of Scientific Explanation and Other Essays in the Philosophy of Science.* New York: Free Press.

Hume, D. [1739] 1888. *A Treatise of Human Nature.* Ed. L. A. Selby-Bigge. Oxford: Clarendon Press.

———. [1748] 1955. *An Inquiry Concerning Human Understanding.* Ed. Charles W. Hendel. Indianapolis: Bobbs-Merrill.

———. [1799] 1970. *Dialogues Concerning Natural Religion*. Ed. Nelson Pike. Indianapolis: Bobbs-Merrill.

Kaplan, D. 1958. "On Significance." Dittoed ms., Carnap papers, University of Pittsburgh Archive of Scientific Philosophy.

———. [1971] 1975. "Homage to Rudolf Carnap." In *Rudolf Carnap, Logical Empiricist,* ed. Jaakko Hintikka, xlvii–xlix. Dordrecht: Reidel.

———. 1975. "Significance and Analyticity: A Comment on Some Recent Proposals of Carnap." In *Rudolf Carnap, Logical Empiricist,* ed. Jaakko Hintikka, 87–94. Dordrecht: Reidel.

Compact Edition of the Oxford English Dictionary: Complete Text Reproduced Micrographically. 1971. Oxford: Oxford University Press.

Parrini, P. 1998. *Knowledge and Reality: An Essay in Positive Philosophy.* Dordrecht: Kluwer Academic Publishers.

Perrin, J. [1913] 1923. *Atoms.* New York: Van Nostrand.

Rees, M. 1997. *Before the Beginning.* Reading, Mass.: Addison-Wesley.

Reichenbach, H. 1938. *Experience and Prediction.* Chicago: University of Chicago Press.

———. 1951. "The Verifiability Theory of Meaning." *Proceedings of the American Academy of Arts and Sciences,* vol. 80, *Contributions to the Analysis and Synthesis of Knowledge,* 46–60. Boston: American Academy of Arts and Sciences.

Salmon, W. C. 1959. "Barker's Theory of the Absolute." *Philosophical Studies* 10:50–53.

———. 1966. "Verifiability and Logic." In *Mind, Matter, and Method: Essays in Honor of Herbert Feigl,* ed. Paul K. Feyerabend and Grover Maxwell, 354–76. Minneapolis: University of Minnesota Press.

———. 1985. "Empiricism: The Key Question." In *The Heritage of Logical Positivism,* ed. Nicholas Rescher, 1–21. Lanham, Md.: University Press of America.

Van Fraassen, B. C. 1980. *The Scientific Image.* Oxford: Clarendon Press.

Index

aesthetics and ethics, 21–22
Allgemeine Erkenntnislehre (Schlick), 7, 43, 170, 350, 353, 367, 368
analytic philosophy, 2, 28–29, 30, 31, 32, 110; linguistic logic and, 38–39
a posteriori statements, 34, 284
a priori knowledge, 6, 17, 292. *See also* relativized a priori; synthetic a priori
atomism, 199, 209, 215, 223; logical, 110, 127, 128
Austrian philosophy, 68, 73–74, 75, 86, 89n20. *See also* Vienna Circle
Avenarius, Richard, 250, 251, 270–71
Avogadro's number, 385
Awodey, S., 3, 57
axiomization, 84, 89n17, 219, 225n26
axioms of constitution, 160, 171, 172
Ayer, A. J., 7, 118n5, 376–78, 384, 385

Bain, Alexander, 236
Basil-Blackwell (publisher), 99
Bauch, Bruno, 17, 31
Bayesian theory, 6, 287, 292, 293, 294–95; confirmation and, 306, 307–8
behaviorism, 97; logical, 5, 263, 265, 266–67, 274, 277
Beiser, Frederick, 168
Bergmann, Gustav, 134, 249
Bernstein, Richard, 343
Black, Max, 133–34
Block, Ned, 266
Bloor, David, 143–52; rule-following and, 143–44, 145, 147; Strong Programme of, 144, 150. *See also* sociology of scientific knowledge (SSK)
Boltzmann, Ludwig, 4, 70, 213, 216, 220, 222, 224nn10, 16; atomism and, 199, 209; on entropy, 211; ideal of, 194; indeterminism and, 201–2, 203, 205–7, 209; Mach and, 196, 197, 212, 224n12; von Mises's critique of, 200
Bolzano, B., 68, 74, 75, 88n2
bootstrapping model, 6, 189n15, 319, 320–21; confirmation theory and, 306–7, 308, 311, 313–18

Brentano, Franz, 74, 88n2, 248, 249
Broad, C. D., 233
Brown, H., 327, 328
Brownian motion, 4, 213, 215–17
Büchner, Ludwig, 235, 236

Callon, Michel, 145, 149–50, 151
Carnap, Rudolph, 2, 24, 29nn1, 7–8, 12, 68, 106 nn4–6, 111, 295, 370n2, 371n4, 384; on causality, 385; on cognitive meaning, 356, 378–79; constitutional system of, 19; debate with Gödel, 3, 57–63, 64; on ethics, 343–44; Hempel and, 112–13, 116, 277, 312, 321, 322; on internal-external distinction, 332–35, 337; Kaplan and, 378–79; Kuhn and, 6, 338; logicism of, 287–93, 294, 296, 298, 307, 322; Marburg School and, 19–20; on mind-body problem, 235, 253–54, 255, 267, 272–76; monism and, 236; on physics, 25–26; Popper's attack on, 304; pragmatics of, 112, 118nn11–12; probability and, 5, 6, 281, 282–83, 284, 305; on rationality, 326–27, 328, 329–39, 342; on syntactical meaning, 36–39, 112; on synthetic a priori, 349; on tolerance, 329, 331–32, 333, 337; on verifiability, 381; on Wittgenstein's *Tractatus*, 33–35, 127–28, 129, 138. *See also* Heidegger-Carnap conflict; *Logische Aufbau der Welt* (Carnap); Neurath-Carnap disputes
Carrier, Martin, 6, 304
Cartan, É., 165–66
Cartesian dualism, 234, 235, 237, 238–39, 241, 267. *See also* psychophysical parallelism
Carus, A. W., 3, 57
Cassirer, Ernst, 3–4, 13–16, 18–29, 163, 201; on expressive function, 24–25; on general covariance, 182–85, 186, 190n29; Heidegger and, 13, 16; mythical view of, 22–23; neo-Kantianism of, 14, 15, 18, 19–20, 21, 73, 87, 204; on relativity theory, 159–62, 180–82, 187, 188nn1–3, 190n26; on synthetic a priori, 353; theory of relativity and, 352; on universal validity, 26–27

categorization paradox, 358–60
causality, 17, 211–13, 248, 385, 386; indeterminist view of, 196–97, 200, 203–4, 205, 208, 210; mind-body duality and, 238–39, 242–43, 256–57
Changing Order (Collins), 143, 147
"chickening out," 134–35, 152
Church, Alonzo, 7, 376–78
Churchland, P. M., 51
Coffa, Alberto, 125, 127, 188*n12*
cognition, 187; coordination as, 168–71, 174, 180
cognitive meaningfulness, 356, 375–86; anti-linguistic turn and, 384–86; Ayer-Church episode and, 376–78; Carnap-Kaplan episode and, 378–79; Hempel's adequacy criterion and, 380–81; self-reflexion of, 381–84; verifiability and, 376, 377–78, 380, 381–82, 383
cognitive significance, 3, 7–8, 264–65, 283, 380. *See also* rationality; reason
cognitive synthesis, 352–56, 364
Cohen, Hermann, 17, 21, 180–81
Collins, Harry, 143–52; rule-following and, 143–44, 145, 147; Winch and, 146–47, 148, 151. *See also* sociology of scientific knowledge (SSK)
communicability, 27, 44–47
comprehensibility, 44, 45
Conant, James, 130, 131, 136, 137, 140, 141
concept, 53–54, 367
confirmation theory, 6, 7, 304–23, 377; auxiliary condition and, 314–15; bootstrap model of, 306–7, 308, 311, 313–18, 319, 320–21; Carnap and, 5, 283, 307, 321, 330–31, 338; Hempel and, 112–13, 116, 305, 308–13, 318–20, 321–22; instantiation and, 309, 314, 315–17, 319, 320; Kuhn's paradigm, 304, 305–6; probabilism in, 307–8. *See also* verifiability
consciousness, 48, 251, 252, 271
consequence conditions, 308, 309–10, 312–13, 321, 322*nn1–5*
consistency condition, 312, 316
constitutional system, 19, 54, 160, 171, 172
content, 44–50; expressibility of, 44–47; intuition and, 47–50; meaning and, 265
Continental philosophy, 28–29, 30, 32, 38–39, 110; neo-Kantianism and, 2, 31–32
conventionalism, 78, 82, 83–84, 214, 220; French, 73, 87, 200, 203
coordinates, relativity of, 163, 178
coordination, 24, 172–73, 179, 352, 367; cognition as, 168–71, 174, 180; constitutional system and, 54, 160, 171, 172; conventionalism and, 220; of equations, 175–76; quantum mechanics and, 221

cosmology, 240–41
covariance, general, 161, 162–67, 178, 182–85, 186, 190*n29*
cowardice, 134–35, 152
credibility, 148–49, 291, 295
critical realism, 243, 250. *See also* realism
Critique of Pure Reason (Kant), 14, 17, 354–55
culture, 21, 22, 25, 26, 147

D'Acconti, Alessandra, 70
Darwinism, 235–36, 241, 248
Dasein, 20
Davos disputation, 13–14
decision theory, 291–92
deduction, transcendental, 354–55
deductive chauvinism, 7, 379
deductive logic, 382. *See also* logic
de Finetti, Bruno, 6, 281, 290–91, 300*nn10–14, 16*, 307; subjectivism of, 292, 293–98, 299
definition, 273, 367, 379
determinism, 198, 208, 220, 223, 245, 248–49. *See also* indeterminism
Diamond, Cora, 130, 131, 134–36, 137–38, 142; on "chickening out," 134–35, 152; on ladder image, 134–35, 139–40, 141
diffeomorphisms, 164
Dilthney, Wilhelm, 20, 32, 40, 247
Dingler, Hugo, 83, 84, 174
dualism, 4–5, 20. *See also* mind-body dualism; psychophysical parallelism
du Bois-Reymond, Emil, 220
Duhem, Pierre, 181, 358, 360, 371*n6*; Vienna Circle and, 69, 70, 83, 84

Ebbinghaus, Hermann, 247, 248
Einstein, Albert, 70, 84, 161, 179, 188*nn1–2, 4*, 216, 244; on general covariance, 162–64; on space and time, 182–83, 186
Einstein theory, 159, 166, 174, 180, 191*n29*. *See also* relativity, general theory of
elucidation notion, 136
empiricism, 330; new, 361–63; nonlinguistic, 7, 8; pragmatic, 111, 112, 113–14, 115, 118; synthetic a priori and, 349–51
energy conservation, law of, 212, 214, 218, 223; psychophysics and, 240, 244, 246
Engelmann, Paul, 73, 129–30
England, 99–101
Enlightenment, 86, 214
entropy, 196, 211
epistemic optimality, 114
epistemology, 8, 16, 175–76, 299; and psychology, 366–70. *See also* knowledge
equisignificance criterion, 380–81
equivalence, 184, 188*n6*, 189*nn16–18*
Erkenntnis and Irrtum, 85, 94, 97–98

Erkenntniskritik (journal), 180–81
ethics, 21–22, 129–30, 343–44
Euclidean geometry, 174–75, 177, 179, 353, 354
Europe, reconstruction of, 105
Exner, Franz Serafin, 4, 202, 203, 220, 221, 224*n*6; indeterminism and, 199, 200, 205, 207–11, 213, 224*nn*15–16, 225*n*17; Planck and, 210–11; on probability, 216, 217; Schrödinger and, 194, 200
experience, 18, 51, 52–53, 251, 265; Carnap on, 268; theory and, 220–21
explication, 112, 114–17, 119*n*19, 148, 149
expressibility, 22, 24–25, 44–47

Fechner, Gustav Theodor, 4–5, 199, 207, 209, 211; psychophysics of, 236, 237–39, 240–42, 244, 254, 257*n*3
Feigl, Herbert, 80, 219, 249, 288, 382; mind-body dispute and, 233–35, 252–53, 254–56, 257, 258*n*22, 259*nn*23, 25, 276
Feyerabend, Paul K., 6, 328, 350, 357, 359, 372*n*6
Forman, P., 198
Frank, Philipp, 4, 67, 70, 71–72, 75, 89*nn*14, 17–18, 200, 225*nn*26–27; Boltzmann and, 194, 197; on causality, 211–13; on experience and theory, 220–21; Hahn and, 79, 81–86; Haller and, 69, 77–78, 85; indeterminism and, 202–3, 211–15, 217, 219; Mach and, 214–15
Frege, Gottlob, 3, 31, 32, 34–35, 36, 131; on logic, 137–38
French conventionalism, 73, 87, 200, 203. *See also* conventionalism
frequency interpretation, 5, 208, 209
frequency theory, 217, 284–87, 295, 299; relative, 288, 289; van Mises on, 282, 284, 294. *See also* probability
Friedman, Michael, 2, 3–4, 13, 29*nn*2–3, 186, 189–90*n*20; on rationality, 326, 332–33, 336, 337, 340, 344*n*2
frugality (inductive simplicity), 256
Fumerton, R. A., 110, 118*n*2
functional dependency, 196, 206, 256
functionalism, 275

Gabriel, Gottfried, 2, 30
Galavotti, Maria Carla, 5, 9*nn*5, 8, 281, 300*nn*
general covariance. *See* covariance, general
General Theory of Knowledge (Schlick), 48, 49, 53, 55*n*1, 250, 276
geometry, 218–20; Euclidean, 174–75, 177, 179, 353, 354
Glymour, Clark, 6, 189*n*15, 306–7, 308, 311, 313–18, 320, 321
Gödel, Kurt, 3, 57–63, 64, 224*n*9
Goldfarb, Warren, 57, 63*n*6, 64, 130, 131, 140

Goodman paradox, 113

Hacker, P. M., 137, 141
Hacking, Ian, 286, 385
Hahn, Hans, 79, 81–86, 203, 225*n*26, 349; Vienna Circle and, 67, 72, 77, 78, 80–81
Haller, Rudolph, 72–73, 77, 85, 203; Neurath and, 68, 69, 72, 88*n*1
Hamann, J. G., 168
Hanle, Paul, 224*n*16
Hanson, Norwood Russell, 6, 350, 357
Hegel, Georg W., 22, 182
Heidegger, Martin, 2, 29*nn*4–6, 135, 168
Heidegger-Carnap conflict, 13, 28, 30–32, 35–36; neo-Kantianism and, 14, 15–16, 20
Heidelberger, Michael, 4, 207, 224*n*13, 233
Heisenberg, Werner Karl, 4, 223
Helmholtz, Hermann von, 24, 245, 353, 371*n*3
Hempel, Carl Gustav, 3, 4, 70, 89*n*8, 109–18, 282, 299*n*1, 367; adequacy condition of, 378; Carnap and, 112–13, 274, 283, 321, 322; on causality, 385; cognitive meaningfulness and, 356; on explanation, 114–17; on hypothesis confirmation, 6, 305, 308–13, 317, 318–20, 321–22, 321–23; logical positivism and, 109–10; on mind-body problem, 5, 263–68, 272, 273, 277; pragmatic empiricism of, 111, 112, 113–14, 115, 118, 119–20*n*21; on verifiability, 376, 382
Hertz, Heinrich, 181, 245
Hesse, Mary, 351, 360–63, 364, 372*n*8
Hilbert, David, 178, 186, 190*n*29, 218; axiomatic method of, 84, 219, 225*n*26; Vienna Circle and, 79, 86, 87, 89*n*17
Höffding, Harald, 244, 249, 258*nn*11–12
Höfler, Alois, 74, 76, 88*n*2, 203, 249, 258*n*19
Hosiasson-Lindenbaum, Janina, 119*n*14
human happiness, 100, 102, 103
Hume, David, 2, 3, 9*n*6, 198, 269, 353, 383; on causation, 322; induction and, 286, 294; Kant and, 168; on types of reason, 375–76, 384
Husserl, Edmund, 2, 17, 20, 31, 32, 248. *See also* Schlick-Husserl opposition
hypothesis confirmation, 327, 338, 377. *See also* confirmation theory; verifiability
hypothetico-deductivism, 52, 308, 317, 319

idealism: logical, 18, 179; transcendental, 19, 21, 271, 354–55
identity theory, 275, 276–77; mind-body dualism and, 239–40, 241, 245, 249, 253, 255–56
indeterminism, 4, 194–223; Boltzmann and, 205–7; Brownian motion and, 213, 215–17; causality and, 196–97, 200, 203–4, 205, 206, 211–13; conventionalism and, 200, 203;

indeterminism *(continued)*,
  Exner's synthesis and, 205, 207–11; Frank and, 202–3, 211–13, 217, 219. *See also* Vienna Circle
induction, rule of, 281, 285, 286, 292, 294
inductive logic, 289–91, 322, 336, 379; rationality and, 328, 329–30, 335–37, 342
instantiation, 6, 113, 275; confirmation and, 309, 314, 315–17, 319
intellectual responsibility, 8, 383
intentionality, 54–55, 185
internal-external distinction, 332–35, 342
International Encyclopedia of Unified Science, 94
intuition, 17, 52, 55, 169, 185, 336–37; content and, 47–50; inductive, 289, 292
invariance groups, 163–64, 165. *See also* covariance, general
irrationalists, 127, 130, 131, 139, 141
Irzik, Gürol, 6, 120*n*22, 325

James, William, 238, 242, 247–48, 254, 258*n*17, 298
Jeffrey, Richard, 114, 118*nn*7, 13, 119*nn*19, 21, 292–93, 299, 300*n*17
Jeffreys, Harold, 287, 288
Jodl, Friedrich, 249, 259*n*20
justification, internal *vs.* external, 332–35, 342

Kant, Immanuel, 3, 34, 50, 176, 222, 357; and Avenarius compared, 251, 270–71; dual origin and, 167–68; Hempel and, 117; intention of, 185; metaphysics and, 14–15; noumenon of, 27–28, 243; a priori of, 349, 350, 352, 354, 360, 361; Vienna Circle and, 67, 75, 76, 78, 85, 86, 88*n*2. *See also* a priori knowledge
Kantianism, 88*n*2, 169, 196–97. *See also* neo-Kantianism
Kaplan, David, 378–79
Kepler's third law of motion, 316, 317
Keynes, John Maynard, 288, 294
Kim, Jaegwon, 5, 9*n*9, 233, 234, 263
kinetic theory, 4, 216
Kitt Peak observatory, 382–83
knowledge, 21, 49–55, 185, 285; Cassirer on, 180–81; experience and, 52–53; intentionality and, 54–55; intuition and, 50–52; judgments and, 16–17; perception and, 49–50, 51, 170; theory of, 367; truth and, 54. *See also* epistemology
Knowledge and Reality (Parrini), 381
Kuhn, Thomas S., 6, 146, 194, 344*n*5, 350, 357, 371*n*6; Hempel and, 115, 117; mind-body problem and, 223*n*1, 234; paradigm of, 147, 148, 304, 305–6, 340, 365–66; on rationality, 326, 327, 328, 335, 338, 339–43

ladder imagery, 134–35, 138–41
Lakatos, Imre, 201, 306, 328, 371*n*6
Lanfredini, Roberta, 2, 43
Lange, Friedrich Albert, 236, 245–46
language: cognitive meaning and, 384–86; expression and, 25; metaphysics and, 35; natural, 22, 23, 380; network model and, 361–62; observational *vs.* theoretical, 356–57, 369; ordinary, 2, 137; paradigm and, 365; scientific, 96; syntax and, 34, 36–39, 57, 112–13; translation and, 272–73; two-language theory, 248–49, 253, 254, 321
Laplace, Pierre-Simon de, 220, 222, 288, 292–93, 294
Lask, Emil, 18, 20
Latour, Bruno, 145, 146, 149–50, 151
Leibniz, G. W., 23–24, 27, 167, 168, 185; mind-body dualism and, 236, 241, 267–68
Le Morvan, P., 322*n*2
Le Roy, Édouard, 82, 84
Leviathan and the Air Pump (Shapin & Shaefer), 146
life philosophy, 31–33, 36, 39, 247
Lindenbaum, Adolf, 119*n*14
linguistic framework, 7, 40, 331, 333–34, 338, 356–57; observation and, 356–57, 369; syntax and, 36–39. *See also* language
local symmetry, 162, 165, 187
Locke, John, 2, 167, 168
logic, 15–16; deductive, 382; life and, 31–32; philosophy as, 137–38; probability and, 283; pure, 17, 18, 20. *See also* inductive logic
"Logical Analysis of Psychology" (Hempel), 263, 268, 272
logical atomism, 110, 127, 128
logical behaviorism, 5, 263, 265, 266–67, 274, 277
Logical Foundations of Probability (Carnap), 289–90, 305, 329
logical idealism, 18, 179
logical positivism, 1, 5, 109–10, 117, 127–28, 277, 328. *See also* positivism
logicism, Carnap and, 287–93, 294, 296, 298, 307, 322. *See also* inductive logic
logico-mathematical rules, 331–32
Logische Aufbau der Welt (Carnap), 19, 20, 25, 283; psychophysics and, 5, 253, 255, 257, 265, 268–69
Lorentz, H. A., 164, 165, 166, 188*n*7, 245, 258*n*14
Loschmidt, Josef, 201, 202, 211
Lotze, Rudolph Hermann, 40*n*3, 248, 258*n*18

Mach, Ernst, 163, 198, 202, 218, 241, 268; on causality, 197, 201, 208, 212, 243; epistemology of, 205–6; Frank and, 214–15; ideal of, 194–95; indeterminism and, 198–99, 202,

203, 207; monism of, 5, 242; Planck's attack on, 83–84, 182, 195–96, 198–99, 214, 224n12; positivism of, 70, 73; psychophysics and, 239–40, 249, 254, 265; unique determination of, 208, 220; Vienna Circle and, 3, 4, 75, 77, 78, 87, 88n2, 203

Marburg School, 16, 18, 19, 21, 39, 186–87, 188n13; relativity theory and, 168, 170, 179, 180

materialism, 235–36, 239, 247

mathematical-physical theory, 23–24, 26

mathematics, 18, 112, 331–32; syntactic interpretation of (SIM), 58–63

Mauthner, E., 139

McDowell, John, 132

meaningful statements, 34, 35, 264–65, 281–82, 376. *See also* cognitive meaningfulness

Meinong, Alexius, 74, 88n2

Menger, Karl, 80

meta-alternation, 149, 150

metaphysics, 14–15, 35–36, 39–40, 96, 132; mind-body problem and, 238, 252, 253; persecution and, 101–4

Mill, John Stuart, 2, 269

mind-body dualism, 4–5, 263–77; Carnap on, 272–76; Hempel on, 263–68; Schlick on, 269–71. *See also* psychophysical parallelism

Minkowski, Hermann, 82, 163, 166, 214

Monatshefte für Mathematik und Physik (magazine), 81–82, 84, 85, 225n22

monism, 5, 236, 242–43, 244

Morris, Charles, 95, 97, 98, 105, 106

mythical thought, 22–23

natural language, 22, 23, 380. *See also* language

natural laws, 207–8, 210, 213

nature, teleological view of, 241, 371–72n6

Naturwissenschaften (magazine), 197–98, 212

neo-Kantianism, 2, 16–22, 31–32, 39, 236; Cassirer and, 14, 15, 18, 19–20, 21, 73, 87, 204; logical empiricism and, 67; psychophysics and, 246, 248–49; relativity theory and, 159, 160–61, 170; Vienna Circle and, 67, 69. *See also* Kant; Marburg School

neopositivism, 127–28. *See also* logical positivism

network model, 360–63, 364–65, 366

Neumann, John von, 244–45

Neurath, Otto, 67, 70–71, 78, 203, 349; Haller and, 68, 69, 72, 88n1; Hempel and, 117; human happiness and, 100, 102, 103; Philosophical Society and, 76, 77; thesis of, 75, 85, 89n20; Vienna Circle and, 79, 88n2, 371n4

Neurath-Carnap disputes, 3, 94–107; England

and, 99–101; metaphysics, persecution and, 101–4; pluralism and, 96, 97, 105. *See also* Unity of Science Movement

Newtonian mechanics, 200, 215, 216, 220, 222, 381; confirmation theory and, 308, 316, 317, 319

Newtonian physics, 21, 166, 174, 177

Nietzsche, Friedrich, 32, 36, 39

Nohl, Herman, 32–33

nonsense, 137–38, 140

Norton, J. D., 162, 178

nothing, 15, 35

noumena, 27–28, 241, 243, 250–51, 252

Oberdan, Thomas, 188n1

objectivity, 26, 83, 149, 286–87, 296, 297, 299; physical, 182–83, 186

"Objectivity, Value Judgment, and Theory Choice" (Kuhn), 340–41

observation, 356–62; network model for, 360–62; strong thesis of, 358–59; theoretical language and, 356–57, 369

ontology, 199, 212, 213, 221–22

Ostwald, Wilhelm, 195, 236

Papago Indians, 382–83

paradigm, 147, 148, 304, 305–6, 340, 365–66

Parrini, Paolo, 1, 2, 7, 8, 349, 381

Peirce, Charles Sanders, 242, 244, 298

perception, 6, 7, 25; knowledge and, 49–50, 51, 170; reason and, 357, 360

Perrin, Jean, 385

persecution, metaphysics and, 101–4

phenomena, 214, 216, 240

phenomenology, 43, 47–48, 53, 265, 268–69

Philosophical Investigations (Wittgenstein), 137, 151–52

Philosophical Society, 76–77

philosophy, 142–43, 330, 360. *See also* analytic philosophy; Continental philosophy; metaphysics

physical objectivity, 182–83, 186. *See also* psychophysical parallelism

physics, 3–4, 25–26, 26, 199, 277, 339; identity view and, 240; natural laws and, 207–8; Newtonian, 21, 166, 174, 177; philosophy and, 360; probability and, 218; relativity theory and, 160, 179

Planck, Max, 4, 184, 204, 210–11, 213, 221, 223n1; on energy conservation, 214, 218; Mach and, 83–84, 182, 195–96, 198–99, 214

Platonism, 37, 102, 103–4, 132

pluralism, 96, 97, 105

Poincaré, Henri, 3, 165, 166, 214; Vienna Circle and, 69, 70, 73, 82, 84

Poisson's theorem, 217, 218
Polyani, M., 344n5
Popper, Karl, 6, 116, 295, 304, 357, 362–63; on rationality, 326, 328, 342
positivism, 19, 78, 264; logical, 1, 5, 109–10, 117, 127–28, 277, 328
posit notion, 285–86
pragmatic empiricism, 111, 112, 113–14, 115, 118
predictability, 97, 297
Principia Mathematica (Russell), 8n4, 18, 19, 20
probabilism, 281–99; Carnap's logicism and, 287–93; confirmation theory and, 307–8; de Finetti's subjectivism and, 293–98, 299; frequency theory and, 284–87, 288, 289, 295, 299; Reichenbach on, 281–87, 298, 299
probability, 5–6, 199, 204, 206–11, 221, 222; calculus of, 208, 216–20; frequency interpretation of, 5, 208, 209; geometry and, 218–20
Prugovečki, E., 165, 166, 188nn6–7
psychological-subjective aspect, 351–52, 368
psychology, 7, 247, 264, 272, 330; epistemology and, 366–70; philosophy of, 5, 9n9
psychophysical parallelism, 5, 233–57, 263–77; Carnap on, 235, 253–54, 272–76; causality and, 238–39, 242–43; critics of, 246–49; Darwinism and, 235–36, 241, 248; Feigl on, 254–56, 257; Hempel on, 5, 263–68, 272, 273, 277; identity view and, 239–40, 241, 245, 249, 253, 255–56; materialism and, 235–36, 239, 247; monism and, 236, 242–43, 244; Schlick on, 234, 249, 250–53, 269–71; Wundt on, 236, 243–44, 248
Putnam, Hilary, 132, 328, 335, 336
Pyrrhonian skepticism, 138

quantum mechanics, 4, 195, 212, 219, 223, 244; Newtonian mechanics and, 200, 220
Quine, W. V. O., 351, 360, 372n9

radical probabilism, 293–98. *See also* probabilism
Ramsey, Frank, 6, 288, 290, 300nn4, 8, 15; subjectivism of, 292, 293, 300n10
rationality, 5, 6, 292, 299, 325–44; Carnap on, 326–27, 328, 329–39; critique of standard, 328–32; inductive logic and, 335–37; Kuhn and, 326, 327, 328, 335, 338, 339–43; relativism and, 332–33, 337, 338, 339–40; scientific method and, 325, 328, 329; scientific theories and, 330–31, 338–39; tolerance and, 329, 331–32, 333, 337. *See also* reason
ravens paradox, 318–20
realism, 37, 149, 196, 209, 287, 288; mind-body problem and, 243, 250, 254, 255; structural, 82
reality, 20, 23, 339; cowardice and, 134–35; rela-

tivity theory and, 176–77; Schlick on, 169–70, 171, 250–52
reason, 6, 15, 28, 176, 292, 361; perception and, 357, 360. *See also* cognitive meaningfulness; rationality
reciprocal effect theory, 243–44
Rees, Martin, 383–84
Reichenbach, Hans, 1, 40n9, 111, 168–79, 360, 370n1–2, 372n7, 380–81; on confirmation, 305, 307; coordination method of, 163, 168–71, 188n13, 189n15, 220; indeterminists and, 203, 204–5; probabilism of, 4, 5, 9n10, 219, 281–87, 289, 298, 372n7; on relativity theory, 159–62, 171–72, 180, 185, 187, 189–90n20, 352; on scientific analysis, 175–78; successive approximation method of, 172–75; on synthetic a priori, 349, 350–51, 353–54, 355, 361, 371n3
Reid, Lynette, 131, 138
Reisch, George, 3, 94
relative frequency, 288, 289. *See also* frequency theory
relativism, 150, 332–33, 337–40, 341
relativity, general theory of, 4, 9n7, 84, 159–91, 204, 215; Cassirer on covariance and, 182–85, 186, 190n29; cognition as coordination and, 168–71, 174, 180; general covariance and, 161, 162–67; Kantian dual origin and, 167–68; Reichenbach-Cassirer debate on, 159–62, 180; scientific analysis and, 175–78; successive approximation and, 172–75
relativized a priori, 160, 161, 171–74, 180, 181, 189n20. *See also* synthetic a priori
representation theorem, 23, 294
Republic (Plato), 102, 103
responsibility, intellectual, 8, 383
reversible-irreversible processes, 210
Rey, Abel, 69, 70, 83, 85
Rickert, Heinrich, 17, 31, 247, 248
Ricketts, Tom, 57, 63–64n6, 130, 137
Riehl, Alois, 243, 249, 252, 253, 254, 258nn9, 20
Rockefeller Foundation, 98
Roth, Paul, 151
rule-following, 143–44, 145, 147
Russell, Bertrand, 2–3, 81, 134, 257n4, 284, 350; logical atomism of, 110, 127, 128; mathematical logic of, 18, 19; monism of, 5, 242
Ryckman, Thomas A., 3, 79, 159
Ryle, Gilbert, 233, 234, 274, 276

Salmon, Wesley, 1, 7, 8, 118nn5–6, 9, 223n, 300nns, 6, 322, 344n1, 375; Bayesian theorem and, 287, 292, 307; on Hempel, 119n6, 19, 21, 120n27; Reichenbach and, 286, 299
Scheffler, I., 358–60

Schilpp, P. A., 331, 335

Schlick, Moritz, 5, 7, 9n7, 55n1, 99, 127, 172, 179, 188nn3, 12–13, 370–71nn2–3, 372nn11–12; on coordination and cognition, 169, 170, 171, 174; indeterminists and, 203, 204; monism and, 236; network model and, 364–65; on psychology and epistemology, 367–69; psychophysics and, 234, 249, 250–53, 255, 258nn21–22, 269–71, 276–77; relativity theory and, 161, 190n20, 352; on synthetic a priori, 349, 350; Vienna Circle and, 67, 71, 75, 78–79, 80, 81, 84, 86, 87

Schlick-Husserl opposition, 2, 43–56; on experience, 52–53; on expressibility of content, 44–47; intentionality and, 54–55; on intuition and content, 47–50; on knowledge and intuition, 50–52

Schrödinger, Erwin, 4, 194, 197, 199–200, 221, 223

Scientific Knowledge (textbook), 145

scientific language, 96. *See also* language

scientific method, 175–78, 286–87, 328, 335

self-applicability, of syntactic view, 61, 62

semantics, 96, 128, 281. *See also* language

"Semantics, Ontology, and Empiricism" (Carnap), 336

sense, and nonsense, 137–38, 140

sense experience, 16–17, 270–71

sensibility, 27, 168, 181, 186–87

Sextus Empiricus, 139

Shaefer, S., 146

Shapin, S., 146

similarity, recognition of, 362–64, 365, 366

skepticism, 138, 142, 143, 147, 149

Smart, J. J. C., 233, 274

Smoluchowski, Marian von, 213, 216

sociology of scientific knowledge (SSK), 143–52; credibility and, 148–49; *Investigations* and, 151–52; meta-alternation principle and, 149, 150; rule-following and, 143–44, 145, 147; TEA-lasers and, 147–48; Winch and, 146–47, 148, 151

sophistry and illusion, 375, 384

Southwest School, 16, 18, 20

space and time, 163, 164, 167, 181, 190n20; physical objectivity and, 182–83, 186. *See also* relativity, general theory

spacial competition, 270–71

special consequence condition, 309–10, 312–13, 321, 322nn1–5. *See also* confirmation theory

Spinoza, Baruch, 241, 244, 246, 252, 267–68

SSK. *See* sociology of scientific knowledge

Stachel, John, 162, 164, 165

statistical ontology, 221–22, 286

Stefan, Josef, 201, 202

Stegmüller, Wolfgang, 111, 118n9

Stern, David G., 3, 8n3, 125

Stöltzner, Michael, 3, 4, 194, 225n27

Strong Programme, 144, 150

strong thesis, 358–59

Structure of Scientific Revolution (Kuhn), 305, 339, 340

Stumpf, Carl, 247, 248

subjectivity, 55, 292; probability and, 288, 293–98, 299; psychological and, 351–52, 368

Substance and Function (Cassirer), 18, 19–20, 21, 182

symbolic meaning, 21, 22–23, 24–25, 26, 28

symbolism, 362, 369–70

syntactical interpretation of mathematics (SIM), 58–64; self-applicability of, 61, 62

syntactical meaning, 34, 36–39, 57, 112–13. *See also* language; semantics

synthetic a priori, 360, 361, 371n3; empiricism and, 349–51; negation of, 349–56, 357; theory of relativity and, 352–54. *See also* relativized a priori

Tarski, Alfred, 3, 62, 112, 255

TEA-lasers, 147–48

theory: choice, 340–41, 342–43; construction, 39; scientific, 79, 330–31, 338–39

theory-change tradition, 305–6

therapeutic reading, of Wittgenstein, 130–32, 136–37, 141–42

thermodynamics, laws of, 195, 196, 209, 213

thing concept, 25, 160

time. *See* space and time

tolerance, 37, 57; rationality and, 329, 331–32, 333, 337

Toulmin, S., 130

Tractatus Logico-Philosophus (Wittgenstein), 3, 73, 125–52; irrationalist reading of, 127, 130, 131, 139, 141; ladder imagery in, 134–35, 138–41; metaphysical reading of, 128–30, 132; philosophical method in, 132–43; reception of, 126–32; sociology of scientific knowledge and, 143–52; therapeutic reading of, 130–32, 136–37, 141–42

transcendental idealism, 19, 21, 271, 354–55. *See also* Kant; neo-Kantianism

transcendental logic, 17

truth, 31, 62, 283, 295, 338; knowledge and, 54

two-language theory, 248–49, 253, 254, 321

Uebel, Thomas E., 3, 4, 67, 88n6, 89n10, 203, 225n22

understanding, 168, 181, 186–87

unique determination, 198, 220

Unity of Science Movement, 94, 95, 97, 99, 101, 104–5, 106nn1, 3, 6

universal validity, 26–27, 28

University of Chicago Press, 95, 97–98

Index

Vaihinger, Hans, 245–46, 252, 258*n15*

validity, 357; objective, 21, 31; universal, 26–27, 28; *vs.* vindication, 334–35, 382. *See also* truth

van Fraassen, Bas, 8, 384

verifiability, 7, 376, 377–78, 380, 381–82, 383

Vienna Circle, 3, 25, 67–89, 127, 128, 198, 206, 219, 371*n4*; conventionalism and, 73, 78, 82, 83–84, 87, 203; Frank-Haller thesis and, 69, 72–73, 85; Hahn-Frank collaboration and, 79, 81–86; Kantian influence and, 67, 75, 76, 78, 85, 86, 88*n2*; master logic texts and, 74–75; mind-body problem and, 234, 255, 264, 272; Neurath-Haller thesis, 68, 69, 72; Schlick and, 71, 75, 78–79, 80, 81, 84, 86, 87. *See also* indeterminism

vindication *vs.* validation, 334–35, 382

von Mises, Richard, 89*n14*, 197, 203, 215–20, 222, 223, 225*n26*; conventionalism and, 200;

frequentism and, 282, 284, 294; probability calculus and, 216–20, 287; Reichenbach and, 4, 5, 205

Wagner, Rudolph, 235

Waismann, Friedrich, 219, 282

Whitehead, Alfred North, 3, 8*n4*, 19

Winch, Peter, 146–47, 148, 151

Wittgenstein, Ludwig, 2, 32, 55*n1*, 110, 255, 350; and mind-body problem, 233, 234. *See also Tractatus Logico-Philosophus*

Wolters, Gereon, 3, 109, 225*n23*

Woolgar, Steven, 145, 149, 150, 151

Wundt, Wilhelm, 236, 243–44, 248

Yearly, Steven, 145, 149, 150

Zimmerman, Robert, 75, 88*n2*

www.ingramcontent.com/pod-product-compliance
Lightning Source LLC
Chambersburg PA
CBHW060750220326
41598CB00022B/2395